Control Systems Safety Evaluation and Reliability
Second Edition

Control Systems Safety Evaluation and Reliability

William M. Goble
Second Edition

ISA–The Instrumentation, Systems, and Automation Society

Notice

The information presented in this publication is for the general education of the reader. Because neither the author nor the publisher have any control over the use of the information by the reader, both the author and the publisher disclaim any and all liability of any kind arising out of such use. The reader is expected to exercise sound professional judgment in using any of the information presented in a particular application.

Additionally, neither the author nor the publisher have investigated or considered the affect of any patents on the ability of the reader to use any of the information in a particular application. The reader is responsible for reviewing any possible patents that may affect any particular use of the information presented.

Any references to commercial products in the work are cited as examples only. Neither the author nor the publisher endorse any referenced commercial product. Any trademarks or tradenames referenced belong to the respective owner of the mark or name. Neither the author nor the publisher make any representation regarding the availability of any referenced commercial product at any time. The manufacturer's instructions on use of any commercial product must be followed at all times, even if in conflict with the information in this publication.

Copyright © 1998   ISA—The Instrumentation, Systems, and Automation Society
                    67 Alexander Drive
                    P.O. Box 12277
                    Research Triangle Park, NC 27709

All rights reserved.

Printed in the United States of America.
10 9 8 7 6 5 4 3

ISBN   1-55617-636-8

No part of this work may be reproduced, stored in a retrieval system, or transmitted in any form or by any means, electronic, mechanical, photocopying, recording or otherwise, without the prior written permission of the publisher.

Library of Congress Cataloging-in-Publication Data

Goble, William M.
    Control systems safety and reliability: techniques and applications / William M. Goble.–2nd ed.
        p. cm. - (Resources for measurement and control series)
    Previous ed. Published under title: Evaluating control systems reliability. 1992.
    Includes index.
    ISBN 1-55617-636-8
    1. Automatic control – Reliability.   I. Goble, William M.
    Evaluating control systems reliability.   II. Title.   III. Series.
TJ213.95.G62   1998
629.8-dc21   97-42557
                                                                       CIP

ISA Resources for Measurement and Control Series (RMC)

- *Measurement and Control Basics*, 2nd Edition (1995)
- *Industrial Level, Pressure, and Density Measurement* (1995)
- *Programmable Controllers*, 3rd Edition (2001)
- *Control Systems Documentation: Applying Symbols and Identification* (1993)
- *Industrial Data Communications: Fundamentals and Applications*, 2nd Edition (1997)
- *Real-Time Control Networks* (1993)

Acknowledgments

This book has been made possible only with the help of many other persons. Early in the process, J. V. Bukowski of Villanova taught a graduate course in reliability engineering where I was introduced to the science. This course and several subsequent tutorial sessions over the years provided the help necessary to get started.

Many others have helped develop the issues important to control system safety and reliability. I want to thank co-workers; John Grebe, John Cusimano, Ted Bell, Ted Tucker, Griff Francis, Dave Johnson, Glenn Bilane, Jim Kinney, and Steve Duff. They have asked penetrating questions, argued key points, made suggestions, and provided solutions to complicated problems. A former boss Bob Adams deserves a special thank you for asking tough questions and demanding that reliability be made a prime consideration in the design of new products.

Fellow members of SP84 have also helped develop the issues. I wish to thank Vic Maggioli, Dimitrios Karydos, Tony Frederickson, Paris Stavrianidis, Paul Gruhn, Aarnout Brombacher, Ad Hamer, Rolf Spiker, Dan Sniezek and Steve Smith. I have learned from our debates.

Several persons made significant improvements to the document as part of the review process. I wish to thank Tom Fisher, John Grebe, Griff Francis, Paul Gruhn, Dan Sniezek, Rainer Faller and Rachel Amkreutz. The comments and questions from these reviewers improved the book considerably. Julia Bukowski from Villanova University and Jan Rouvroye of Eindhoven University deserve a special thank you for their comprehensive and detail review. Iwan van Beurden of Eindhoven University also deserves a special thank you for a detail review and check of the examples and exercise answers. I also wish to thank Rick Allen, a good friend, who reviewed the draft and tried to teach the rules of grammar and punctuation.

Finally, I wish thank my wife Sandy and my daughters Tyree and Emily for their patience and help. Everyone helped proofread, type, and check math. While the specific help was greatly appreciated, it is the encouragement and support for which I am truly thankful.

Contents

PREFACE   xiii

ABOUT THE BOOK   xv

ABOUT THE AUTHOR   xvii

Chapter 1   INTRODUCTION   1
Control System Safety and Reliability, 1
Reliability Engineering, 5

Chapter 2   UNDERSTANDING RANDOM EVENTS   9
Random Variables, 9
Mean, 18
Variance, 21
Common Distributions, 22

Chapter 3   FAILURES: STRESS VERSUS STRENGTH   31
Failures, 31
Failure Categorization, 31
Failure Sources, 37
Stress-Strength, 45
Measuring Strength, 52

Chapter 4   RELIABILITY AND SAFETY   59
Reliability Definitions, 59
The Constant Failure Rate, 73
Safety Terminology, 78

Chapter 5 FAILURE MODES AND EFFECTS ANALYSIS 89
Introduction, 89
FMEA Procedure, 89
FMEA Limitations, 90
FMEA Format, 90
Failure Modes, Effects, and Diagnostic Analysis (FMEDA), 93

Chapter 6 FAULT TREE ANALYSIS 101
Introduction, 101
Fault Tree Method, 101
Fault Tree Symbols, 103
Qualitative Fault Tree Analysis, 106
Quantitative Fault Tree Analysis, 106

Chapter 7 NETWORK MODELING 117
Reliability Networks, 117
Quantitative Network Evaluation, 134

Chapter 8 MARKOV MODELING 147
Repairable Systems, 147
Solving Markov Models, 149
Discrete-Time Markov Modeling, 151

Chapter 9 DIAGNOSTICS 187
Improving Safety and MTTF, 187
Measuring Diagnostic Coverage, 200
Diagnostic Techniques, 202
System Coverage, 209
Fault Injection Testing, 210

Chapter 10 COMMON CAUSE 213
Common-Cause Failures, 213
Common-Cause Modeling, 217
Common-Cause Avoidance, 223
Estimating the Beta Factor, 225
Including Common Cause in System Models, 227

Chapter 11 SOFTWARE RELIABILITY 237
Software Failures, 237
Stress-Strength View of Software Failures, 240
Software Complexity, 243
Software Reliability Modeling, 253
Software Reliability Model Assumptions, 263

Chapter 12 MODELING DETAIL 271
Key Issues, 271
Probability Approximations, 272
Multiple Failure Modes, 285
On-line Diagnostics, 292

Diagnostics and Common Cause, 298
Comparing the Techniques, 303

Chapter 13 RELIABILITY AND SAFETY MODEL CONSTRUCTION 307
System Model Development, 307
Sensors and Final Elements, 331

Chapter 14 SYSTEM ARCHITECTURES 335
Introduction, 335
Controller Construction, 335
System Configurations, 341
1oo1: Single-Channel System, 342
1oo2: Dual-Channel System, 348
2oo2: Dual-Channel System, 353
1oo1D: Dual-Channel System, 358
2oo3: Triple Controller System, 364
2oo2D Architecture, 376
1oo2D Architecture, 381
1oo2D Architecture with Comparison, 389
Comparing Architectures, 394

Chapter 15 SAFETY INSTRUMENTED SYSTEMS 397
Introduction, 397
Risk Cost, 397
Risk Reduction, 398
SIS Architectures, 404

Chapter 16 LIFE CYCLE COSTING 423
The Language of Money, 423
Procurement Costs, 425
Operating Costs, 426
Cost of System Failure, 427
Time Value of Money, 433
SIS Life-Cycle Cost, 439

APPENDIX A STANDARD NORMAL DISTRIBUTION TABLE 443

APPENDIX B MATRIX MATH 447

APPENDIX C PROBABILITY THEORY 455

APPENDIX D RELIABILITY PARAMETERS 475

APPENDIX E CONTINUOUS-TIME MARKOV MODELING 481

APPENDIX F ANSWERS TO EXERCISES 495

INDEX 513

Preface

The ability to numerically evaluate control system design parameters, like safety and reliability, have always been important in order to balance the tradeoffs between cost, performance and maintenance in control system design. However, there is more involved than just economics. Proper protection of personnel and the environment have become the issue. Increasingly, quantitative analysis of safety and reliability is becoming essential as international regulations require justified and measured safety protection performance.

ISA's S84.01 standard defines quantitative performance levels for safety instrumented systems(SIS). New IEC safety standards and the industry specific companion standards do the same. In general these standards are not prescriptive, they do not say exactly how to design the system. Instead, they advise the quantitative safety measurements that must be met and the designer considers various design alternatives to see which design meets the targets.

This general approach is very consistent with those who work to economically optimize their designs. Design constraints must be balanced in order to provide the optimal design. The ultimate economic success of the process is affected by all of the design constraints. True design optimization requires that alternative designs be evaluated in the context of the constraints. Numeric targets and methods to quantitatively evaluate safety and reliability are the tools needed to include this dimension in the optimization process.

Like many areas of engineering, it must realized that system safety and reliability cannot be quantified with total certainty at the present time. Different assumptions are made in order to simplify the problem. Failure rate data, the primary input required for most methods, is not precisely specified or readily available. Precise failure rate data requires an extensive life test where operational conditions match expected usage.

Several factors prevent this testing. First of all, current control system components from quality vendors have achieved a general level of reliability that allows them to operate for many, many years. Precise life testing requires that units be operated until failure. The time required for this testing is far beyond the usefulness of the data (components are obsolete before the test is complete). Secondly, operational conditions vary significantly between control system installations. One site may have failure rates that are much higher than another site. Lastly, variations in usage will affect reliability of a component. This is especially true when design faults exist in a product. Design faults are probable in the complex components used in today's systems. Design faults, "bugs," are almost expected in complicated software.

In spite of the limitations of variability, imprecision, simplified assumptions, and different methods: rapid progress is being made in the area of safety and reliability evaluation. ISA standards committees are working in different areas of this field. SP84 has a committee working on methods of calculating system reliability. Several methods that utilize the tools covered in this book are proposed.

Software reliability has been the subject of intense research for over a decade. These efforts are beginning to show some results. This is important to the subject of control systems because of the explosive growth of software within these systems. Although software engineering techniques have provided better design fault avoidance methods, the growth has outstripped the improvements. Software reliability may well be the control system reliability crisis of the future.

Safety and reliability are important design constraints for control systems. When those involved in the system design share common vocabulary, understand evaluation methods, include all site variables and understand how to evaluate reliable software; then safety and reliability can become true design parameters. This is the goal.

William M. Goble
Perkasie, PA
January 1998

About the Book

The material for this book came from several sources. Over the years, the author has written many papers on safety and reliability topics. The preparation, presentation, and feedback from these papers has contributed greatly. Notes from the graduate course "SYS506, Engineering Reliability" taught by the author at the University of Pennsylvania and ISA's "ES35, Evaluating System Safety and Reliability" have provided the necessary organization and body. Student questions provided insight into material that needed more explanation.

Work assignments and engaging debate held in conjunction with ISA's SP84 committee also provided major input. The notation and methodology of ISA TR84.02 draft documentation was used. Research and debate held at Eindhoven University of Technology, Eindhoven, Netherlands has been a significant input. Other information came from corporate experiences in the development and manufacture of safety critical equipment.

The book is intended to serve a wide variety of users. Professional control system designers will be able to obtain an introduction and detailed background necessary to understand how to meet important new safety regulations and reliability engineering topics. This can help allow proper evaluation of control system components, evaluation of various system architectures, better communications with vendors, and means to increase accuracy of life cycle cost estimates. The book should also provide a breadth of information for reliability and safety engineers who would like information from the control systems industry.

The book is also intended to provide an introductory text for college courses in the reliability field. Many examples and detailed explanations should fulfill this need. The practical presentation may help introduce the issues of "engineering judgment," the difference between theory and real implementation. A final objective of the book is to provide background to those professionals who wish to quantitatively evaluate safety instrumented systems.

The book is organized in two sections. The first part of the book provides basic foundation material. Probability, statistics, reliability theory definitions, and basic reliability modeling techniques are included in this category. The second part of the book deals specifically with advanced topics relevant to safety instrumented systems and control systems. Readers currently knowledgeable in reliability theory may wish to skim part one to obtain notation background.

The science of reliability engineering continues to advance. Improvements in our modeling techniques will continue. Readers are advised to stay in touch via ISA's magazines, transactions, and technical proceedings.

About the Author

Dr. William M. Goble has more than 25 years of experience in analog and digital electronic circuit design, software development, engineering management and marketing. He is currently Principal Partner with exida.com, a knowledge company focused on automation safety and reliability.

He holds a B.S. in electrical engineering from Penn State and an M.S. in electrical engineering from Villanova. He has a PhD from Eindhoven University of Technology in Eindhoven, Netherlands from the Department of Mechanical Reliability and did research in methods of modeling the safety and reliability of programmable electronic systems. He is a Professional Engineer in the state of Pennsylvania and holds a Certified Functional Safety Expert certificate from TÜV.

He is a well known speaker and also teaches courses and consults on various reliability and safety engineering topics. He has authored or co-authored many technical papers and magazine articles primarily on software and hardware safety and reliability, quality improvement and quantitative modeling.

He serves on the adjunct faculty of the University of Pennsylvania teaching graduate courses in engineering reliability. He has developed and teaches courses on safety and reliability for ISA—The Instrumentation, Systems, and Automation Society where he is a Fellow member. He is also a member of IEEE and the ISA SP84 Committee.

1 Introduction

Control System Safety and Reliability

Safety and reliability have become essential parameters of automatic control systems design. Economic benefits of a safe and reliable system include less lost production, higher product quality, reduced maintenance costs, and lower risk costs. Additional benefits include regulatory compliance, the ability to schedule maintenance, and many others — including peace of mind and the satisfaction of a job well done.

Given the importance of safety and reliability, how are they achieved? How are they measured? A number of fundamental concepts are used to achieve high reliability and high safety: high-strength design, fault-tolerant design, on-line failure diagnostics, and high common-cause strength. All these important concepts will be developed in later chapters of the book.

Reliability and safety are measured using a number of well-defined parameters, including reliability, availability, MTTF (mean time to failure), RRF (risk reduction factor), PFD (probability of failure on demand), safety availability, and others. These terms have been developed over the last 50 years or so by the reliability and safety engineering community.

BPCS and SIS

Within the control system, the safety protection equipment is frequently separated from the control equipment. The control equipment is called the basic process control system (BPCS), and the protection equipment is called the safety instrumented system (SIS) (Ref. 1). The BPCS reads process sensors, does control and sequencing calculations, and commands actuation devices (typically valves or motors). The SIS reads sensors, does calculation and logic required to identify potentially dangerous conditions, and generates outputs to actuators designed to mitigate the dangerous situation. An SIS may protect personnel, equipment, the environment, or any combination of the three. Figure 1-1 shows these two categories of equipment.

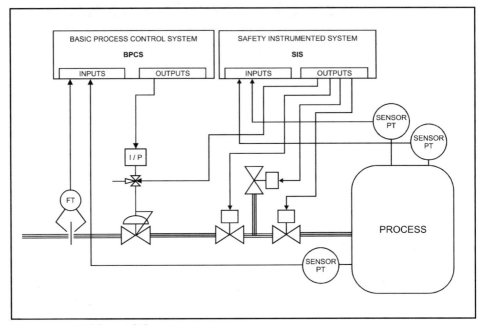

Figure 1-1. BPCS and SIS.

Within each category is a different emphasis from the reliability and safety perspective. The emphasis within the BPCS is on availability, the probability of successful operation at any moment in time. Availability is a measure of uptime. In the BPCS the maximum uptime is desired, since an unscheduled outage is a very expensive event in many processes. Failure modes (safe or dangerous) are a consideration, but availability comes first. The emphasis within the SIS is opposite. It must be designed to fail-safe. Availability is an important consideration, but safety must come first.

Standards

Several standards play an important role in the safety and reliability evaluation of control systems. Standards provide checklists of issues that should be addressed in qualitative evaluation, as well as performance measures against which reliability and safety calculations can be compared. Standards also provide explanation and examples of how systems can be designed to maximize safety and reliability.

ISA's S84.01 standard (Ref. 1), *Applications of Safety Instrumented Systems for the Process Industries*, provides guidance in a number of important areas. It describes the boundaries of the SIS and the BPCS, and lists definitions of terms and acronyms. The standard provides quantitative target goals for various levels of safety in a chart called "safety integrity levels." This important concept divides safety needs of a process into order-of-magnitude levels. S84.01 describes three safety integrity levels, as shown in Figure 1-2.

SAFETY INTEGRITY LEVEL	SAFETY AVAILABILITY	PROBABILITY OF FAILURE ON DEMAND (PFD)
3	0.999 - 0.9999	0.001 - 0.0001
2	0.99 - 0.999	0.01 - 0.001
1	0.9 - 0.99	0.1 - 0.01

Figure 1-2. Safety Integrity Levels from ISA S84.01.

S84.01 also describes the important concept of a "safety life cycle," a systematic design process that begins with conceptual process design and ends with SIS decommissioning. A simplified version of the safety life cycle chart is shown in Figure 1-3. The complete chart is presented in the standard. Note that safety requirements are verified in the sixth step. This verification can be done in a quantitative manner. The various reliability and safety metrics are calculated and compared with targets defined in the safety requirements. This book provides tools to accomplish that task.

The S84.01 standard includes many informative annexes that describe important design concepts and provide example system solutions. Example methods for determining the safety integrity level of a process are also provided. This standard is highly recommended to anyone involved in the design, evaluation, and operation of safety instrumented systems.

Another recent standard is the IEC61508 document (Ref. 2), a seven-part standard that describes safety instrumented systems from a general perspective. It also describes the safety integrity level concept, with four levels defined as shown in Figure 1-4. This standard is quite detailed, with specific equipment requirements for each safety integrity level.

Introduction

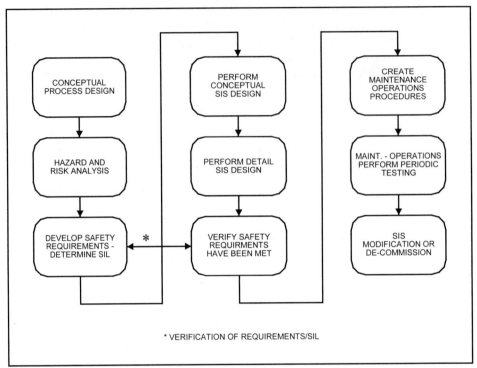

Figure 1-3. Simplified Safety Life Cycle.

SAFETY INTEGRITY LEVEL	PROBABILITY OF FAILURE ON DEMAND (PFD)	RISK REDUCTION FACTOR (RRF)
4	< 0.0001	> 10,000
3	0.001 - 0.0001	1,000 - 10,000
2	0.01 - 0.001	100 - 1,000
1	0.1 - 0.01	10 - 100

Figure 1-4. Safety Integrity Levels from IEC61508.

Reliability Engineering

The science of reliability engineering has developed a number of qualitative techniques that allow an engineer to understand system operation in the presence of a component failure. These techniques include failure modes and effects analysis (FMEA), qualitative fault tree analysis (FTA), and hazops (hazard and operational analysis). Other techniques based on probability theory and statistics allow the control engineer to quantitatively evaluate reliability and safety of control system designs. Reliability networks (sometimes called reliability block diagrams) and fault trees use combinational probability to evaluate system level probability of success, probability of safe failure, or probability of dangerous failure. Markov models, another popular technique, show system success and failure via circles called states. These techniques will be covered.

Life-cycle cost modeling may be the most useful technique of all to answer questions of cost and justification. Using this analysis tool, the output of a reliability analysis in the language of statistics is converted to the clearly understood language of money. It is frequently quite surprising how much money can be saved using reliable and safe equipment.

Reliability and safety engineering is built on a foundation of probability and statistics. However, a successful control system reliability evaluation depends just as much on control and safety systems knowledge. This knowledge includes an understanding of the components used in these systems, the component failure modes and their effect on the system, and the system failure modes and failure stress sources present in the system environment. Thus, logic, systems engineering, and some mathematics are combined to complete the tool set needed for reliability and safety evaluation. Real-world factors, including on-line diagnostic capability, repair times, software failures, human failures, common-cause failures, failure modes, and time-dependent failure rates, must be addressed in a complete analysis.

Perspective

The field of reliability and safety engineering is relatively new, with significant research having been driven by military needs in the mid-1940s. Introductory work in hardware reliability was done in conjunction with the German V2 rocket program, where innovations such as the 2oo3 voting scheme were invented. Human reliability research began with American studies done on radar operators and gunners during World War II. Military systems were among the first to reach complexity levels at which reliability engineering became important. Methods were needed to answer important questions, such as "Which configuration is more reliable on an airplane: four small engines or two large engines?"

Control systems and safety protection systems have also followed an evolutionary path toward greater complexity. Early control systems were simple; push buttons and solenoids, sight gages, thermometers, and dipsticks were typical control tools. Later, single-loop pneumatic controllers dominated. Most of these machines were not only inherently reliable, but many failed in "graceful," predictable ways. With a pneumatic system, when the air tubes leaked, the output went down. When the air filter clogged, the output went to zero. Diagnostics seemed inherent. When the hissing noise changed, a good technician could locate the problem. Safety protective systems were built from relays and sensing switches. With the addition of safety springs and special contacts, these devices would virtually always fail with the contacts open. Again, they were simple devices that were inherently reliable, with predictable (mostly) fail-safe failure modes.

The inevitable need for better processes eventually pushed control systems to a level of complexity at which sophisticated electronics became the optimal solution for control and safety protection. "Distributed" microcomputer-based controllers, introduced in the mid-1970s, offered economics, flexibility, and glamour. The level of complexity in control systems kept increasing, until programmable electronic systems became the standard. Systems today utilize a hierarchical collection of computers of all sizes — from microcomputer-based sensors to worldwide computer communication networks. Industrial control and safety instrumented systems are now among the most complex systems anywhere. These complex systems are the type that can benefit most from reliability engineering.

Control systems designers need answers to their questions. "Which control architecture gives the best reliability for the application?" "What combination of systems gives the lowest cost of ownership for the next five years?" "Should I use a personal computer to control our reactor?" "What architecture is needed to meet SIL3 safety requirements?" These questions are answered using quantitative reliability and safety analysis.

Qualitative versus Quantitative

Some experienced control system engineers have a healthy skepticism regarding quantitative safety and reliability engineering. This may come from the old quotation, "There are lies, damned lies, and statistics." Quantitative evaluation does utilize some statistical methods. Consequently, there will be uncertainty in results, as well as variation in actual results obtained from actual systems. Or, the controversy may stem from another famous quotation, "Garbage in, garbage out." Poor failure rate estimates and poor simplification assumptions can ruin the results of any reliability and safety evaluation. Good qualitative reliability engineering should be used to prevent "garbage" from entering the evaluation. Qualitative engineering provides the foundation for all quantitative work.

Quantitative safety and reliability evaluation is a growing science, with knowledge and techniques evolving each year. Despite variation and uncertainty, quantitative techniques can be valuable. Lord Kelvin stated, "... but when you cannot express it with numbers, your knowledge is of a meager and unsatisfactory kind; it may be the beginning of knowledge, but you have scarcely, in your thoughts, advanced to the stage of science, whatever the matter may be." These words apply to control systems safety and reliability.

Quantitative safety and reliability evaluation can add great depth and insight (Ref. 3). Sometimes intuition can be deceiving. After all, it was once intuitively obvious that the world was flat. Many aspects of probability and reliability can appear counterintuitive. The quantitative reliability evaluation either verifies the qualitative evaluation or adds substantially to it. Therein lies its value.

EXERCISES

1.1 Are methods used to determine safety integrity levels of a process presented in S84.01?

1.2 Is availability a measure of the uptime of a system?

1.3 Are safety integrity levels defined by order-of-magnitude quantitative numbers?

1.4 Can quantitative evaluation techniques be used to verify safety integrity requirements?

1.5 Should quantitative techniques be used exclusively to verify safety integrity?

REFERENCES AND BIBLIOGRAPHY

1. S84.01, *Applications of Safety Instrumented Systems for the Process Industries*, ISA—The Instrumentation, Systems, and Automation Society, 1996.

2. IEC61508, *Functional Safety — Safety Related Systems*, 1997.

3. Gruhn, P., "Safety System Risk Analysis Needs to Be More Quantitative," *Control*, Chicago: Putnam, August 1991.

4. Coppola, A., "Reliability Engineering of Electronic Equipment: A Historical Perspective," *IEEE Transactions of Reliability*, April 1984.

5. Barlow, R. E., "Mathematical Theory of Reliability: A Historical Perspective," *IEEE Transactions of Reliability*, April 1984.

2
Understanding Random Events

Random Variables

The concept of a random variable seems easy to understand, and yet many questions and statements indicate misunderstanding. A manager reads that, on average, an industrial boiler explodes every 15 years and knows that the unit in his plant has been running 14 years. He calls a safety engineer to determine how to avoid the explosion next year. This is clearly a misunderstanding. Boiler explosions and many other types of failure events are classified as random because only chances and averages can be predicted, not precise events. Predictions are based on statistical data gathered from a large number of sites. These statistical techniques are used because they offer the best information obtainable. A failure event cannot be precisely predicted.

The process of failure is like many other processes that have variations in outcome. Such variations cannot be predicted by substituting variables into a formula, because the exact formula or the variables involved may not be completely understood. These processes are called "random processes," primarily because they are not well characterized.

Some random variables can have only certain values. These random variables are called "discrete" random variables. Other variables can have a numerical result anywhere within a range. These are called "continuous" random variables. Statistics are used to gain some knowledge about these random variables and the processes that produce them.

Statistics

Statistics are usually based on data samples. Consider the case of a researcher who wants to understand how a computer program is being used. The researcher calls six computer program users at each of

20 different locations and asks exactly which program function is being used at that moment. The program functions are categorized as follows:

Category 1 — Editing functions (cut, copy, and paste)

Category 2 — Input functions (data entry)

Category 3 — Output functions (printing and formatting)

Category 4 — Disk functions

Category 5 — Check functions (spelling and grammar)

The results of the survey (sample data) are presented in Table 2-1. This is a list of data values.

	User 1	User 2	User 3	User 4	User 5	User 6
Site 1	1	2	3	2	4	1
Site 2	3	3	3	2	2	1
Site 3	2	2	1	3	3	2
Site 4	1	3	2	2	2	2
Site 5	2	1	4	3	3	2
Site 6	2	2	3	2	3	2
Site 7	1	2	2	3	2	1
Site 8	2	2	2	2	3	2
Site 9	3	3	2	2	2	4
Site 10	2	2	2	2	2	2
Site 11	2	5	2	2	3	2
Site 12	3	2	2	4	2	2
Site 13	5	2	1	2	2	3
Site 14	2	2	2	3	2	2
Site 15	3	2	2	4	2	2
Site 16	2	2	2	3	2	5
Site 17	2	3	1	2	2	3
Site 18	1	2	2	2	2	2
Site 19	2	2	3	2	3	1
Site 20	2	2	2	2	1	2

Table 2-1. Computer Program Function Usage

Histogram

One of the more common ways to organize data is the histogram. A histogram is a graph with data values on the horizontal axis and the quantity of samples for each value on the vertical axis. A histogram of data for Table 2-1 is shown in Figure 2-1.

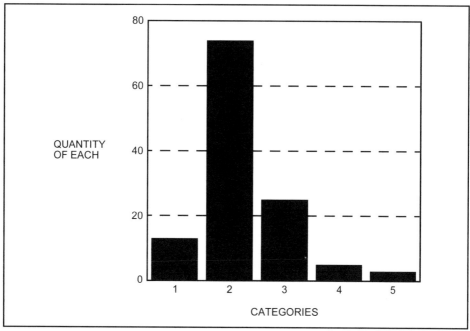

Figure 2-1. Computer Program Usage Histogram.

> **EXAMPLE 2-1**
>
> **Problem:** Assume that the computer usage survey results of Figure 2-1 are representative for all users. If another call is made to a user, what is the probability that the user will be using a function in category 5?
>
> **Solution:** The histogram shows that three answers from the total of 120 were within category 5. Therefore, the chances of getting an answer in category 5 are $\frac{3}{120}$, which equals 2.5%.

Probability Density Function

A probability density function (pdf) relates the value of a random variable with the probability of getting that value (or value range). For discrete random variables, the probability of getting each result is provided in a pdf. For continuous random variables, a pdf provides the probability of getting a random variable value from a range. The random variable values typically form the horizontal axis, and probability numbers (a range of 0 to 1) form the vertical axis.

A probability density function has the following properties:

$$f(x) \geq 0 \text{ for all } x \tag{2-1}$$

and

$$\sum_{i=1}^{n} P(x_i) = 1 \qquad (2\text{-}2)$$

for discrete random variables, or

$$\int_{-\infty}^{+\infty} f(x)dx = 1 \qquad (2\text{-}3)$$

for continuous random variables.

Figure 2-2 shows a pdf for a fair coin toss (a process with a discrete output). The chart shows that the probability of getting "heads" as a result is one-half. The probability of getting "tails" as a result is also one-half. Notice that all probabilities total one.

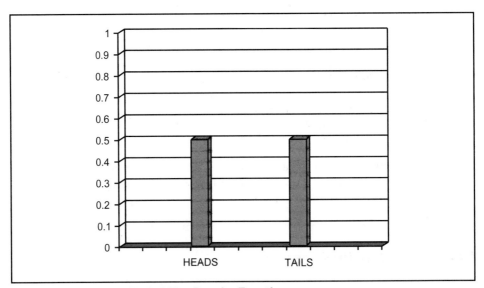

Figure 2-2. Coin Toss Probability Density Function.

Figure 2-3 shows a discrete pdf for the toss of a pair of fair dice. There are 36 possible outcomes. The probability of getting a result of seven is $\frac{6}{36}$, because there are six ways to get a seven. The probability of getting a result of two is $\frac{1}{36}$, because only one combination of the dice will give that result. Again, the probabilities total one.

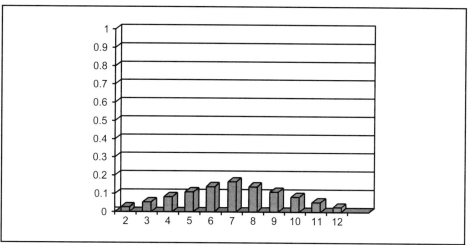

Figure 2-3. Dice Toss Probability Density Function.

For a continuous random variable, the pdf again shows probability as a function of random variable. But with a continuous random variable, the interest is in the probability of getting a process outcome within an interval, not a particular value. Figure 2-4 shows a continuous pdf for a process that can have an outcome between 100 and 110. All outcomes are equally likely.

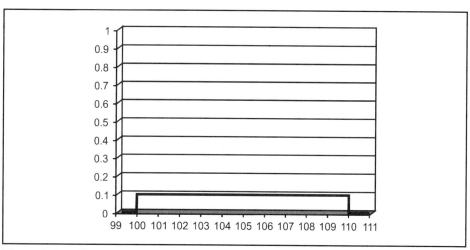

Figure 2-4. Uniform Distribution Probability Density Function.

Such distributions are called "uniform" distributions. The probability of getting an outcome within an interval is proportional to the length of the interval. For example, the probability of getting a result between 101.0 and 101.2 is 0.02 (0.2 interval length × 0.1 probability). As the length of the interval is reduced, the probability of getting a result within the interval drops. The probability of getting an exact particular value is zero.

For a continuous pdf, the probability of getting an outcome in the range from a to b equals

$$P(a \leq X \leq b) = \int_a^b f(x)dx \qquad (2\text{-}4)$$

EXAMPLE 2-2

Problem: A uniform pdf over the interval 0 to 1,000 for the random variable X is given by $f(x) = 0.001$, if $0 \leq X \leq 1000$, and 0 otherwise. What is the probability of an outcome in the range of 110 to 120?

Solution: Using Equation 2-4,

$$P(110 \leq X \leq 120) = \int_{110}^{120} 0.001\, dx = 0.001x \Big|_{110}^{120}$$

$$= 0.001(120 - 110) = 0.01$$

EXAMPLE 2-3

Problem: The pdf for the random variable T is given by

$$f(t) = kT,$$

if $0 \leq t \leq 4$;

0 otherwise.

What is the value of k?

Solution: Using Equation 2-3 and substituting for $f(t)$:

$$\int_0^4 kt\, dt = 1$$

which equals

$$\frac{kt^2}{2}\Big|_0^4 = 1$$

Evaluating,

$$\frac{16k}{2} = 1, \therefore k = 0.125$$

EXAMPLE 2-4

Problem: A random variable T has a pdf of $f(t) = 0.01e^{-0.01t}$ if $t \geq 0$, and 0 if $t < 0$. What is the probability of getting a result when T is between 0 and 50?

Solution: Using Equation 2-4,

$$P(0 \leq t \leq 50) = \int_{t=0}^{t=50} 0.01 e^{-0.01t} \, dt$$

Integrating by substitution ($u = -0.01t$), we obtain

$$P(0 \leq t \leq 50) = \int_{t=0}^{t=50} 0.01 e^{-0.01t} \, dt$$

$$= \int_{u=0}^{u=-0.5} 0.01 e^u \frac{du}{-0.01}$$

Simplifying and converting back to t,

$$\int_{u=0}^{u=-0.5} e^u \, du = -e^u \Big|_0^{-0.5} = -e^{-0.01t} \Big|_0^{50}$$

Evaluating,

$$= [-e^{-0.01(50)}] - [-e^{-0.01(0)}]$$

$$= -0.60653 + 1 = 0.39347$$

Cumulative Distribution Function

A cumulative distribution function (cdf) is defined by

$$F(x_n) = \sum_{i=1}^{J} P(x_i) \qquad (2\text{-}5)$$

for discrete random variables, or

$$F(x) = \int_{-\infty}^{x} f(x) \, dx \qquad (2\text{-}6)$$

for continuous random variables. A cdf is often represented by the capital letter F. The cumulative density function represents the cumulative probability of getting a result in the interval from $-\infty$ to the present value of x. It equals the area under the pdf curve accumulated up to the present value.

A cdf has the following properties:

$$F(\infty) = 1, F(-\infty) = 0, \text{ and } \frac{dF}{dx} \geq 0$$

The probability of a random process result within an interval can be obtained using a cdf.

It can be shown that

$$P(a \leq X \leq b) = \int_a^b f(x)dx = F(b) - F(a) \qquad (2\text{-}7)$$

This means that the area under the pdf between a and b is simply the cdf area to the left of b minus the cdf area to the left of a.

EXAMPLE 2-5

Problem: Calculate and graph the discrete cdf for a dice toss process where the pdf is as shown in Figure 2-3.

Solution: A cdf is the sum of previous probabilities at each discrete variable. Table 2-2 shows the sum of the probabilities for each result. The cdf is plotted in Figure 2-5.

Result	Probability	Cumulative Probability
2	0.0278	0.0278
3	0.0556	0.0834
4	0.0834	0.1668
5	0.1111	0.2779
6	0.1388	0.4167
7	0.1666	0.5833
8	0.1388	0.7221
9	0.1111	0.8332
10	0.0834	0.9166
11	0.0556	0.9722
12	0.0278	1.000

Table 2-2. Sum of Probabilities (Example 2-4)

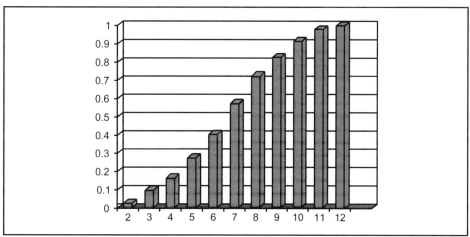

Figure 2-5. Dice Toss Cumulative Distribution Function.

EXAMPLE 2-6

Problem: Calculate and plot the cdf for the pdf shown in Figure 2-4.

Solution: During the interval from 100 to 110, the pdf equals a constant, 0.1. The pdf equals zero for other values of x. Therefore, using Equation 2-6,

$$F(x) = \int_{-\infty}^{x} 0.1\, dx = 0.1x \text{ for } 100 \leq x \leq 110$$

The cdf is zero for values of x less than 100. The cdf is one for values of x greater than 110. This cdf is plotted in Figure 2-6.

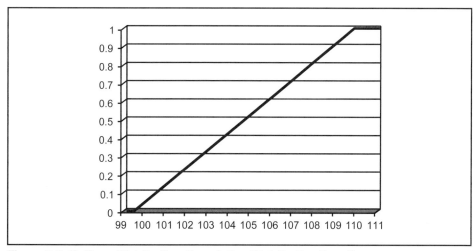

Figure 2-6. Uniform Distribution Cumulative Distribution Function.

Mean

The average value of a random variable is called the "mean" or "expected value" of a distribution. For discrete random variables, the equation for mean, E, is

$$E(x_i) = \bar{x} = \sum_{i=1}^{n} x_i P(x_i) \qquad (2\text{-}8)$$

The x_i in this equation refers to each discrete variable outcome. The n refers to the number of different possible outcomes. $P(x_i)$ represents the probability of each discrete variable outcome.

EXAMPLE 2-7

Problem: Calculate the mean value of a dice toss process.

Solution: Using Equation 2-8,

$$E(x) = 2\left(\frac{1}{36}\right) + 3\left(\frac{2}{36}\right) + 4\left(\frac{3}{36}\right) + 5\left(\frac{4}{36}\right)$$
$$+ 6\left(\frac{5}{36}\right) + 7\left(\frac{6}{36}\right) + 8\left(\frac{5}{36}\right) + 9\left(\frac{4}{36}\right)$$
$$+ 10\left(\frac{3}{36}\right) + 11\left(\frac{2}{36}\right) + 12\left(\frac{1}{36}\right)$$
$$= \frac{252}{36} = 7$$

Another more common form of "mean" is given by Equation 2-9:

$$\bar{x} = \frac{\sum_{i=1}^{n} x_i}{n} \qquad (2\text{-}9)$$

where x_i refers to each data sample in a collection of n samples. This is called the "sample mean." Given a large number of data samples, Equation 2-9 will yield results similar to Equation 2-8. As the number of samples grows larger, the results obtained from Equation 2-9 will get closer to the results obtained from Equation 2-8. When a large number of samples are used, the probabilities are inherently included in the calculation.

Understanding Random Events

> **EXAMPLE 2-8**
>
> **Problem:** A pair of fair dice is tossed 10 times. The results are 7, 2, 4, 6, 10, 12, 3, 5, 4, and 2. What is the sample mean?
>
> **Solution:** Using Equation 2-9, the samples are added and divided by the number of samples. The answer is 5.5. Compare this to the answer of seven obtained in Example 2-7. The difference of 1.5 is large but not unreasonable since the sample size of 10 is small.

> **EXAMPLE 2-9**
>
> **Problem:** The dice toss experiment is continued. The next 90 samples are 6, 8, 5, 7, 9, 4, 7, 11, 3, 6, 10, 7, 4, 6, 4, 8, 8, 5, 9, 3, 2, 11, 9, 8, 2, 9, 5, 11, 12, 8, 9, 7, 10, 6, 6, 8, 10, 7, 8, 11, 6, 7, 9, 8, 7, 8, 7, 10, 7, 6, 8, 7, 9, 11, 7, 7, 12, 7, 7, 9, 5, 6, 3, 7, 6, 10, 8, 7, 8, 6, 10, 7, 11, 6, 9, 11, 5, 8, 7, 4, 6, 8, 9, 7, 10, 9, 11, 7, 4, and 7. What is the sample mean?
>
> **Solution:** Using Equation 2-9, the samples are added and divided by the number of samples. The total of all 100 samples (including those from Example 2-8) is 725. Dividing this by the number of samples yields a sample mean result of 7.25. This result is closer to the distribution mean obtained in Example 2-7.

For continuous random variables, the equation for mean value is

$$E(x) = \int_{-\infty}^{\infty} x f(x) dx \qquad (2\text{-}10)$$

> **EXAMPLE 2-10**
>
> **Problem:** The pdf for a random variable is given by $f(t) = 1/8t$, if $0 \leq t \leq 4$; otherwise, it is equal to zero. What is the mean?
>
> **Solution:** Using Equation 2-10,
>
> $$E(t) = \int_0^4 t \frac{1}{8} t \, dt$$

> **EXAMPLE 2-10 (continued)**
>
> Integrating,
>
> $$E(t) = \frac{1}{8}\frac{t^3}{3}\Big|_0^4 = \frac{64}{24}$$
>
> The mean represents the "center of mass" of a distribution. It takes into account different densities. This example shows this attribute. The distribution spans values from zero to four. The mean is $\frac{64}{24}$ (2.667). This value is greater than the middle value of two.

Median

The median value of a sample data set is a measure of center. The median value is defined as the data sample where there are an equal number of samples of greater value and lesser value. It is the "middle" value in an ordered set. If there are two middle values, it is the value halfway between those two values.

Knowing both the mean and the median allows one to evaluate symmetry. In a perfectly symmetric pdf, like a normal distribution, the mean and the median are the same. In an asymmetric pdf, like the exponential, the mean and the median are different.

> **EXAMPLE 2-11**
>
> **Problem:** Find the median value of a dice toss process.
>
> **Solution:** The set of possible values are 2, 3, 4, 5, 6, 7, 8, 9, 10, 11, and 12. Since there are 11 values, the middle value would be the sixth. Counting shows the median value to be seven.

> **EXAMPLE 2-12**
>
> **Problem:** Find the median value of the sample data presented in Example 2-8.
>
> **Solution:** Sorting the 10 values provides 2, 2, 3, 4, 4, 5, 6, 7, 10, and 12. Two middle values are present, a four and a five. Therefore, the median is halfway between these two at 4.5.

Variance

The mean is a measure of the "center of mass" of a distribution. Knowing the center is valuable, but more must be known. Another important characteristic of a distribution is spread, or the amount of variation. A process under control will be consistent and will have little variation. A measure of variation is important for control purposes (Refs. 1 and 2).

One way to measure variation is to subtract values from the mean. When this is done, the question "How far from the center are the data?" can be answered mathematically. For each discrete data point, calculate

$$(x_i - \bar{x}) \tag{2-11}$$

In order to calculate variance, these differences are squared to obtain all positive numbers and then averaged. When there is a given discrete probability distribution, the formula for variance is

$$\sigma^2(x_i) = \sum_{i=1}^{n} (x_i - \bar{x})^2 P(x_i) \tag{2-12}$$

where x_i refers to each discrete outcome and $P(x_i)$ refers to the probability of realizing each discrete outcome. If a set of sample data is given, the formula for sample variance is

$$s^2(x_i) = \frac{\sum_{i=1}^{n}(x_i - \bar{x})^2}{n} \tag{2-13}$$

where x_i refers to each piece of data in the set. The size of the data set is given by n. As in the case of the sample mean, the sample variance will provide a result similar to the actual variance as the sample size grows larger. However, there is no guarantee that these numbers will be equal. For this reason, the sample variance is sometimes called an "estimator" variance.

For continuous random variables, the variance is given by

$$\sigma^2(x) = \int_{-\infty}^{\infty} (x - \bar{x})^2 f(x) dx \tag{2-14}$$

Standard Deviation

The formula for standard deviation is

$$\sigma(x) = \sqrt{\sigma^2(x)} \qquad (2\text{-}15)$$

Standard deviation is often assigned the lowercase Greek letter sigma (σ). It is a measure of "spread" commonly associated with the normal distribution.

EXAMPLE 2-13

Problem: What is the variance of a dice toss process?

Solution: In Example 2-6, the mean of a dice toss process was calculated as seven. Using Equation 2-12,

$$\sigma^2(x_i) = (2-7)^2 \frac{1}{36} + (3-7)^2 \frac{2}{36} + (4-7)^2 \frac{3}{36} + (5-7)^2 \frac{4}{36}$$
$$+ (6-7)^2 \frac{5}{36} + (7-7)^2 \frac{6}{36} + (8-7)^2 \frac{5}{36} + (9-7)^2 \frac{4}{36}$$
$$+ (10-7)^2 \frac{3}{36} + (11-7)^2 \frac{2}{36} + (12-7)^2 \frac{1}{36}$$
$$= 5.834$$

EXAMPLE 2-14

Problem: What is the sample variance of the dice toss data presented in Example 2-9?

Solution: Using Equation 2-13, the sample mean of 7.25 is subtracted from each data sample. This difference is squared. The squared differences are summed and divided by 100 on a PC spreadsheet. The answer is 5.9075. This result is comparable to the value obtained in Example 2-13.

Common Distributions

Several well-known distributions play an important role in reliability engineering. The exponential distribution is widely used to represent the probability of component failure. This is a direct result of a constant failure rate assumption. The normal distribution is used in many areas of science. In reliability engineering it is used to represent strength distributions. It is also used to represent stress. The lognormal distribution is used to model repair probabilities.

Exponential Distribution

The exponential distribution is commonly used in the field of reliability. In its general form it is written

$$f(x) = ke^{-kx}, \text{ for } x \geq 0; \qquad (2\text{-}16)$$
$$= 0, \text{ for } x < 0$$

The cumulative distribution function is

$$F(x) = 0, \text{ for } x < 0; \quad = 1 - e^{-kx}, \text{ for } x \geq 0 \qquad (2\text{-}17)$$

The equation is valid for values of $k > 0$. The cdf will reach the value of one at $x = \infty$ (which is expected). Figure 2-7 shows a plot of an exponential distribution pdf and cdf where $k = 0.6$. Note: $\text{pdf}(x) = d[\text{cdf}(x)]/dx$.

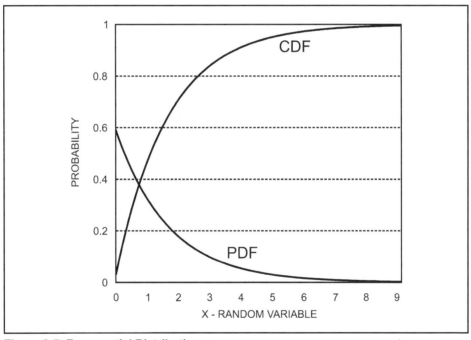

Figure 2-7. Exponential Distribution.

Normal Distribution

The normal distribution is the most well-known and widely used probability distribution in general science fields. It is so well known because it applies (or seems to apply) to so many processes. In reliability engineering, it primarily applies to measurements of product strength and external stress.

The pdf is perfectly symmetric about the mean. The "spread" is measured by variance. The larger the value, the flatter the distribution. Figure 2-8 shows a normal distribution pdf and cdf where the mean equals four and the standard deviation equals one. Because the pdf is perfectly symmetric, the cdf always equals one-half at the mean value of the pdf. The pdf is given by

$$f(x) = \frac{1}{\sigma\sqrt{2\pi}} e^{\left[\frac{-(x-\mu)^2}{2\sigma^2}\right]} \qquad (2\text{-}18)$$

where μ is the mean value and σ is the standard deviation. The cdf is given by

$$F(x) = \frac{1}{\sigma\sqrt{2\pi}} \int_{-\infty}^{x} e^{\left[\frac{-(x-\mu)^2}{2\sigma^2}\right]} \qquad (2\text{-}19)$$

There is no closed-form solution for Equation 2-19; that is, a formula cannot be written in the variable x that will give the value for $F(x)$. Fortunately, simple numerical techniques are available and standard tables exist to provide needed answers.

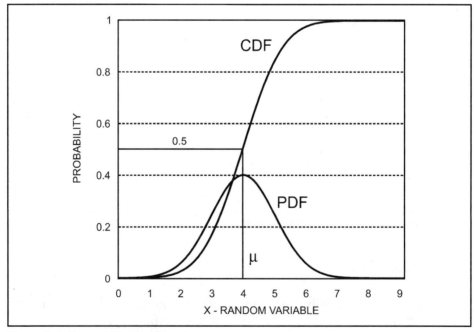

Figure 2-8. Normal Distribution.

Standard Normal Distribution

If

$$z = \frac{(x - \mu)}{\sigma} \quad (2\text{-}20)$$

is substituted into the normal pdf function,

$$f(z) = \frac{1}{\sqrt{2\pi}} e^{-\left(\frac{z^2}{2}\right)} \quad (2\text{-}21)$$

This is a normal distribution that has a mean value of zero and a standard deviation of one. Tables can be generated (Appendix A and Ref. 2, Table 1.1.12) showing the numerical values of the cumulative distribution function. These are done for different values of z. Any normal distribution with any particular σ and μ can be translated into a standard normal distribution by scaling its variable x into z using Equation 2-20. Through the use of these techniques, numerical probabilities can be obtained for any range of values for any normal distribution.

EXAMPLE 2-15

Problem: A control system will fail if the ambient temperature exceeds 70°C. The ambient temperature follows a normal distribution with a mean of 40°C and a standard deviation of 10°C. What is the probability of failure?

Solution: The probability of getting a temperature above 70°C must be found. Using Equation 2-20,

$$z = \frac{(70 - 40)}{10} = 3$$

Checking the chart in Appendix A for a value of 3.00, the left column (0.00 adder) across from the three, a value of 0.99865 is found. This means that 99.865% of the area under the pdf occurs before the temperature of 70°C.

$$P(T \leq 70) = 0.99865$$

To obtain a probability for temperatures greater than 70°C, use the rule of complementary events (see Appendix C). Since the system will fail as soon as the temperature goes above 70°C, this is the probability of failure.

$$P(T > 70) = 1 - 0.99865 = 0.00135$$

This is shown in Figure 2-9.

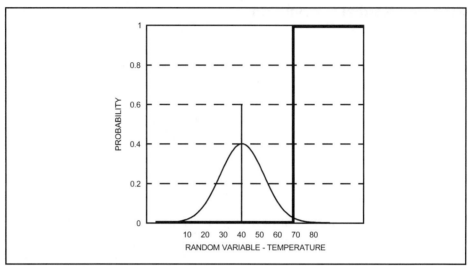

Figure 2-9. Example 2-15 Distribution.

Lognormal Distribution

The random variable X has a lognormal pdf if $\ln X$ has a normal distribution. Thus, the lognormal distribution is related to the normal distribution and also has two parameters.

The lognormal distribution has been used to model probability of completion for many human activities, including "repair time." It has also been used in reliability engineering to model uncertainty in failure rate information. Figure 2-10 shows the pdf and cdf of a lognormal distribution with a μ of one-tenth and a σ of one. It should be noted that these are not the mean and standard deviations of the distribution.

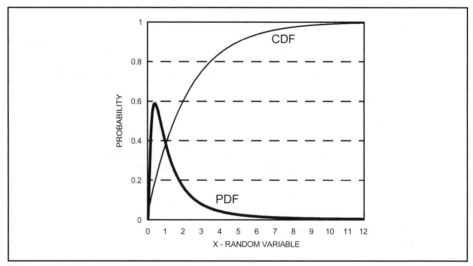

Figure 2-10. Lognormal Distribution.

The pdf is given by

$$f(t) = \frac{1}{t\sigma\sqrt{2\pi}} e^{\left[\frac{-(\ln t - \mu)^2}{(2\sigma^2)}\right]}, \text{ for } t \geq 0; \quad (2\text{-}22)$$

$$= 0, \text{ for } t < 0$$

EXERCISES

2.1 A failure log is provided. Create a histogram of failure categories: software failure, hardware failure, operations failure, and maintenance failure.

Symptom	Failure Category
1. PC crashed	Software
2. Fuse blown	Hardware
3. Wire shorted during repair	Maintenance
4. Program stopped during use	Software
5. Unplugged wrong module	Maintenance
6. Invalid display	Software
7. Scan time overrun during game	Software
8. Water in cabinet	Maintenance
9. Wrong command	Operations
10. General protection fault	Software
11. Cursor disappeared	Software
12. Coffee spilled	Hardware
13. Dust on thermocouple terminal	Hardware
14. Memory-full error	Software
15. Mouse nest between circuits	Hardware
16. PLC crashed after download	Software
17. Output switch shorted	Hardware
18. LAN addresses duplicated	Maintenance
19. Computer memory chip failed	Hardware
20. Power supply failed	Hardware
21. Sensor switch shorted	Hardware
22. Valve stuck open	Hardware
23. Bypass wire left in place	Maintenance
24. PC virus caused crash	Software
25. Hard disk failure	Hardware

FAILURE LOG: Ten sites

Draw a Venn diagram (see Appendix C for explanation) of the failure sources.

2.2   Based on the histogram of Exercise 2.1, what is the chance of the next failure being a software failure?

2.3   A discrete random variable has the possible outcomes and probabilities listed below.

Outcome	Probability
1	0.1
2	0.2
3	0.5
4	0.1
5	0.1

Plot the pdf and the cdf.

2.4   What is the mean in Exercise 2.3?

2.5   What is the variance in Exercise 2.3?

2.6   The following set of numbers is provided from an accelerated life test in which power supplies were run until failure (days until failure). What is the mean of these data?

| 75 | 95 | 110 | 112 | 121 | 125 | 140 | 174 | 183 |

| 228 | 250 | 342 | 554 | 671 | 823 | 1065 | 1289 | 1346 |

What is the average failure time in hours? What is the median failure time in hours?

2.7   A controller will fail when the temperature gets above 80°C. The ambient temperature follows a normal distribution, with a mean of 40°C and a standard deviation of 10°C. What is the probability of failure?

REFERENCES AND BIBLIOGRAPHY

1. Goble, W. M., and Tucker, T. W., "Distributed Statistical Process Control — A High Performance Solution for Product Quality," *Proceedings of the Third Multi-National Instrumentation Conference*, Miconex '88, Beijing, China, 1988.

2. Avallone, E. A., and Baumeister III, T. (eds.), *Mark's Standard Handbook for Mechanical Engineers*, 9th Ed., New York: McGraw-Hill, 1986.

3. Goble, E. R., "Probability of Penny Flipping," *Pennridge High School Science Fair Report*, 1991.

4. Haeussler, E. F., Jr., and Paul, R. S., *Introductory Mathematical Analysis*, Englewood Cliffs, NJ: Prentice-Hall, 1987.

5. Dovich, R. A., *Reliability Statistics*, Milwaukee: ASQC Quality Press, 1990.

6. Neter, J., Wasserman, W., and Whitmore, G. A., *Applied Statistics*, Boston: Allyn and Bacon, 1978.

7. Marsh, C. E., and Tucker, T. W., "Application of SPC Techniques to Batch Units," *ISA Transactions*, Vol. 30, No. 1, 1991.

8. Mamzic, C. L., *Statistical Process Control*, Research Triangle Park, NC: ISA—The Instrumentation, Systems, and Automation Society, 1996.

3
Failures: Stress versus Strength

Failures

A failure occurs when a device (a system, a unit, a module, or a component) fails to perform its intended function. Control systems fail. Everyone understands that this potentially expensive and potentially dangerous event can occur. To prevent control system failures, the causes of failure are studied. When failure causes are understood, system designs can be improved.

To obtain sufficient depth of understanding, all levels in the system must be examined. For safety and reliability analysis purposes, four levels are defined. A "system" is built from "units." If the system is redundant, multiple units are used. Nonredundant systems use a single unit. Units are built from "modules," and modules are built from "components" (see Figure 3-1).

Many real control systems are constructed in just this manner. Although system construction could be defined in other ways, these levels are optimal for use in safety and reliability analysis, especially the analysis of fault-tolerant systems. These terms will be used throughout the book.

Failure Categorization

A group of control system suppliers and users once brainstormed at an ISA standards meeting on the subject of failures. A listing of failure sources and failure types resulted:

- Humidity
- Software bugs
- Temperature
- Power glitches

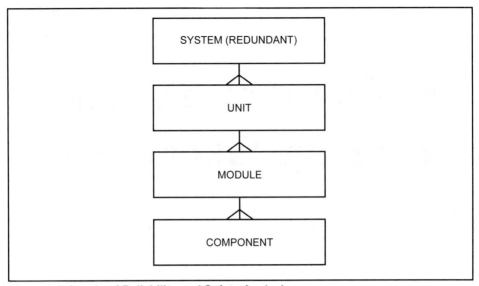

Figure 3-1. Levels of Reliability and Safety Analysis.

- System design errors
- Electrostatic discharge (ESD)
- Battery wearout
- Broken wires
- Corrosion-induced open circuits
- Random component failures
- Repairman error
- Radio frequency interference (RFI)
- Operator closing wrong switch
- Vibration
- Improper grounding
- Wrong configuration loaded
- Incorrect replacement
- Component made incorrectly
- Wrong software version installed

This list is quite diverse and includes a number of different failure types, as well as failure sources, which need to be sorted out. Many different factors stress a system and cause failure. Failures are categorized so that failure data can be organized in a consistent way in order to learn about better system design.

In the study of system failures, two categories emerge: physical failures and functional failures (Figure 3-2). Physical failures are often called random failures, and in many cases only physical failures are considered true failures. A physical failure happens when a component or components within a module fail. Physical failures are relatively well understood. Databases are kept on physical failures by the military (MIL-STD-217, Ref. 1) and by industry groups (Refs. 2-5). Life test results by component manufacturers provide physical failure data on new components (Refs. 6 and 7).

Other failures are called functional failures (or systematic failures). A functional failure occurs when the system is capable of operating but does not perform its intended function. An example of this is the common "software crash." A piece of data is entered into a personal computer and, when the return key is pressed, the computer simply quits. The computer has failed to accomplish its intended function, but no component has failed — no physical failure has occurred. The computer restarts properly after the "reset" button is pressed. The failure may or may not even appear repeatable depending on the data entered.

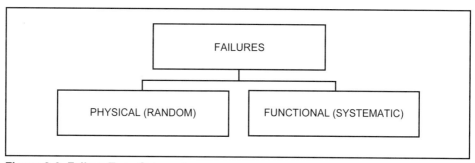

Figure 3-2. Failure Type Categories.

Most functional failures are a result of design faults. Functional failures may be permanent or may be transient in nature. Failure rate data for functional failures are currently difficult to obtain; however, future possibilities exist. Uncensored failure logs from system sites have been suggested as a good source for these data. Newer computer-based systems have automatic diagnostic logs that record error events as a function of time. These promise to provide additional information about functional failure rates in the future.

Both physical and functional failures have attributes that are important to control system safety and reliability analysis. This failure information is needed to help determine how to prevent future failures. Information is recorded about the failure source and the effect on the controller function. The term "failure source" is used to represent the cause of a failure — the "cause of death." This information should include all the elements that have stressed the system. The failure source may be something that causes an increase in "stress," or it may be something that causes a decrease in "strength." Table 3-1 shows the failure classifications of our previous failure source/failure type list.

Item	Class
Humidity	Failure source
Software bugs	Functional failure
Temperature	Failure source
Power glitches	Failure source
System design errors	Functional failure
Electrostatic discharge (ESD)	Failure source
Battery wearout	Physical failure
Broken wires	Physical failure
Corrosion-induced open circuits	Physical failure
Random component failures	Physical failure
Repairman error	Failure source
Radio frequency interference (RFI)	Failure source
Operator closing wrong switch	Functional failure
Vibration	Failure source
Improper grounding	Failure source
Wrong configuration loaded	Functional failure
Incorrect replacement	Functional failure
Component made incorrectly	Failure source
Wrong software version installed	Functional failure

Table 3-1. Classification Listing

Physical Failures

A physical failure (sometimes called a random failure) is almost always permanent and attributable to some component or module. For example, a system that consists of a single-board controller module fails. The controller output de-energizes and no longer supplies current to a solenoid valve. The controller diagnostics identify a bad output transistor component. An "autopsy" of the output transistor shows that it would not conduct current, and that it failed with an open circuit.

The failure occurred because a thin bond wire inside the transistor melted. Plant operations reports a nearby lightning strike just before the controller module failed. Lightning causes an impulse of electrical stress that can exceed the strength of a transistor. It should be noted that lightning is considered a random event. Many failure classification schemes use the term "random failure" instead of "physical failure" because the stress events are generally random.

For this failure:

Title	Module de-energized
Root cause	Output transistor failed; melted bond wire
Failure type	Physical
Primary stressor	Electrical surge
Secondary stressors	Temperature

In another example of a physical failure, a system failed because a power supply module failed with no output. An examination of the power supply showed that it had an open electrolytic capacitor (a capacitor with a value far below its rating, which is virtually the same as an open circuit). The capacitor autopsy showed that the electrolyte had evaporated. Without electrolyte, these capacitors cannot hold a charge and become open circuits.

Evaporation is a process that occurs over time. Eventually, all electrolyte in an electrolytic capacitor will evaporate and the capacitor will wear out. Electrolyte evaporation can be viewed as a strength reduction process. In time, the strength approaches normal stress levels and the component fails. For this failure:

Title	Power supply failed
Root cause	Electrolytic capacitor failed (evaporation)
Failure type	Physical
Primary Stressor	Electrical voltage, current
Secondary Stressors	Temperature, weak case design

Functional Failures

If all the physical components of a system are working and yet the system fails to perform its function, that failure is classified as "functional." Functional failures have the same attributes as physical failures. Each failure has a failure source and an effect on the controller function. The failure sources are almost always design faults, although sometimes a maintenance or an installation error causes a functional failure.

The exact source of a functional failure can be obscure. Often the failure will occur when the system is asked to perform some unusual function, or perhaps the system receives some combination of input data that was never tested. Some of the most obscure failures involve combinations of stored information, elapsed time, input data, and function performed. The ability to resist such stress may be considered a measure of "strength" in the design.

For example, a computer-based control system de-energized its outputs whenever a tuning parameter was set to a value below 0.0001 and the "ramping" option was enabled. The problem was traced to a design fault in the software that occurred only when a particular program path was executed by the computer. As you might guess, that particular sequence of computer program is executed only when the tuning parameter value is below 0.0001 and the ramping option enabled. The testing done by the software designers did not include that particular combination. For this failure:

Title	Output module de-energized
Root cause	Software fault
Failure type	Functional
Primary stressor	Parameter below 0.0001
Secondary stressors	Ramping option enabled

In another example, a control system did not properly annunciate (notify operator about) a high-pressure prealarm. An expensive rupture disk on a tank blew open as a result. The failure autopsy showed that a two-wire (4-20 mA) pressure transmitter could not generate a full-scale signal. The current loop ran out of compliance voltage because wire resistance in the current loop was too high. The system designer specified 28-gage wire instead of the proper lower-resistance 16-gage wire. This is a functional failure since no physical component failed. For this failure:

Title	Prealarm annunciation failure
Root cause	Wrong wire size
Failure type	Functional
Primary stressor	Design fault
Secondary stressors	Electrical current

In a well-publicized functional failure of the U.S. telephone network (Ref. 8), 20 million customers lost service for several hours. When the number of messages (the stress) received by a packet switch exceeded the quantity that could be stored in a memory buffer (the strength), the switch would shut itself down instead of discarding excess messages. Messages were then routed to other switches that became overloaded, cascading the problem across wide regions. The problem was traced to a line of software code. The software was changed to improve the design.

For this failure:

Title	Telephone network failed
Root cause	Software fault; memory buffer too small
Failure type	Functional
Primary stressor	Number of messages
Secondary stressors	None identified

Failure Sources

Many different factors can cause failure, individually or in combination. It is important to keep a broad, open perspective.

Consider humidity. Does the presence of humidity cause failures? No, failures are not normally the direct result of humidity. The specifications of many electronic controller modules state that they will work when the humidity is between 10% and 90%, noncondensing. However, humidity does accelerate the chemical process of corrosion. A badly corroded electrical contact will eventually fail. Is humidity the cause of the failure? No, corrosion is the source of stress that causes the failure. Humidity merely accelerated the process. But consider the case in which the humidity *exceeds the rated specification*. Water is dripping into the electronic controller and it fails; in this case, "humidity" is clearly the failure source. The issues can be confusing. A categorization scheme for failure sources will help clear the picture.

Failure sources are both internal and external to the system (see Figure 3-3). Internal failure sources typically result in decreased strength, and include design faults (product) and manufacturing faults (process). The faults can occur at any level: component, module, unit, or system. External failure sources increase stress, and include environmental sources, maintenance faults, and operational faults.

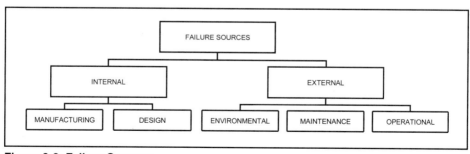

Figure 3-3. Failure Sources.

Internal — Design

Design faults, which can cause system failures, are a major source of functional failures. Many case histories are recorded (Ref. 9). In some cases the designer did not understand how the system would be used. Sometimes the designer did not anticipate all possible input conditions. Different designers working on different parts of the system may not completely understand the whole system. These symptoms may apply to component design, module design, unit design, or system design.

If the designer does not understand the environment (stress levels) and the intended usage of the system, design faults (insufficient strength) are more likely. Consider the example of a microcomputer-based controller module. Occasionally the controller module stops working — it freezes. After a short time, an independent "watchdog" timer circuit within the module resets the microprocessor. The module then restarts normal operation. The failure is transient. Obviously, no physical failure has occurred.

After extensive troubleshooting, the autopsy report shows that at high temperatures a logic gate occasionally responds to a noise pulse and clocks the microprocessor too rapidly. The microprocessor misinterprets an instruction and stops operating. Although environmental noise (a random electrical stress) triggers the failure, a design error has caused this functional failure. The designer did not understand the environment and intended usage of the logic circuit. The design was too weak, with not enough safety factor. For this failure:

Title	Transient controller module failure
Root cause	Hardware design fault; gate noise
Failure type	Functional
Primary stressor	Electrical noise
Secondary stressors	Temperature

It is difficult for designers to anticipate all possible input conditions. In a well-publicized case, an early computer system (NORAD, North American Air Defense, October 5, 1960), designed to detect missiles heading for the United States, signaled an alarm. The system had been in operation less than a month. It was a false alarm; the BMEWS radar in Thule, Greenland, had detected the rising moon. The designers did not anticipate this input. Reports indicate that this design was "strengthened" shortly after the failure occurred via a redesign. This is characteristic of design faults that are properly repaired: strength increases.

This failure indicates:

Title	False alarm on moon
Root cause	Software fault; unanticipated input
Failure type	Functional
Primary stressor	Moon rising signal from sensor
Secondary stressors	None identified

The primary defense against a design fault is a careful review by qualified experts (Refs. 10-12). Various design review techniques are in use, including an "active review" process where a second party (not the designer) describes the design to a group of reviewers. Other techniques involve the use of experienced personnel who specifically look for design faults.

Internal — Manufacture

Manufacturing faults occur when some step in the process of manufacturing a component, module, unit, or system is done incorrectly. For example, a controller module failed when it no longer supplied current to a valve positioner. It was discovered that a resistor in the output circuit had a corroded wire that no longer conducted electricity. This was a surprise since the resistor wire is coated to protect against corrosion. Closer examination showed that the coating was not applied correctly when the resistor was made. Small voids in the coating occurred because the coating machine was not cleaned on schedule. A manufacturing fault of this type lowers the strength of a device (component, module, unit, or system). For this failure:

Title	Output module failed
Root cause	Corroded wire on resistor component
Failure type	Physical
Primary stressor	Corrosive atmosphere
Secondary stressors	Temperature, humidity, strength decreased by coating voids

Many manufacturing faults reduce product strength; they make the product weaker, more susceptible to failure stress sources. This significantly shortens product life. Fortunately, many of these faults can be detected by "accelerated testing" techniques. Such techniques include the use of elevated temperatures, temperature cycles (rapid hot and cold), elevated voltages, and/or high humidity. Manufacturers can use accelerated stress testing, often called "burn-in," on newly manufactured products. The technique will often cause the manufacturing faults to quickly become failures. This process serves to "weed out" many manufacturing defects, rapidly decreasing the failure rate due to manufacturing faults.

Other examples of manufacturing defects include bad solder joints on printed circuit boards, a component installed backwards, a missing part, or an overtorqued bolt. At the lowest level this can even include defects in raw material. An example is a crystalline imperfection in a silicon wafer used to make an integrated circuit.

External — Environmental

Control systems are used in industrial environments where many factors are present that help cause failures (Figure 3-4). Temperature can cause direct failures. For example, a PC-based industrial computer module that performs a data logging function is mounted in a shed. One hot day the temperature inside the shed reaches 65°C. The jacket of the floppy disk melts and jams the disk drive. For this failure:

Title	PC controller floppy disk fails
Root cause	Floppy disk got too hot
Failure type	Physical
Primary stressor	Temperature
Secondary stressors	None identified

Temperature also helps speed along many chemical processes that cause failure, including corrosion, diffusion, oxidation, and evaporation.

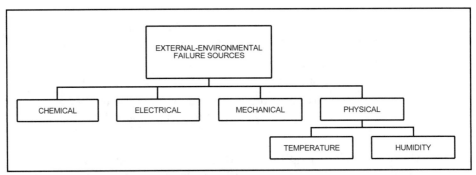

Figure 3-4. External Environmental Failure Sources.

Many other industrial failure sources are present. Large electrical currents are being switched, generating wideband electrical noise. Mechanical processes such as mixing, loading, and stamping cause shock and vibration. Consider another example: A controller module mounted near a turbine had all of its calibration data stored in a programmable memory that was mounted in an integrated circuit socket. After several months of operation, the controller failed. Its outputs went dead. An examination of the failed controller module showed that the memory had vibrated out of the socket. The controller computer was programmed to safely de-energize the outputs if calibration data were lost.

As the memory chip vibrated out of its socket, the strength of the electrical contact decreased. The strength of the contact (the mechanical contact force) does vary from unit to unit; in the presence of vibration the average mechanical contact force decreases. Whenever stress remains a constant and strength decreases, the rate at which failures occur will increase. For this failure:

Title	Controller module failed
Root cause	Memory chip socket failed
Failure type	Physical
Primary stressor	Vibration
Secondary stressors	Temperature

Environmental failure sources are external to the product. Chemical, electrical, and mechanical processes are all present. The failure rates caused by environmental sources vary with the product design. Some products are stronger and are designed to withstand higher environmental stress; they do not fail as frequently.

For example, failures in electronic equipment have been traced to corrosion within the module. Electrical contacts can no longer conduct current because insulating oxides have formed. Stronger designs that utilize contact lubricants (a protective sealant) can withstand this chemical process for much longer periods of time. Contact lubricants form a thin sealant layer around contacts, thus preventing chemical corrosion.

Module input/output circuits are susceptible to electrical overstress because they typically are connected via cable to field sensors. Cables pass through areas of high electrical stress. Voltage spikes can easily exceed the ratings of input amplifiers. Unless these failure stress sources are thwarted, they will cause failures. A stronger design can utilize special protective components to absorb energy, thereby limiting voltages and currents.

Mechanical vibration can fracture metal components, especially at resonant frequencies. Stronger designs use bracing to withstand the stress of this failure source.

It could be argued that all environmental failures are really system-level design errors, since the designer did not use a product with enough strength. The distinction can be arbitrary. Normally, if the design strength is greater than the anticipated environmental stress, the failure is an environmental failure, not a design failure.

Temperature and humidity are environmental failure sources that affect failure rates in two ways:

1. They can cause direct failure when their magnitude exceeds design limits. For example, an integrated circuit mounted in a plastic package works well even when hot (some are rated to junction temperatures of 150°C). But when the temperature exceeds the glass transition point of the plastic (approximately 180°C), the component fails. In another example, the keyboard for a personal computer worked well in a control room even through the humid days of summer. However, it failed immediately when a glass of water tipped into it. For this failure:

Title	Telephone network failed
Root cause	Keyboard failed (water spill)
Failure type	Physical
Primary stressor	Water on keyboard
Secondary stressors	None identified

2. Temperature and humidity can also accelerate other failure sources. Many chemical and physical processes operate as an exponential function of temperature. The hotter it gets, the faster the process proceeds. Consider another case history in which a temperature controller was mounted near a glass furnace with an average ambient temperature of 70°C. It failed after one year of operation with its outputs at full scale. An autopsy of the controller showed that an electrically programmable logic array no longer contained its "program" and did not perform the correct logic functions. The component is programmed with an electrical charge that is trapped within the device; the device does "leak," and its strength decreases. Over a period a time, the charge will dissipate and the component will fail. This time period is expected to be greater than 10 years when the component is operated at 50°C. At the higher temperature, the charge leakage is accelerated, and the expected life is drastically reduced. For this failure:

Title	Temperature controller failed
Root cause	Logic component lost program
Failure type	Physical
Primary stressor	Temperature
Secondary stressors	None identified

Humidity also affects other failure sources. The presence of humidity may accelerate corrosion. Humidity may also provide a conduction path that allows electrostatic discharge to cause damage.

External — Maintenance Faults

Incorrect repair procedures can cause failures. For example, a controller module in a dual fault-tolerant system fails. The system continues to operate because its redundant backup module is now controlling. While installing the replacement controller module, the repair technician plugs the output switch cable into the communication socket, shorting the system output. For this failure:

Title	Redundant controller failed
Root cause	Cable plugged into wrong socket
Failure type	Physical
Primary stressor	Maintenance error (cable)
Secondary stressors	None identified

Since maintenance activities are a human reliability problem, they are affected by many variables: complexity, system repairability (foolproofing), condition of the repair technician, training, familiarity with the system, and time pressure. Like operational failure rates, maintenance failure rates are difficult to estimate. Failure rates should be estimated on a site basis by reviewing historical plant maintenance logs. The factors cited above should be taken into account and used as "adjustment factors" based on experience.

External — Operational Faults

Any system that allows a human operator to command system operation can fail due to an incorrect command. For example, an operator, intending to enable the remote set point of a control loop, selects the wrong point on his CRT station. This incorrect control loop is instead switched from automatic to manual.

For this example:

Title	Wrong loop set
Root cause	Operator selected wrong loop
Failure type	Functional
Primary stressor	Incorrect operation
Secondary stressors	None identified

Such errors do happen. This author once made a similar operator error while solo-piloting an aircraft. On final approach to a night landing, instead of turning on the landing lights, he turned off the engine! Fortunately, the landing, without lights or engine, was accomplished.

Operator errors are listed in accident reports in many industries, yet some say there is no such thing as an operator error. Charles Perrow states in his book, *Normal Accidents: Living with High-Risk Technology* (Ref. 13), that complex and coupled systems are the real cause of the problem. Certainly, operator faults are affected by system complexity, design of the man/machine interface (foolproofing), familiarity with the system, operator condition (fatigue, boredom, etc.), and reliability of the instrumentation. Operator faults can even be induced by production pressures. J. M. Juran calls these "management-induced operator errors" (Ref. 14). Production pressures and system complexity do contribute to operator errors. But, like the pilot error described above, sometimes the operator just does the wrong thing.

The many factors involved in operator errors make it difficult to estimate operator failure rates. Theoretical models for quantitative analysis do exist (Ref. 15). Failure rate estimates can be done on a plant site basis by reviewing historical plant failure logs. Failure rate is calculated by dividing the number of failures by the operating hours. Relative system complexity can be estimated and used as an "adjustment factor."

Failure Databases

Many companies now keep detailed records of process control equipment failures (Ref. 16). This information promises to be a valuable source of better reliability and safety statistics in the future.

Several important pieces of information should be kept in these databases: equipment type, manufacturer, model number, serial number, failure type (physical or functional), primary and secondary stressors (failure sources), time in operation, time since last inspection, and repair time required. While all this information may be impossible to obtain, the future value is there.

Some companies have used these data to correlate failures with plant conditions. Failure prevention actions can often be defined by inspection of these failure data.

Other benefits of a good failure database include reduced false alarms, proper identification of which equipment actually failed, increased warranty coverage, and decreased inspection time in safety-critical systems.

Stress-Strength

Failures occur when some form of "stress" exceeds the associated strength of the product. A. C. Brombacher explains this in detail in his book, *Reliability by Design: CAE Techniques for Electronic Components and Systems* (Ref. 17). The concept is easy to understand and is used extensively in mechanical engineering when choosing the size and type of material to be used for components. The mechanical engineer is dealing with a stress consisting of a physical force. The associated mechanical strength is the ability of the component to resist the force.

In the safety and reliability analysis of control systems, many other types of stress are present. Brombacher explains that stress is composed of a combination of stressors, which can take many forms: chemical corrosion, electrical voltage/current, mechanical vibration/shock, temperature, humidity, and even human error. All external failure sources are stressors.

The strength of a particular device (component, module, unit, or system) is a result of its design and manufacture. The choice of various design parameters and protection components dictates the initial strength of a product. If the manufacturing process identically duplicates the product, the strength of each device will be the same, as designed. Defects in raw materials and the manufacturing process will reduce strength in a variable way. Internal failure sources affect strength.

Stress

Stress varies with time. Consider stress caused by temperature. Temperature changes somewhat periodically on a daily basis, and typically varies with the season. Thus, another periodic variation is added to the daily variation. Other, more random factors affect temperature, such as wind, clouds, and local heat sources. Some temperature sources such as fire or explosion generate temperatures that have an impulse function. When temperature is viewed as a whole, the stress level can only be characterized as "random."

Many independent parameters affect temperature. Each of these sources will affect temperature in some variable manner that is characterized by a probability density function (pdf). Given a large number of independent sources, the central limit theorem would tell that a pdf for ambient temperature is a normal distribution. The probability of any particular temperature-level range depends on the normal distribution characterized by its mean and its standard deviation (Figure 3-5).

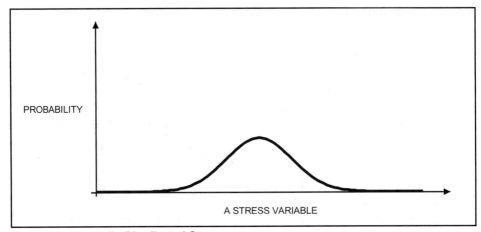

Figure 3-5. Normally Distributed Stress.

When evaluating stress levels that cause failures, the probability of getting a stress level lower than a particular value must be determined. This is represented by an inverse distribution function (Figure 3-6).

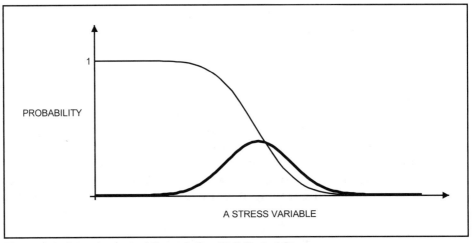

Figure 3-6. Complement of Cumulative Distributed Stress.

Strength

The "strength" of a product is the result of its design and manufacture. The product is designed to meet a certain strength level over its predicted life. The manufacturing process is designed to replicate the product. This applies from the component level on up. If the manufacturing process is perfect, the strength will always be the same. This is illustrated in Figure 3-7.

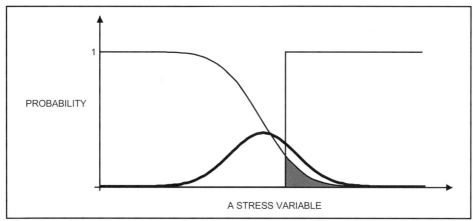

Figure 3-7. Identical Strength cdf Versus Inverse Stress cdf.

The probability of getting a stress less than a particular value is plotted along with a step function which shows a cumulative distribution function (cdf) for strength. The step occurs at the known strength of the devices. The crosshatched area on the figure represents the probability of failure (stress is greater than strength).

In some cases the area representing probability of failure can be calculated. Remember that a particular module failure occurs when any one of a number of stress levels exceeds the associated strength level. Stress level is defined as the variable x, and strength level as the variable y. The excess of strength over stress is represented by the variable w:

$$w = y - x \qquad (3\text{-}1)$$

The module succeeds when $w > 0$ and fails when $w \leq 0$. The failure probability is represented by the area where the stress-strength curves interfere. This area is reduced when more "safety factor" (the difference between mean strength and mean stress) is used by the designer.

EXAMPLE 3-1

Problem: A module will fail when the temperature of its power transistor exceeds the transistor's strength of 90°C. The power transistor operates at a temperature that is 30°C hotter than the ambient temperature. The ambient temperature in a large number of installations over a long period of time is characterized by a normal distribution with a mean of 40°C and a standard deviation of 10°C. What is the probability of module failure?

> **EXAMPLE 3-1 (continued)**
>
> **Solution:** How much of the area under the normal curve that represents stress (ambient temperature) exceeds the strength (90°C) of the transistor? First, relate the transistor strength to ambient temperature. Since the transistor operates thirty degrees hotter than ambient,
>
> $$90 - 30 = 60 = \text{maximum ambient temperature}$$
>
> The normal distribution is converted to standard form using Equation 2-20,
>
> $$z = \frac{60 - 40}{10} = 2$$
>
> Using Appendix A at $z = 2$, a number of 0.97725 is obtained. This means that 97.725% of the area under this normal distribution is below a temperature of 60°C. The area beyond the 60°C point is
>
> $$100\% - 97.725\% = 2.275\%$$
>
> The probability of module failure due to excessive temperature stress is 0.02275. This is shown in Figure 3-8.

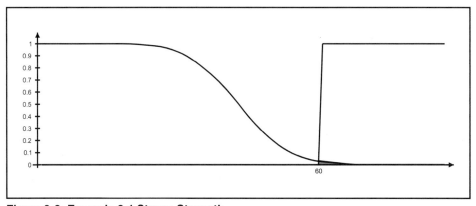

Figure 3-8. Example 3-1 Stress-Strength.

A simulation can show how many modules fail as a function of time given a particular stress-strength relationship. A simulation was done using a normally distributed stress and a constant strength. Choosing values (Figure 3-9) similar to previous work done at Eindhoven University in the Netherlands, the results are shown in Figure 3-10. Within the limits of the simulation, a relatively constant number of modules would fail as a function of time.

Failures: Stress versus Strength 49

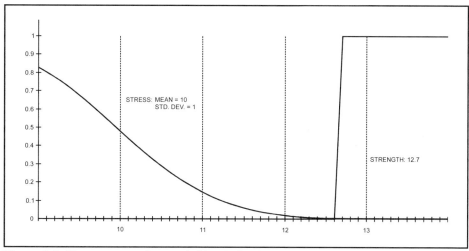

Figure 3-9. Stress versus Strength Simulation Values-Identical Strength.

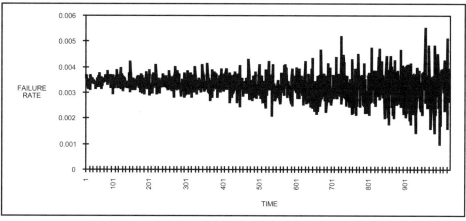

Figure 3-10. Failure Rate with Time for Identical Constant Strength.

Strength Varies

Actual manufacturing processes are not ideal, and newly manufactured products are not identical in strength. The raw materials used to make components vary from batch to batch; the component manufacturing process varies; and the module-level manufacturing process can differ in any number of places. The degree of quality control can vary. The result of this variation in strength is some sort of probability density distribution. Intuition, supported by the central limit theorem, would indicate another normal distribution.

Figure 3-11 shows this stress-strength situation. Again, the probability of failure can be calculated from a stress-strength relationship, although in this case numerical integration techniques must be used (Ref. 18). A simulation of this stress-strength relationship shows interesting results (Figure 3-12). Fewer and fewer modules fail with time. This occurs because the weaker devices fail quickly and are removed from the population. Remaining units are stronger and last longer.

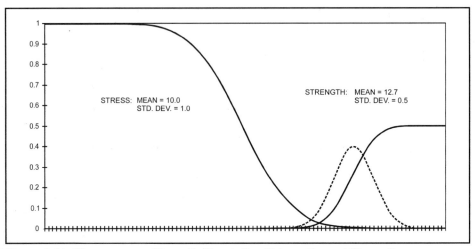

Figure 3-11. Stress versus Strength Simulation Values-Normal Strength.

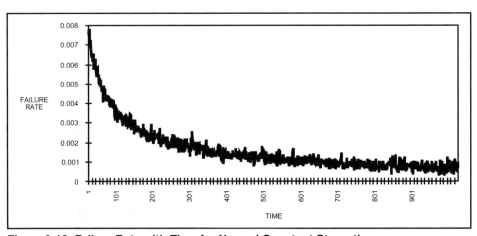

Figure 3-12. Failure Rate with Time for Normal Constant Strength.

Strength Decreases

Strength changes with time. Unfortunately, many strength factors seem to decrease as a product is used. As strength decreases, the likelihood of failure increases, as does the rate at which failures occur. Although there are rare circumstances where product strength increases with time, a vast majority of strength factors decrease. Decreasing strength is shown in

Failures: Stress versus Strength

Figures 3-13 and 3-14. A simulation of this stress-strength relationship in shown in Figure 3-15. The number of failures per unit of time decreases and then increases. This unfortunate reality is known as "wearout."

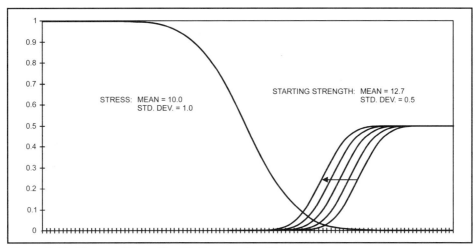

Figure 3-13. Stress - Strength with Decreasing Strength.

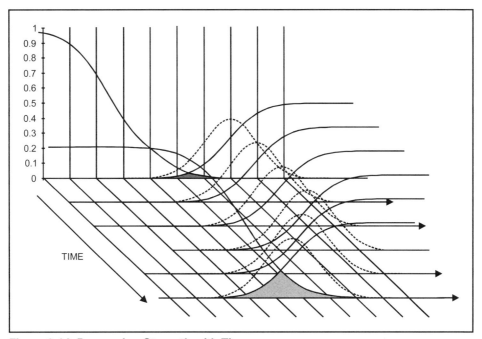

Figure 3-14. Decreasing Strength with Time.

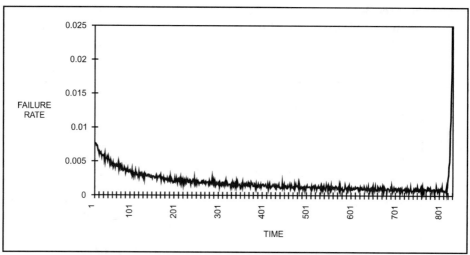

Figure 3-15. Failure Rate with Time When Strength Decreases.

Measuring Strength

Strength is measured by testing a device (a component, a module, a unit, or a system) until it fails (or at least until the limits of the test equipment have been reached). When several units are actually tested to destruction, a good indication of strength is obtained.

Many industrial stressors have been characterized by international standards. These include temperature, humidity, corrosive atmospheres, electromagnetic radiation, electrical surge, electrostatic discharge, shock, and vibration. Each stressor is defined, typically by "levels," in a standard.

Temperature

Temperature stress testing is delineated in the IEC68-2 standard (Ref. 19). Different temperature levels for operation and storage are defined, as are a series of tests under various conditions. The most common tests for operational temperature stress are 68-2-1 Test Ab, 68-2-2 Test Bb, and 68-2-14 Tests Na and Nb. A typical industrial specification might read "Operating Temperature: 0 to 60°C, tested per IEC68-2."

Although the standards do not require operation beyond the test limits, testing to destruction should be done. A high-strength industrial controller should successfully operate at least 30°C beyond the specification. High strength is designed into industrial electronic modules by adding large metal heat sinks and heat sink blocks on critical components such as microprocessors and power output drivers.

Humidity

Humidity testing is also done per the IEC68-2 standard. The most common testing is done using 68-2-3 Test Ca for operational humidity and 68-2-30 Test Dd for cycling humidity. Common specifications for industrial devices are 5 to 95% noncondensing for operation and 0 to 100% condensing for storage. Control modules with plastic spray coating and contact lubricant will resist humidity to much higher stress levels.

Mechanical Shock/Vibration

Mechanical shock and vibration stressors can be quite destructive to control systems, especially when they are installed near rotating equipment such as pumps and compressors. Testing is typically done by vibrating the equipment at various frequencies between 10 and 150 Hz per IEC68-2-3 Test Fc. Although displacement and acceleration are specified for different portions of the frequency spectrum, it is useful to think of the stressor levels in G's. IEC68-2-27 Test Ea defines how mechanical shocks are applied.

A typical specification for industrial control equipment would be 2G vibration and 15G shock. High-strength equipment may be rated to 4G vibration and 30G shock. Control systems that feature bolt-down connectors, screw-in modules, rigid printed circuit mounting, and mechanical bracing will operate successfully to higher stress levels.

Corrosive Atmospheres

Failures due to corrosive atmospheres are common in control systems, especially in the chemical, oil and gas, and paper industries. Some of the corrosive gases attack the copper in electronic assemblies. ISA's S71.04 standard (Ref. 20) specifies several levels of stress. Testing can be done to verify the ability of a system to withstand corrosive atmospheres over a period of time.

High-strength modules should withstand a class G3 atmosphere for more than 10 years. Plastic coating of electrical assemblies, contact lubricant, and a physical cover over electronic modules will increase strength against corrosion stressors.

Electromagnetic Fields

The IEC1000-4 series of standards (Ref. 21) specify levels of various electric stressors. IEC1000-4-3 specifies test methods for radiated electromagnetic fields (EMF), and IEC1000-4-6 specifies test methods to verify strength against conducted EMFs.

Industrial control equipment is typically specified to withstand radiated EMFs (sometimes called radio frequency interference, or RFI) at levels of 10 v/m over frequencies between 80 and 1,000 MHz. There should be enough strength to withstand 10 V conducted disturbances over a 150 kHz to 80 Mhz range. Module shielding and EMF filtering will add strength to a module.

Electrostatic Discharge

IEC1000-4-2 explains how to determine the strength against electrostatic discharges. An electrostatic discharge can occur when two dissimilar materials are rubbed together. When an operator gets out of a chair or a person removes a coat, an electrostatic discharge may result.

Equipment that will withstand a 15 kv air discharge and an 8 kv contact discharge is suited for use in an industrial environment. High strength can be provided by grounded conductive (metal) enclosures that provide a short path for electrostatic discharge.

Electrical Surge and Fast Transients

Electrical stress in the form of transient voltages and currents can come from secondary effects of lightning, motor starters, relay contacts, and even arcing contacts (welding machines). These surge levels are specified in IEC1000-4-5. Electrical fast transients are specified in IEC1000-4-4.

Surge and fast-transient levels specified for power lines are higher than for signal lines, as higher levels of voltage surge are more likely on power systems. Surge protection networks and inductive filters on signal cables offer high strength against these stressors.

Highly Accelerated Life Testing

One method developed (Ref. 22) to identify weaknesses in the design of a module is known as highly accelerated life testing (HALT). In this test, an electronic assembly is subjected to random three-axis vibration while being simultaneously heated and cooled at a rapid rate (from –90 to 150°C). The objective is to increase stress levels until failure occurs. The failure is identified and if possible, the assembly is strengthened. Figure 3-16 shows a module undergoing HALT testing.

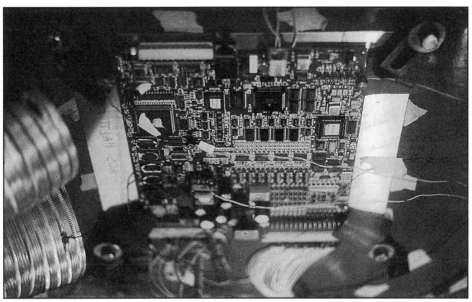
Figure 3-16. Controller Module Under HALT Test.

EXERCISES

3.1 What failure source has caused the most failures in your plant?

3.2 Can software design faults cause failures of computer-based controllers?

3.3 What failure sources are primarily responsible for "infant mortality" failures?

3.4 What method can be used to cause manufacturing faults to become failures in a newly manufactured product?

3.5 How is a wearout mechanism related to the strength of a product?

3.6 The input module to a process control system will withstand 2000 v without failure. If the voltage applied to the input exceeds 2000 v, the module will fail. Input voltage transients for many installations over a period of time are characterized by a normal distribution with a mean of 1500 v and a standard deviation of 200 v. What is the probability of module failure due to input transient electrical overstress?

3.7 A stress is characterized as a normal distribution with a mean of 10 and a standard deviation of 1. If devices will fail when subjected to stresses above 12.7, what is the probability of failure?

3.8 Is highly accelerated life testing a method used by some manufacturers to increase the strength of an electronic design?

3.9 Contact lubricant is a fluid used to seal electrical connectors and provides strength against what stressors?

REFERENCES

1. MIL-STD-217F, Reliability Prediction of Electronic Equipment, New York: Rome Laboratory, AFSC, 1985.

2. *Guidelines for Process Equipment Reliability Data, with Data Tables*, New York: American Institute of Chemical Engineers, 1989.

3. Klinger, D. J., Nakada, Y., and Menendez, M. A., *AT&T Reliability Manual*, Chapter 4, New York: Van Nostrand Reinhold, 1990.

4. IEEE 500, Component Reliability Data, New York: IEEE.

5. Fardip database, Tonbridge, Kent, TECHNIS, 1993.

6. *Reliability Audit Report*, Phoenix: Motorola DMTG Reliability DATA Management Group, 1990.

7. *Master Selection Guide*, Carrollton, TX: Texas Instruments Inc., 1990.

8. Watson, G. F., "Three Little Bits Breed a Big, Bad Bug," *IEEE Spectrum*, May 1992.

9. Kletz, T. A., *What Went Wrong? Case Histories of Process Plant Disasters*, Houston: Gulf Publishing, 1988.

10. Fagan, M. E., "Advances with Inspections," *IEEE Transactions of Software Engineering*, July 1986.

11. Parnas, D. L., and Weiss, D. M., "Active Design Reviews: Principles and Practices," *Proceedings of the Eighth International Conference on Software Engineering*, Washington, DC: IEEE Computer Society Press, 1985.

12. Bukowski, J. V., Goble, W. M., and Johnson, D. A., "Software Reliability Feedback: A Physics of Failure Approach," *Proceedings of the Annual Reliability and Maintainability Symposium*, New York: IEEE, 1992.

13. Perrow, C., *Normal Accidents: Living with High-Risk Technology*, New York: Basic Books, 1984.

14. Juran, J. M. (ed.), *Quality Control Handbook*, 3rd Ed., New York: McGraw-Hill, 1974.

15. Dougherty, E. M., Jr., and Fragola, J. R., *Human Reliability Analysis*, New York: John Wiley & Sons, 1988.

16. Kenderdine, R. S., "Implementing a Computerized Maintenance Management System," *Chemical Engineering*, 1997.

17. Brombacher, A. C., *Reliability by Design: CAE Techniques for Electronic Components and Systems*, New York: John Wiley & Sons, 1992.

18. Bukowski, J. V., and Lele, A., "The Case for Architecture Specific Common Cause Failure Rates and How They Affect System Performance," *Proceedings of the Annual Reliability and Maintainability Symposium*, New York: IEEE, 1997.

19. IEC68-2, "Environmental Testing."

20. ISA S71.04, *Environmental Conditions for Process Measurement and Control Systems: Airborne Contaminants*, Research Triangle Park, NC: ISA—The Instrumentation, Systems, and Automation Society, 1985.

21. IEC1000-4, "Electromagnetic Compatibility for Industrial-Process Measurement and Control Equipment: Electrical Fast Transients/Bursts," 1988.

22. Hobbs, G.K., "Development of Stress Screens," *Proceedings of the Annual Reliability and Maintainability Symposium*, New York: IEEE, 1987.

BIBLIOGRAPHY

1. Brombacher, A. C., Houtermans, M. J. M., Rouvroye, J. L., Spiker, R. Th. E., "Systematic Failures in Safety Systems: How to Analyze; How to Optimize," *Advances in Instrumentation and Control*, October 1996.

2. Brombacher, A. C., and Steinz, H. C., "Analysing the "Strength" of a Safeguarding System," *Hydrocarbon Technology International*, London: Sterling Publications, 1988.

3. Brombacher, A. C., Vos, R., Fennema, P. H., and van't Loo, J., "The robust design of electronic systems," ASQC Quality in Electronics Conference, San Jose: American Society for Quality Control, 1989.

4. Des Plas, E. P., "Reliability in the Manufacturing Cycle," *Proceedings of the Annual Reliability and Maintainability Symposium*, New York: IEEE, 1986.

5. Klyatis, L. M., "One Strategy of Accelerated Testing Technique," *Proceedings of the Annual Reliability and Maintainability Symposium*, New York: IEEE, 1997.

6. Nelson, W., *Accelerated Testing*, New York: John Wiley & Sons, 1990.

7. Ramerirez, M. A., "Preventing Maintenance Induced Errors," *Proceedings of the Annual Reliability and Maintainability Symposium*, New York: IEEE, 1985.

8. Van Geest, D. C. L., Hoeksma, R. H., Brombacher, A. C., and Herrmann, O. E., "Integration of Physical Reliability Knowledge into the Design of VLSI Circuits," *Proceedings of the International Reliability Physics Symposium*, New York: IEEE, 1993.

9. Walinga, J. S., Brombacher, A. C., and Fennema, P. H., "Why Do Circuits Really Fail?" *Proceedings of the IAESTED Symposium*, Lugano: IAESTED, 1989.

4
Reliability and Safety

Reliability Definitions

The term "random variable" is well understood in the field of statistics. It is the independent variable — the variable being studied. Samples of the random variable are taken, and statistics are computed about that variable in order to learn how to predict its future behavior. In reliability engineering, the primary random variable is T: time to failure, or failure time. Reliability engineers gather data about when and how things fail. This information is used to gain insight into future performance of system designs.

For example, Table 4-1 records the time to failure for 10 modules that were life tested. The sample average (or mean) of the failure times can be calculated. For this test, the sample mean time to failure (MTTF) is calculated to be 3,248 hr. This measurement provides some information about the future performance of similar modules. Several different terms are used to describe predicted system performance. The terms are defined to provide different kinds of information engineers need.

Module	Time to Fail, in Hours
1	2,327
2	4,016
3	4,521
4	3,176
5	70
6	3,842
7	3,154
8	2,017
9	5,143
10	4,215

Table 4-1. Life Test Results for 10 Modules

59

Reliability

Reliability is a measure of success. Reliability generally is defined as "the probability that a device will perform its intended function when required to do so if operated within its specified design limits." This definition includes four important aspects:

1. The device's "intended function" must be known.
2. "When the device is required to function" must be judged.
3. "Satisfactory performance" must be determined.
4. The "specified design limits" must be known.

All four aspects must be addressed when defining a situation to be a success or a failure.

Mathematically, reliability (R) has a precise definition: "The probability that a device will be successful in the time interval from zero to t." In terms of the random variable T,

$$R(t) = P(T > t) \tag{4-1}$$

Reliability equals the probability that T, failure time, is greater than t, operating time interval.

Consider a newly manufactured and successfully tested component. It operates properly when put into service at time $t = 0$. As the time interval increases, it becomes less likely that the component will remain successful. Since the component will eventually fail, the probability of success for an infinite time interval is zero. Thus, all reliability functions start at a probability of one and decrease to a probability of zero (Figure 4-1).

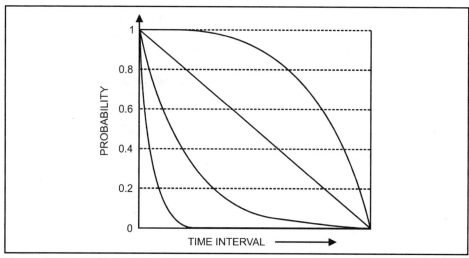

Figure 4-1. Reliability Functions.

Reliability is a function of operating time interval. A statement such as "System reliability is 0.95" is meaningless because the time interval is not known. The statement "The reliability equals 0.98 for a mission time of 100 hr" makes perfect sense.

Reliability is a relatively stringent measure that is usually applied to situations such as aircraft flights and space missions where no repair is possible. In these circumstances, the system must operate continuously without failure. Industrial control systems are usually repairable. For these repairable systems, other standards of measure — such as availability or MTTF — are more relevant.

Unreliability

Unreliability, $F(t)$, a measure of failure, is defined as "the probability that a device will fail in the time interval from zero to t."

$$F(t) = P(T \le t) \tag{4-2}$$

In terms of the random variable T, unreliability equals the probability that failure time will be less than or equal to operating time. Since any component must be either successful or failed, $F(t)$ is the one complement of $R(t)$.

$$F(t) = 1 - R(t) \tag{4-3}$$

$F(t)$ is a cumulative distribution function. It begins with a probability of zero and increases to a probability of one.

Availability

Reliability is a measure that requires system success for an entire time interval. No failures (and subsequent repairs) are allowed. This measurement is insufficient for engineers who need to know the average chance of success when repairs are possible. Another measure of system success for repairable systems is required (Figure 4-2). That metric is availability. Availability is defined as "the probability that a device is successful at time t." No time interval is involved. If a system is operating, it is available. It does not matter whether it has failed in the past and has been repaired, or has been operating continuously from time $t = 0$ without failure. Availability is a measure of "uptime" in a system, unit, or module.

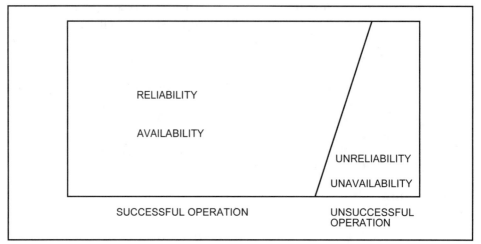

Figure 4-2. Successful - Unsuccessful Operation.

Availability and reliability are different. Reliability is always a function of operating time interval and failure rates. It is a measurement that starts at one and goes to zero as the time interval gets longer. Availability is a function of failure rates, repair rates, and operating time. But availability results will reach a steady-state value as a function of operating time interval. Steady-state availability is a function only of failure rates and repair rates. A plot of availability versus reliability for a single repairable module is shown in Figure 4-3.

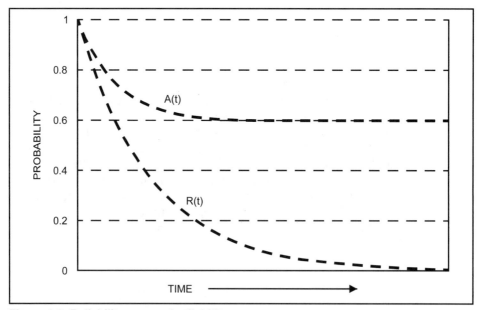

Figure 4-3. Reliability versus Availability.

Reliability and Safety

Unavailability

Unavailability, a measure of failure, is used primarily for repairable systems. It is defined as "the probability that a device is not successful (is failed) at time t." Unavailability is the one complement of availability; therefore,

$$U(t) = 1 - A(t) \tag{4-4}$$

EXAMPLE 4-1

Problem: A controller has an availability of 0.99. What is the unavailability?

Solution: Using Equation 4-4, Unavailability = 1 − 0.99 = 0.01.

Probability of Failure

The probability of failure during any interval of operating time is given by a probability density function (pdf). Such probability density functions are defined as

$$f(t) = \frac{dF(t)}{dt} \tag{4-5}$$

The pdf can be mathematically described in terms of the random variable T:

$$\lim_{\Delta t \to 0} P(t < T \le t + \Delta t) \tag{4-6}$$

This can be interpreted as the probability that the failure time, T, will occur between the operating time, t, and the next interval of operation, $t + \Delta t$.

The probability of failure function can provide failure probabilities for any time interval. The probability of failure between the operating hours of 2,000 and 2,200, for example, is

$$P(2000, 2200) = \int_{2000}^{2200} f(t)dt \tag{4-7}$$

> **EXAMPLE 4-2**
>
> **Problem:** A valve positioner has an exponential probability density of failure function (constant failure rate) of
>
> $$f(t) = 0.0002e^{-0.0002t}$$
>
> What is the probability that it will fail after the warranty (6 months, 4,380 hr) and before plant shutdown (12 months, 8760 hr)?
>
> **Solution:** Using Equation 4-7,
>
> $$P(4380, 8760) = \int_{4380}^{8760} 0.0002e^{-0.0002t} dt$$
>
> This evaluates to
>
> $$P(4380, 8760) = -e^{-0.0002 \times 8760} - (-e^{-0.0002 \times 4380})$$
>
> Calculating the result:
>
> $$P(4380, 8760) = -0.1734 + 0.4165 = 0.243$$
>
> This result states that the probability of failure during the interval from 4380 hr to 8760 hr is 24.3%.

Mean Time to Failure (MTTF)

One of the most widely used reliability parameters is the MTTF. Unfortunately, it is also sometimes misused and misinterpreted as "guaranteed minimum life." Consider Table 4-1. The MTTF of 3,248 hr was calculated using a simple average technique. One of the modules failed after only 70 hr, however.

MTTF is merely the "expected" failure time. It is defined from the statistical definition of expected value (Equation 2-10). Using the random variable failure time, t, and substituting the probability density function $f(t)$:

$$E(t) = \int_0^{+\infty} tf(t)dt \qquad (4\text{-}8)$$

Reliability and Safety

Substituting

$$f(t) = -\frac{d[R(t)]}{dt} \tag{4-9}$$

into the expected value formula yields

$$E(t) = -\int_0^{+\infty} t\, d[R(t)]$$

Integrating by parts, this equals

$$E(T) = [-tR(t)]_0^\infty - \left[-\int_0^{+\infty} R(t)dt\right]$$

The first term equals zero at both limits. This leaves the second term, which equals

$$MTTF = E(T) = \int_0^{+\infty} R(t)dt \tag{4-10}$$

Thus, in reliability theory, the definition of MTTF is the definite integral evaluation of the reliability function. Note that the definition of MTTF is *not* related to the inverse of failure rate.

$$MTTF \neq \frac{1}{\lambda} \text{ by definition} \tag{4-11}$$

Note: The formula MTTF = $1/\lambda$ is valid for single components with a constant failure rate or a series of components, all with constant failure rates. See the section "The Constant Failure Rate" later in this chapter.

Mean Time to Repair (MTTR)

MTTR is the "expected value" of the random variable repair time (or time to repair), not failure time. The definition includes the time required to detect that a failure has occurred, as well as the time required to make the repair once the failure has been detected and identified. Like MTTF, MTTR is an average value. The term MTTR applies only to repairable devices. Figure 4-4 shows that MTTF represents the average time required to move from successful operation to unsuccessful operation. MTTR is the average time required to move from unsuccessful operation to successful operation. Mean dead time (MDT) is another commonly used term which means the same as MTTR.

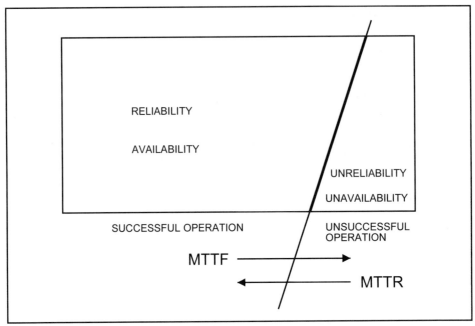

Figure 4-4. MTTF and MTTR in Operation.

Mean Time between Failures (MTBF)

MTBF is a term that applies only to repairable systems. Like MTTF and MTTR, it is an average value, but it is the time between failures. This implies that a component has failed and then has been repaired. Mathematically,

$$\text{MTBF} = \text{MTTF} + \text{MTTR} \tag{4-12}$$

Figure 4-5 shows a graphical representation of MTTF, MTTR, and MTBF.

The MTBF term has been misused. Since MTTR is usually much smaller than MTTF, MTBF is approximately equal to MTTF. The term MTBF is often substituted for MTTF, which applies to both repairable systems and nonrepairable systems.

EXAMPLE 4-3

Problem: A power supply module is potted in epoxy. The manufacturer quotes "MTBF equals 50,000 hr." Is this term being used correctly?

Solution: This power supply is potted in epoxy. Although it can be argued that it is possible to repair the unit, it is generally not practical to consider repair. The term MTBF applies to repairable equipment and has been misused in this case.

Reliability and Safety 67

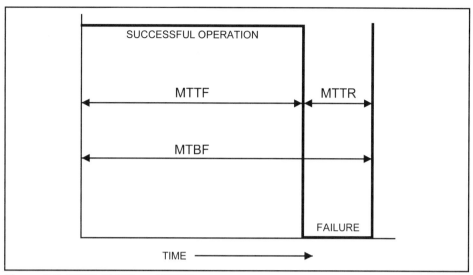

Figure 4-5. MTTF, MTTR and MTBF.

EXAMPLE 4-4

Problem: An industrial I/O module contains 16 circuit assemblies mounted in sockets. The manufacturer claims that the module has an MTBF of 20,000 hr. Is the term MTBF being used correctly?

Solution: The module has been designed to be repairable. The term MTBF is appropriate, although the term MTTF can be more precise since average failure time is the variable of interest.

EXAMPLE 4-5

Problem: An industrial I/O module has an MTTF of 87,600 hr. When the module fails, it takes an average of 2 hr to repair. What is the MTBF?

Solution: Using Equation 4-12, the MTBF = 87,602 hr. The MTBF is approximately equal to the MTTF.

EXAMPLE 4-6

Problem: An industrial I/O module has an MTTF of 87,400 hr. When the module fails, it takes an average of 400 hr to repair. What is the MTBF?

Solution: Using Equation 4-12, the MTBF = 87,800 hr. It is interesting to note that compared to the module of Example 4-5, this module will fail sooner because it has a lower MTTF. Using a larger MTBF number as a positive indicator would be misleading. MTTF is a more precise term than MTBF for the measurement of successful operation.

Failure Rate

Instantaneous failure rate, often called "hazard rate" by reliability engineers, is a commonly used measure of reliability that gives the number of failures per unit time from a quantity of components exposed to failure.

$$\lambda(t) = \frac{\text{Failures per Unit Time}}{\text{Quantity Exposed}} \qquad (4\text{-}13)$$

Failure rate has units of inverse time. It is common practice to use units of "failures per billion (10^9) hours." This failure unit is known as FIT. For example, an integrated circuit will experience seven failures per billion operating hours at 25°C, and thus has a failure rate of seven FITs.

EXAMPLE 4-7

Problem: Three hundred industrial I/O modules have been operating in a plant for seven years. Five failures have occurred. What is the average failure rate for this group of modules?

Solution: Using Equation 4-13, the average failure rate is $5/(300 \times 7 \times 8760)$ = 0.000000271798 = 272 FITS.

The failure rate function is related to the other reliability functions. It can be shown that

$$\lambda(t) = \frac{f(t)}{R(t)} \qquad (4\text{-}14)$$

Time-Dependent Failure Rates

If a collection of 50 modules were subjected to life testing, the time when any unit fails could be recorded. An extremely stressful environment could be created to accelerate failures. A check to see how many modules fail each week may show that the percentage of failures decreases, increases, or stays constant.

Consider the failure log for a highly accelerated life test (HALT) as shown in Table 4-2. The number of failures decreases during the first few weeks. The number then remains relatively constant for many weeks. Toward the end of the test, the number begins to increase.

"Failure rate" is calculated in column four and equals the number of failures divided by the number of module hours (surviving modules times hours) in each weekly period. The failure rate also decreases at first, then remains basically constant, and finally increases. These changes in the failure rate of a module are typically due to several factors, including variations in strength and strength degradation with time.

Week	Number Surviving at Start of Week	Failures	Failure Rate
1	50	9	0.0011
2	41	5	0.0007
3	36	3	0.0005
4	33	2	0.0004
5	31	2	0.0004
6	29	1	0.0002
7	28	2	0.0004
8	26	1	0.0002
9	25	1	0.0002
10	24	0	0.0000
11	24	2	0.0005
12	22	1	0.0002
13	21	1	0.0003
14	20	0	0.0000
15	20	1	0.0003
16	19	0	0.0000
17	19	1	0.0003
18	18	1	0.0003
19	17	0	0.0000
20	17	1	0.0003
21	16	1	0.0004
22	15	0	0.0000
23	15	1	0.0004
24	14	0	0.0000
25	14	1	0.0004
26	13	0	0.0000
27	13	1	0.0005
28	12	0	0.0000
29	12	1	0.0005
30	11	0	0.0000
31	11	1	0.0005
32	10	1	0.0006
33	9	1	0.0007
34	8	1	0.0007
35	7	1	0.0008
36	6	1	0.0010
37	5	2	0.0024
38	3	3	0.0059

Table 4-2. Life Test Data

Decreasing Failure Rate

A decreasing failure rate is characteristic of a "fault removal process." Consider a collection of modules in which a portion of the modules have manufacturing faults. The entire collection of modules is placed on an accelerated life test. Since manufacturing faults reduce the strength of a module, the modules with faults will fail in a relatively short period of time. Failed modules are removed from the test. After a period of time, no modules that have manufacturing faults remain in the collection. The failure rate due to manufacturing faults will have dropped to zero (Figure 4-6).

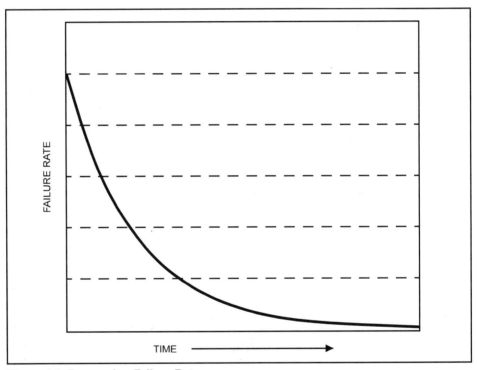

Figure 4-6. Decreasing Failure Rate.

Constant

If failures in a collection of modules are due to stresses from the environment and the strength is relatively constant, the failure rate for a large collection tends to be constant (refer back to Figure 3-10, which shows an approximately constant failure rate). The stresses are of many different types. They appear to come at random times and appear to have random magnitudes. In some cases, specific reasons why a stress is generated might be determined. For the most part, however, the specific stress events that cause failure are not understood or recorded and are treated as random (in time).

Increasing Failure Rate: Wearout

Some components have consumable resources. When the resource is gone, the component is worn out. A battery is an obvious example of a component with a consumable resource. A motor bearing is another example. In the case of the battery, a chemical "wearout" occurs. In the case of the motor bearing, a mechanical wearout occurs. Consumption (or wear) occurs in these two components primarily as a function of use. Other components have wearout mechanisms that are independent of use. Electrolyte within an electrolytic capacitor will evaporate. The chemical wearout process occurs even if the component is not being used.

Components wear at different rates. Imagine a life test of 100 fan motors in which the motor bearings wore at exactly the same rate; one day all the motors would fail at the exact same instant. Because components do not wear at the same rate, they do not fail at the same time. However, as a group of components approach wearout, the failure rate increases.

The process of wear can be thought of as a gradual reduction in strength; eventually, the strength drops below that required for normal use. Reductions in strength occur in other ways. Some failures are caused only by repeated stress events. Each stress event lowers the component strength. Electrostatic discharge (ESD) is an example of this. Each time a component receives an ESD strike, it is damaged and its strength is reduced. After some number of ESD hits, the component strength drops below that required for normal operation and the component fails. Any strength reduction process will result in an increasing failure rate (refer back to Figure 3-15, where the failure rate increases rapidly at the end).

Bathtub Curve

A group of modules will likely be susceptible to many kinds of stresses: chemical, mechanical, electrical, and physical. The strength factors as initially manufactured will vary and strength will change at different rates as a function of time. When the failure rates due to these different failure sources are superimposed, the well-known "bathtub curve" results.

The failure rate of the module collection decreases in early life. The failure rate will be relatively constant after the manufacturing defects are removed. This failure rate can be very low if the module has few design faults and high strength. As physical resources on the module are consumed or if other strength reduction factors occur, the failure rate increases (sometimes rapidly), completing the right side of the "bathtub" (Figure 4-7).

There are many variations of the bathtub curve. In some cases no wearout region exists. Some modules rarely have manufacturing defects that are not detected during the manufacturing test process, and these modules do not have a decreasing failure rate region.

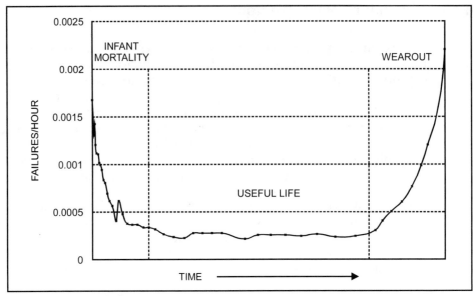

Figure 4-7. Bathtub Curve.

Another variation in failure rate curves is known as the "roller coaster curve." Several different kinds of manufacturing faults are present in a collection of modules. Each manufacturing fault takes a different amount of time to cause failures. Many of the faults are removed by inspection/ test procedures. When the failure rates are modified by inspection procedures, a curve results that looks like part of a roller coaster (Figure 4-8).

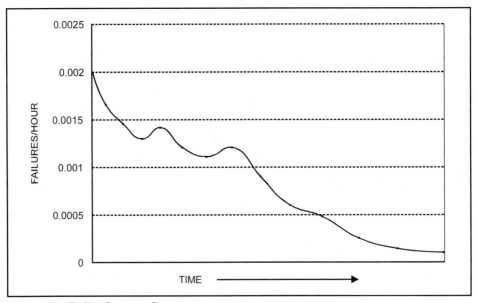

Figure 4-8. Roller Coaster Curve.

Reliability and Safety

The Constant Failure Rate

A useful probability density function in the field of reliability engineering is the exponential. For this distribution:

$$f(t) = \lambda e^{-\lambda t} \tag{4-15}$$

By integrating this function it can be determined that

$$F(t) = 1 - e^{-\lambda t} \tag{4-16}$$

and

$$R(t) = e^{-\lambda t} \tag{4-17}$$

The failure rate equals

$$\lambda(t) = \frac{f(t)}{R(t)} = \frac{\lambda e^{-\lambda t}}{e^{-\lambda t}} = \lambda$$

In other words, a collection of components that have an exponentially decreasing probability of failure will have a constant failure rate. The constant failure rate is characteristic of many kinds of products. As seen in the stress-strength simulations, this is characteristic of a rather uniform constant strength and a random stress. Other products typically exhibit a decreasing failure rate. In these cases, however, the constant failure rate represents a worst-case assumption and can still be used.

The MTTF of a component with an exponential probability density function can be derived from the definition of MTTF:

$$MTTF = \int_0^{+\infty} R(t)\,dt$$

Substituting the exponential reliability function:

$$MTTF = \int_0^{+\infty} e^{-\lambda t}\,dt$$

and integrating,

$$MTTF = -\frac{1}{\lambda}[e^{-\lambda t}]_0^{\infty}$$

When the exponential is evaluated, the value at $t = \infty$ is zero and the value at $t = 0$ is one. Substituting these results, we have a solution:

$$MTTF = -\frac{1}{\lambda}[0 - 1] = \frac{1}{\lambda} \qquad (4\text{-}18)$$

Equation 4-18 is valid for single components with an exponential pdf or a series system composed of components which all have an exponential pdf.

EXAMPLE 4-8

Problem: A motor has a constant failure rate of 150 FITs. What is the motor reliability for a mission time of 1,000 hr?

Solution: Values are substituted in Equation 4-17, including the failure rate of 150/1,000,000,000 and the time interval of 1,000.

$$\begin{aligned} R(1000) &= e^{-0.00000015 \times 1000} \\ &= e^{-0.00015} \\ &= 0.99985 \end{aligned}$$

EXAMPLE 4-9

Problem: Calculate the MTTF of the motor from Example 4-8.

Solution: Since the motor has an exponential pdf, Equation 4-18 can be used. Therefore:

$$\begin{aligned} MTTF &= \frac{1}{0.00000015} \\ &= 6{,}666{,}666.7 \text{ hr} \end{aligned}$$

EXAMPLE 4-10

Problem: Field failure records from an industrial I/O module indicate it has an average failure rate of 272 FITs. What is the module reliability for a mission time of 8,760 hr (one year)?

Solution: Values are substituted in Equation 4-17, including the failure rate of 272/1,000,000,000 and the time interval of 8,760.

$$\begin{aligned} R(t) &= e^{-0.27 \times 10^{-6} \times 8760} \\ &= 0.9976 \end{aligned}$$

> **EXAMPLE 4-11**
>
> **Problem:** Calculate the MTTF of the module from Example 4-10.
>
> **Solution:** Since the module has an exponential pdf, Equation 4-18 can be used. Therefore:
>
> $$MTTF = \frac{1}{0.27 \times 10^{-6}} = 3{,}676{,}470 \text{ hr}$$

A Useful Approximation

Mathematics has shown how certain functions can be approximated by a series of other functions. One of these approximations can be useful in reliability engineering. For all values of x, it can be shown (Ref. 14) that

$$e^x = 1 + x + \frac{x^2}{2!} + \frac{x^3}{3!} + \frac{x^4}{4!} + \dots \qquad (4\text{-}19)$$

For a sufficiently small value of x, the exponential can be approximated with

$$e^x = 1 + x$$

A rearrangement yields

$$1 - e^x \approx -x$$

Substituting gives the result:

$$1 - e^{-\lambda t} \approx \lambda t$$

A single component (or series of components) with an exponential pdf has a failure probability that equals

$$P(\text{failure}) = 1 - e^{-\lambda t} \approx \lambda t \qquad (4\text{-}20)$$

when λt is small. Thus, this probability can be approximated by substituting $\lambda \times t$. This can save engineering time. Be careful, however. The approximation degrades with higher values of failure rates and interval times. Remember, this is not a fundamental formula — only an approximation.

EXAMPLE 4-12

Problem: A component has an exponential probability of failure. The failure rate (lambda) is 0.00001 failures per hour. We wish to calculate the probability of failure for a 1,000 hr mission. If we use the approximation, what is the error?

Solution: Using Equation 4-20, the probability of failure can be approximated.

$$P(\text{failure}) = 0.00001 \times 1000$$
$$= 0.01$$

To check the error, the full formula must be used.

$$F(1000) = 1 - e^{-01}$$
$$= 0.00995017$$

In this case, the error is only 0.00004983375. Note that the approximation gave a pessimistic result. This is on the conservative side when dealing with probabilities of failure.

Availability: Constant Failure Rate Components

Remember that reliability and availability are different. Availability defines the chances of success at any moment in time. Availability represents percentage uptime of a device. Availability over a long period of time, or steady-state availability, is usually the desired measurement. For long-term conditions it is assumed that the "repair rate" is constant and equals 1/MTTR. Repair rate is frequently represented by the lowercase Greek letter mu (μ). The formula is:

$$\mu = \frac{1}{\text{MTTR}} \qquad (4\text{-}21)$$

EXAMPLE 4-13

Problem: Diagnostics detect when a failure occurs within 3 sec. The average repair time is 4 hr. Assume a constant repair rate. What is the repair rate?

Solution: Using Equation 4-21,

$$\mu = \frac{1}{4} = 0.25$$

Reliability and Safety

For single components with constant failure and constant repair rates, availability can be calculated using this formula (see Chapter 8 for more detail):

$$A = \frac{\mu}{\lambda + \mu} \qquad (4\text{-}22)$$

Substituting Equations 4-18 and 4-21 into 4-22 yields:

$$A = \frac{MTTF}{MTTF + MTTR} \qquad (4\text{-}23)$$

This is a common formula for availability. Note that it has been derived for a single component with a constant failure rate and a constant repair rate. It will be shown later in this book that it also applies to series systems; however, note that *it does not apply* to many types of systems, including parallel systems (i.e., systems with redundancy).

EXAMPLE 4-14

Problem: An industrial control module has an MTTF rating of 25 years. Assuming a constant failure rate, what is its failure rate?

Solution: Using Equation 4-18,

$$\begin{aligned}
\lambda &= \frac{1}{MTTF} \\
&= \frac{1}{(25 \times 8760)} \\
&= 0.457 \times 10^{-5} \text{ failures per hr}
\end{aligned}$$

EXAMPLE 4-15

Problem: An industrial control module has an MTTF rating of 25 years. Assuming a constant failure rate, what is the reliability for a one-year period of operation?

Solution: Using Equation 4-16 with l = 0.000004566 failures per hour (4,566 FITs):

$$R(t) = e^{-0.457 \times 10^{-5} \times 8760}$$

$$= 0.9607$$

The reliability can also be calculated using the approximation method: unreliability = λt = 0.04, and reliability = 1 − 0.04 = 0.96, which yields a quicker and more pessimistic answer.

> **EXAMPLE 4-16**
>
> **Problem:** An industrial control module has an MTTF rating of 25 years. Assuming a constant failure rate and a constant repair rate of 0.25 (4 hr MTTR), what is the steady-state availability?
>
> **Solution:** Using Equation 4-23,
>
> $$A = \frac{MTTF}{MTTF + MTTR}$$
>
> $$= 0.99998$$

> **EXAMPLE 4-17**
>
> **Problem:** An industrial I/O module has an MTTF rating of 100 years. Assuming a constant failure rate and an MTTR of 4 hr, what is the reliability for a one-year period of operation and the availability?
>
> **Solution:** The constant failure rate is 0.000001142 (1,142 FITs); the reliability is 0.99; the availability is 0.999995. Note that availability is higher than reliability, as might be expected when looking at Figure 4-3.

Safety Terminology

When evaluating system safety, an engineer must examine more than the probability of successful operation. Failure modes of the system must be reviewed. The normal metrics of reliability, availability, and MTTF only suggest a measure of success. Additional metrics to measure safety include probability of failure on demand (PFD), risk reduction factor (RRF), safety availability, and mean time to fail dangerously (MTTFD). Other related terms are probability of failing safely (PFS), mean time to fail safely (MTTFS), and diagnostic coverage. These terms are especially useful when combined with the other reliability engineering terms.

Failure Modes

Failure modes must be considered in systems designed for safety protection applications. Two failure modes are important: safe and dangerous. ISA's S84.01 defines "safe state" as "state(s) in which the equipment under control, or process, shall attain after proper operation of the SIS." In a majority of the most critical applications, designers choose a de-energized condition as the safe state. A system designed for safety protection applications should de-energize its outputs to achieve a safe state. These systems are called "normally energized."

Reliability and Safety

When a normally energized system is operating successfully (Figure 4-9), the input circuits read the sensors, perform calculation functions, and generate outputs. Input switches are normally energized to indicate a safe condition. Output circuits supply energy to a load (typically a valve). The sensor switch opens (de-energizes) to indicate a potentially dangerous condition. If the logic solver (typically a safety PLC) is programmed to recognize the sensor input as a potentially dangerous condition, it will de-energize its output(s). This action is designed to mitigate the danger.

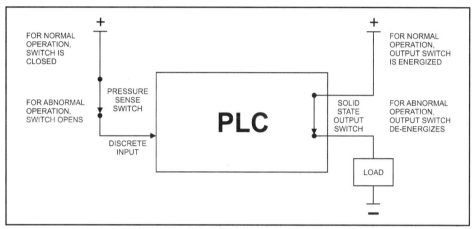

Figure 4-9. Successful Operation - Normally Energized System.

A safe failure in such a system (Figure 4-10) happens when the output de-energizes even though there is no potentially dangerous condition. This is frequently called a "false trip." This can happen in many different ways. Input circuits can fail in such a way that the logic solver thinks a sensor reads danger when it does not. The logic solver itself can miscalculate and command the output to de-energize. Output circuits can fail open-circuit. Many of the components within an SIS can fail in a mode that will cause the system to fail safely.

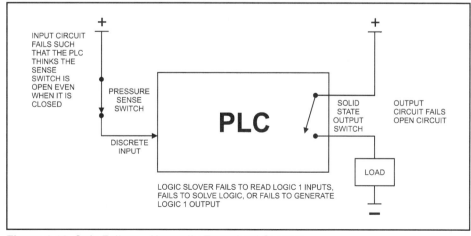

Figure 4-10. Safe Failure - Normally Energized System.

Dangerous failures are defined as failures that prevent the SIS from responding to a potentially dangerous condition known as a "demand." Figure 4-11 shows this situation. Many component failures might cause dangerous system failure, especially if a system is not designed for high safety. A "purpose-built" or safety PLC is specifically designed to avoid this failure mode using a number of design techniques.

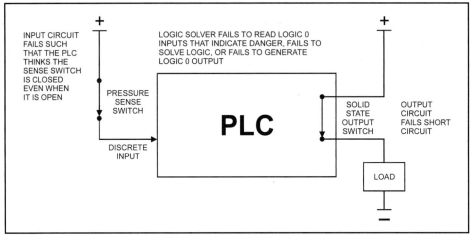

Figure 4-11. Dangerous Failure - Normally Energized System.

PFS/PFD

There is a probability that a normally energized SIS will fail with its outputs de-energized. This is called probability of failing safely (PFS). There is also a probability that the system will fail with its outputs energized. This is called probability of failure on demand (PFD). The latter term refers to the fact that when a safety protection system is failed dangerously, it will *not* respond when a demand (a potentially dangerous condition) occurs. Figure 4-12 shows the relationship of safe and dangerous failure modes to overall system operation.

Unreliability has been defined as the probability that a device is not successful during an interval of time. Unreliability includes all failure modes; therefore:

$$F(t) = PFS(t) + PFD(t) \qquad (4\text{-}24)$$

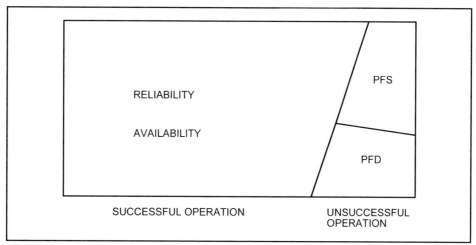

Figure 4-12. Failure Modes.

PFD average (PFDavg) is a term used to describe the average probability of failure on demand. Since PFD (like unreliability) increases with time, the average value over a period of time is calculated (Figure 4-13).

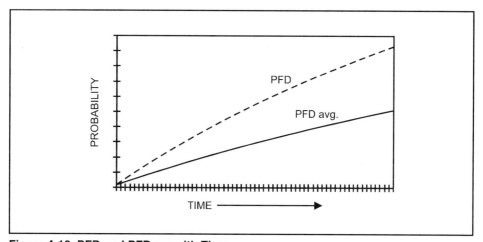

Figure 4-13. PFD and PFDavg with Time.

Unavailability has been defined as the probability that a device is not successful at any moment in time. Unavailability includes all failure modes. Therefore, for repairable systems:

$$U(t) = PFS(t) + PFD(t) \tag{4-25}$$

and

$$A(t) = 1 - [PFS(t) + PFD(t)] \tag{4-26}$$

This applies for time-dependent calculations or steady-state calculations.

Safety Availability

Safety availability, which is different from availability, is defined as the percentage of time that an SIS is able to perform its protective function when the process is operating. Since the process is not operating when shut down, the area in Figure 4-12 marked PFS is excluded from consideration. This situation is shown in Figure 4-14. For this definition, safety availability equals 1 − PFD.

$$SA(t) = 1 - PFD(t) \qquad (4\text{-}27)$$

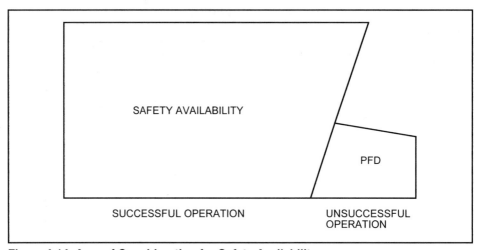

Figure 4-14. Area of Consideration for Safety Availability.

EXAMPLE 4-18

Problem: A steady-state PFS value of 0.001 and a steady-state PFD value of 0.0001 have been calculated for an SIS. What is the availability and the safety availability?

Solution: Using Equation 4-26,

$$A = 1 - (0.001 + 0.0001) = 0.9989$$

Using Equation 4-27,

$$SA = 1 - 0.0001 = 0.9999$$

MTTFS/MTTFD

MTTF describes the average amount of time until a system fails. The definition includes all failure modes. Frequently, engineers need to calculate the average amount of time until the system fails in one mode or another. In the case of SISs, the mean time to fail safely (MTTFS) and the mean time to fail dangerously (MTTFD) are of interest (Figure 4-15).

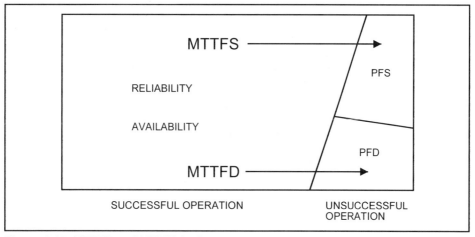

Figure 4-15. MTTFD and MTTFS in Operation.

The exact definition of mean time to failure in any mode must be explained. Consider the failure times shown in Table 4-3. These are identified with failure modes in Figure 4-16. Assume that one PLC is being measured. It starts operating at time $t = 0$ and fails dangerously after 2,327 hr. It is then repaired and operates another 4,016 hr before it fails safely. After repair it operates another 4,521 hr before failing dangerously again. So far the system has failed dangerously twice: the first time after operating 2,327 hr; the second time after operating 8,537 hr (4016 + 4521). In industrial control systems, the measure of interest is the average operating time between failures of a particular mode, ignoring all other modes.

Failure	Time to Fail, in Hours	Mode
1	2,327	Dangerous
2	4,016	Safe
3	4,521	Dangerous
4	3,176	Safe
5	70	Safe
6	3,842	Safe
7	3,154	Safe
8	2,017	Dangerous
9	5,143	Safe
10	4,215	Dangerous

Table 4-3. PLC System Failure Times

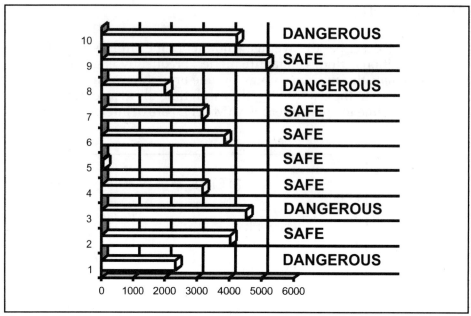

Figure 4-16. Failure Times with Failure Mode.

EXAMPLE 4-19

Problem: A PLC has measured failure data from Table 4-3. What is the MTTFD?

Solution: Four dangerous failures are recorded (Figure 4-17). The total operating times are:

  First dangerous failure   2,327 hr

  Second dangerous failure   8,537 hr (4016 + 4521)

  Third dangerous failure   12,259 hr (3176 + 70 + 3842 + 3154 + 2017)

  Fourth dangerous failure   9,358 hr (5143 + 4215)

The average of these four values is 8,120 hr. Another way to calculate the MTTFD would be to note that the total operating hours is 32,481 divided by four dangerous failures.

Reliability and Safety 85

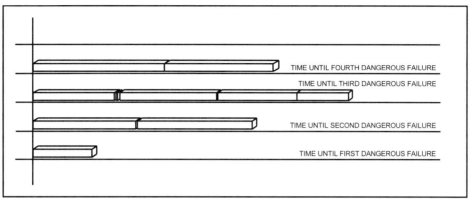

Figure 4-17. Operating Times until Dangerous Failure.

Diagnostic Coverage

The ability to detect a failure is an important feature in any control or safety system. This feature can be used to reduce repair times and to control operation of several fault-tolerant architectures. The measure of this ability is known as the diagnostic coverage factor, C. The diagnostic coverage factor measures the probability that a failure will be detected, given that it occurs. Diagnostic coverage is calculated by adding the failure rates of detected failures and dividing by the total failure rate. As an example, consider a system of 10 components. The failure rates and detection abilities are as follows:

Component	Failures/hr	Detection Ability
1	0.00991	Detected
2	0.00001	Not detected
3	0.00001	Not detected
4	0.00001	Not detected
5	0.00001	Not detected
6	0.00001	Not detected
7	0.00001	Not detected
8	0.00001	Not detected
9	0.00001	Not detected
10	0.00001	Not detected

Although only one component failure out of a possible 10 is detected, the coverage factor is 0.991 (or 99.1%). For this example, the detected failure rate is 0.00991. This number is divided by the total failure rate of 0.01. The coverage factor is not 0.1, as might be assumed by dividing the number of detected failures by the total number of known possible failures.

In control system reliability and safety analysis, it is generally necessary to define the coverage factor for safe failures and the coverage factor for dangerous failures. The superscript S is used for the safe coverage factor, C^S. The superscript D is used for the dangerous coverage factor, C^D. The evaluation of PFS and PFD will be affected by each different coverage factor.

In some fault-tolerant architectures, two additional coverage factor designations may be required. Detection of component failures is done by two different techniques, classified as reference or comparison. Reference diagnostics can be done by a single unit. The coverage factor of reference diagnostics will vary widely depending on the implementation, with most results ranging from 0.0 to 0.999. Comparison diagnostics will require two or more units. The coverage factor depends on implementation, but results are generally good, with most results ranging from 0.9 to 0.999.

Reference diagnostics utilize the predetermined characteristics of a successfully operating unit. Comparisons are made between actual measured parameters and the predetermined values for these parameters. Measurements of voltages, currents, signal timing, signal sequence, and temperature are utilized to accurately diagnose component failures. Advanced reference diagnostics include digital signatures and frequency domain analysis.

Comparison diagnostic techniques depend on comparing data between two or more operating units within a system. The coverage factor will vary since there are trade-offs between the amount of data compared and the diagnostic coverage effectiveness.

In some fault-tolerant architectures, the diagnostic coverage factor changes when the system is operating in a degraded condition. The appropriate coverage factor must be used in reliability and safety analysis depending on the operational state of the system. When fault-tolerant systems that normally compare degrade to a single operational unit, the comparison must stop and the reference coverage factor must be used. The following coverage factor notation is defined:

> C — Coverage due to either reference diagnostics, comparison diagnostics, or a combination of both.
>
> C_1 — Coverage due to single-unit reference diagnostics.
>
> C_2 — Coverage due to comparison diagnostics, where C always equals the higher of the two values.

EXERCISES

4.1 Which term requires successful operation for an interval of time, reliability or availability?

4.2 Which term is more applicable to repairable systems, reliability or availability?

4.3 Unavailability for a system is given as 0.001. What is the availability?

4.4 When does the formula MTTF = $1/\lambda$ apply?

4.5 Availability of the process control system is quoted as 99.9%. What is the unavailability?

4.6 A control module has an MTTF of 60 years. Assuming a constant failure rate, what is the failure rate?

4.7 A control module has an MTTF of 60 years. It has an average repair time of 8 hr. What is the steady-state availability?

4.8 A control module has an MTTF of 60 years. What is the reliability of this module for a time period of six months?

4.9 An SIS has a PFS of 0.002 and a PFD of 0.001. What is the availability?

4.10 An SIS has a PFS of 0.002 and a PFD of 0.001. What is the safety availability?

REFERENCES AND BIBLIOGRAPHY

1. Billinton, R., and Allan, R. N., *Reliability Evaluation of Engineering Systems: Concepts and Techniques*, New York: Plenum Press, 1983.

2. Des Plas, E. P., "Reliability in the Manufacturing Cycle," *Proceedings of the Annual Reliability and Maintainability Symposium*, New York: IEEE, 1986.

3. Wong, K. L., "Off the Bathtub onto the Roller-Coaster Curve," *Proceedings of the Annual Reliability and Maintainability Symposium*, New York: IEEE, 1988.

4. Wong, K. L., "The Physical Basis for the Roller-Coaster Hazard Rate Curve For Electronics," *Quality and Reliability Engineering International*, Vol. 7, No. 6, 1991.

5. O'Connor, P. D. T., *Practical Reliability Engineering*, Milwaukee: ASQC Quality Press, 1985.

6. Wong, K. L., "A New Framework for Electronic Assembly/System Reliability Prediction," *Proceedings of the Annual Reliability and Maintainability Symposium*, New York: IEEE, 1997.

7. Grebe, J. C., "How to Increase Control System Reliability," *Control Engineering*, February 1995.

5
Failure Modes and Effects Analysis

Introduction

A failure modes and effects analysis (FMEA) is a systematic technique that is designed to identify problems. It is a "bottom-up" method that starts with a detailed list of all components within the system. An entire system can be analyzed one component at a time. Alternatively, the system can be hierarchically divided into subsystems and modules as required. The FMEA can be done on each grouping in the hierarchy. The primary reference for the FMEA technique is MIL-STD-1629 (Ref. 1).

FMEA Procedure

The minimum steps required in the FMEA process are simple:

1. List all components.

2. For each component, list all failure modes.

3. For each component/failure mode, list the effect on the next higher level.

4. For each component/failure mode, list the severity of effect.

An FMEA can be very effective in identifying critical failures within a system. One of the primary reasons for doing this is so that the system design can be changed to eliminate critical failures. For this reason, the best possible time to do an FMEA is during the design phase of a project. The FMEA should be done while design changes can still be made without disrupting the entire project. Ideally, the completed FMEA will have no critical failures identified. All will have been designed out!

An FMEA also provides important documentation input to the reliability and safety evaluation. The various failure modes in components or modules that must be modeled at the next higher level are identified.

FMEA Limitations

The FMEA technique does not identify combinations of failures or common-cause failures that cause critical problems. Since each component is reviewed individually, combinations are not addressed. In fault-tolerant systems, common-cause failures are rarely identified since they require more than one component failure.

Operational and maintenance failures are also likely to be missed during the FMEA unless the reviewer is skilled in human reliability analysis and recognizes component failure modes due to human interaction. In general, the skill and persistence of the reviewer is important to the quality of an FMEA. All failure modes of components must be known, or else they will not be included.

FMEA Format

An FMEA is documented in a tabular format as shown in Table 5-1. Computer spreadsheets are ideal tools for this tabular format. Each column in the table has a specific definition.

Column 1 describes the name of the device under review. Depending on the scope of the FMEA, this could be a component, a module, or a unit. Column 2 is available to list the part number or the code number of the device under review. Column 3 describes the function of the component. A good functional description of each component can effectively help to document system operation.

Column 4 describes the failure modes of the component. One row is typically used for each component failure mode. Examples of component failure modes include fail short, fail open, drift, stuck at one, stuck at zero, etc., for electronic components. Mechanical switch failure modes might include stuck open, stuck closed, contact weld, ground short, etc. References 2 and 3 provide database listings of failure modes for possible system components. Column 5 describes the cause of the failure mode of column 4. Generally, this is used to list the primary "stress" causing the failure — for example, heat, chemical corrosion, dust, electrical overload, RFI, human operational error, etc. (see Chapter 3).

Column 6 describes how this component failure mode affects the component function of the module (or subsystem). Column 7 lists how this component failure mode affects the next system sublevel. In safety evaluations, this column is used to indicate safe versus dangerous failures. Depending on the scope of the FMEA, it is possible to consider all levels of the system (component, module, unit, or system). Frequently, an FMEA at the component level describes the effect at the module level and perhaps the effect at the unit level. Unless the FMEA is being done for a specific system architecture, another FMEA is done for higher levels.

Column 8 lists the failure rate of the particular component failure mode. The use of this column is optional when FMEAs are being done for qualitative purposes. When quantitative failure rates are desired and specific data for the application are not available, failure rates and failure mode percentages are available from handbooks (Refs. 2-5).

Lastly, column 9 is reserved for comments and relevant information. This area gives the reviewer an opportunity to suggest improvements in design, methods to increase component strength (against the perceived stress), or perhaps needed user documentation considerations.

EXAMPLE 5-1

Problem: Figure 5-1 shows a simplified reactor with an emergency cooling system. The system consists of a gravity-feed water tank, a control valve (VALVE1), a cooling jacket around the reactor, a cooling jacket drain pipe, a temperature-sensing switch (TSW1), and a power supply. In normal operation, the temperature-sensing switch is closed because reactor temperature is below a dangerous limit. Electrical current flows from the power supply through the valve and the temperature-sensing switch. This electrical current (energy) keeps the valve closed. If the temperature inside the reactor gets too hot, the temperature-sensing switch opens. This stops the flow of electrical current, and the control valve opens. Cooling water flows from the tank, through the valve, through the cooling jacket, and through the jacket drain pipe. This water flow cools the reactor, lowering its temperature. Perform an FMEA for this system.

Solution: The FMEA procedure requires creation of a list with all failure modes for each of the system components, then filling out each relevant column for each failure in the list. Table 5-1 shows the results of this system-level FMEA. The FMEA has identified six critical items that should be reviewed to determine the need for correction. We could consider installing two temperature switches and wiring them in series. This would eliminate a critical failure when one temperature-sensing switch fails shorted. Or, we could install two drain pipes and pipe them in parallel to prevent a single clogged drain from causing a critical failure. A level sensor on the water tank could warn of insufficient water level. Many other possible design changes could be made to mitigate the critical failures or to reduce the number of false trips. Did the FMEA reviewer miss any failure modes of any components?

Table 5-1. FMEA for Example 5-1

1 Name	2 Code	3 Function	4 Mode	5 Cause	6 Effect	7 Criticality	8 Failure Rate	9 Remarks
Cooling Tank		Water storage	Leak Plugged outlet	Corrosion Dirt	Lost water No water	Dangerous Dangerous		Consider design change to detect Second outlet?
Valve	VALVE1	Open for coolant	Jam closed Fail open Coil open Coil short	Dirt, corr. Corr., power Elec. surge Corr., wire	No water False trip False trip False trip	Dangerous Safe Safe Safe		Second valve?
Jacket		Path for coolant	Leak Clog	Dirt, corr.	None No water	None Dangerous		Small flow in normal operation?
Drain pipe		Path for coolant	Clog	Dirt, corr.	No water	Dangerous		
Temp. switch	TSW1	Sense overtemp.	Short Open	Elec. surge	No cooling False trip	Dangerous Safe		Two switches?
Power supply	PS1	Energy for valve	Short Open	Maint. error Many	False trip False trip	Safe Safe		

Failure Modes and Effects Analysis

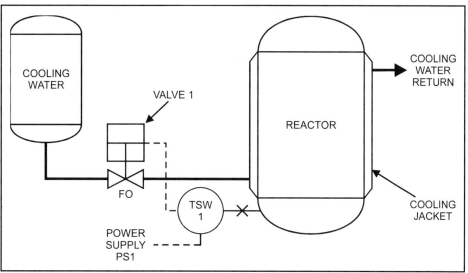

Figure 5-1. Cooling System.

EXAMPLE 5-2

Problem: A high-pressure protection system is shown in Figure 5-2. Three pressure switches sense pressure in a reaction tank. Each pressure switch has two contact outputs, and the six contacts are wired in a two-out-of-three voting circuit. This system is normally energized. When all components are operating properly, excessively high pressure in the tank opens the contacts on all switches. The electrical circuit de-energizes and two valves close, which cuts fuel flow to a burner. This reduces pressure in the tank to a safe value. Perform an FMEA for the system.

Solution: The FMEA is shown in Table 5-2. Notice that no overall critical effects were discovered. Did the FMEA reviewer find everything?

Failure Modes, Effects, and Diagnostic Analysis (FMEDA)

The FMEA approach can be extended to include an evaluation of the diagnostic ability of a circuit, module, unit, or system. Additional columns are added to the chart as shown in Table 5-3.

The 10th column is an extension to the standard for the purpose of identifying that this component failure is detectable by on-line diagnostics. A number "1" is entered to designate that this failure mode is detectable. A number "0" is entered if the failure mode is not detectable. Column 11 is an extension to the standard used to identify the diagnostic used to detect the failure. The name of the diagnostic should be listed. Perhaps the error code generated or the diagnostic function could also be listed.

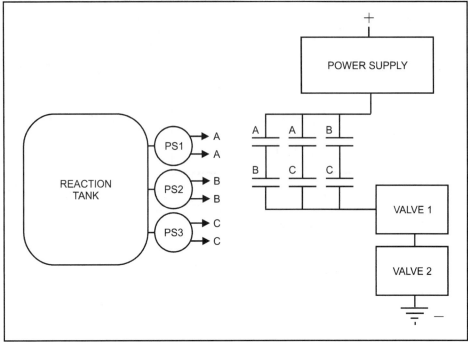

Figure 5-2. Pressure Protection System.

Column 12 is used to numerically identify failure mode. A "1" is entered for safe failure modes. A "0" is entered for dangerous failure modes. The number is used in spreadsheets to calculate the various failure rate categories. The safe detected failure rate is listed in column 13. This number can be calculated using the previously entered values if a spreadsheet is used for the table. It is obtained by multiplying the failure rate (column 8) by the failure mode number (column 12) and the detectability (column 10).

The safe undetected failure rate is shown in column 14. This number is calculated by multiplying the failure rate (column 8) by the failure mode number (column 12) and one minus the detectability (column 10). Column 15 lists the dangerous detected failure rate. It is obtained by multiplying the failure rate (column 8) by one minus the failure mode number (column 12) and the detectability (column 10). Column 16 shows the calculated failure rate of dangerous undetected failures. It is obtained by multiplying the failure rate (column 8) by one minus the failure mode number (column 12) and one minus the detectability (column 10).

Failure Modes and Effects Analysis

1 Name	2 Code	3 Function	4 Mode	5 Cause	6 Effect	7 Criticality	8 Failure Rate	9 Remarks
Valve	VALVE1	Cuts fuel flow when closed	Jam open Fail closed Coil open Coil short	Dirt, corrosion Corrosion, power Elec. surge Corrosion, wire	May not cut off fuel False trip False trip False trip	Dangerous Safe Safe Safe		Second valve still cuts fuel
Valve	VALVE2	Cuts fuel flow when closed	Jam open Fail closed Coil open Coil short	Dirt, corrosion Corrosion, power Elec. surge Corrosion, wire	May not cut off fuel False trip False trip False trip	Dangerous Safe Safe Safe		Second valve still cuts fuel
Pressure switch	PS1	Sense overpressure	Short Open	Power surge Many	No false trip, no danger False trip	Safe Safe		Redundant switches still work Detected?
Pressure switch	PS1	Sense overpressure	Short Open	Power surge Many	False trip	Safe Safe		Redundant switches still work Detected?
Pressure switch	PS1	Sense overpressure	Short Open	Power surge	False trip	Safe Safe		Redundant switches still work Detected?
Power supply		Energy for valve	Short Open	Maintenance error Many	False trip False trip	Safe Safe		Fail safe Fail safe

Table 5-2. FMEA for Pressure Protection System

Conventional Input Circuit

The FMEDA analysis is very useful for evaluation of electronic circuits. Consider the PLC input circuit of Figure 5-3. This design is optimized for low cost. In normal operation, an AC voltage is applied to the circuit terminals on the left of the drawing. A filter circuit and a voltage divider supply reduced voltage to diodes D1 and D2. These diodes limit the voltage to the opto-coupler circuit OC1. The output of the opto-coupler conducts when AC voltage is present and the microcomputer reads a 5-v signal across R4. When the AC input voltage goes to low (zero), the voltage across R4 goes to zero.

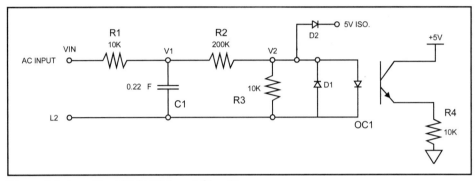

Figure 5-3. Low Cost PLC AC Input Circuit.

The FMEDA for this circuit is shown in Table 5-3. Component failure rates must be listed for each component failure mode since diagnostic coverage calculations are done based on a weighted average of failure rates (see Chapter 3).

This circuit has no intrinsic diagnostic capability. The microprocessor (or other diagnostic circuit) has no way to determine whether the reading is due to a sensor input or a failed component. This type of circuit is not appropriate for use in safety PLCs or high-availability systems.

Critical Input Circuit

Another AC input circuit in shown in Figure 5-4. This circuit was designed for high diagnostic coverage. Very high levels of diagnostics are needed for high-safety and high-availability systems. The circuit uses two sets of opto-couplers. The microprocessor that reads the inputs can read both opto-couplers. Under normal operation, both readings should be the same. In addition, readings must be taken four times per AC voltage cycle. This allows the microprocessor to read a dynamic signal. When all components are operating properly, a logic 1 is a series of pulses. This circuit design is biased toward fail-safe with a normally energized input sensor. Table 5-4 shows the FMEDA.

Failure Modes and Effects Analysis

1	2	3	4	5	6	7	8	9	10	11	12	13	14	15	16
Name	Code	Function	Mode	Cause	Effect	Criticality	Failure Rate	Remarks	Detectability	Diagnostics	Mode	SD	SU	DD	DU
R1-1K	4555-1	Input filter	Short		No filter	Safe	0.5		0		1	0	0.5	0	0
			Open	Open contact	Read 0	Safe	1		0		1	0	1	0	0
			Drift low		Low filter	Safe	0.01	Too low	0		1	0	0.01	0	0
			Drift high			Safe	0.01	Too high	0		1	0	0.01	0	0
C1-0.18	4600-2	Input filter	Short	Overvoltage	No input	Safe	4		0		1	0	4	0	0
			Open	Open contact	No filter	Safe	2		0		1	0	2	0	0
R2-200K	4555-200	I limit	Short		Blow in	Dangerous	0.5	Damage	0		0	0	0	0	0.5
			Open	Open contact		Safe	1		0		1	0	1	0	0
			Drift low		Hi I	Dangerous	0.01	Damage	0		0	0	0	0	0.01
			Drift high			Safe	0.01	Too high	0		1	0	0.01	0	0
R3-10K	4555-10	Pull down	Short		Read 0	Safe	0.5		0		1	0	0.5	0	0
			Open	Open contact		Dangerous	1		0		0	0	0	0	1
			Drift low	Low (1)	Low filter	Safe	0.01	Too low	0		1	0	0.01	0	0
			Drift high			Safe	0.01	Too high	0		0	0	0	0	0.01
D1	4800-4	Voltage clip	Short	(2)	Short in	Safe	4		0		1	0	4	0	0
			Open	Open contact	No prot.	Dangerous	4		0		0	0	0	0	4
D2	4800-4	Voltage limit	Short		Jam in 1	Dangerous	4		0		0	0	0	0	4
			Open		Overvoltage	Dangerous	4	Damage	0		0	0	0	0	4
OC1	5100-6	Isolator	Diode short	Wearout (3)	Read 0	Safe	45	No light	0		1	0	45	0	0
			Diode open	No contact	Read 0	Safe	2	No light	0		1	0	2	0	0
			Tran. short		Jam in 1	Dangerous	10		0		0	0	0	0	10
			Tran. open		Read 0	Safe	10		0		1	0	10	0	0
			Total failure rate				93.56								
			Total safe failure rate				93.56			C Safe	0.00				
			Total dangerous failure rate				70.04			C Dang.	0.00				
			Safe detected failure rate				23.52			C Total	0.00				
			Safe undetected failure rate				0								
			Dangerous detected failure rate				70.04								
			Dangerous undetected failure rate				0								
							23.52						0.04	0	23.5

(10^{-9} failure per hour)

1. Low threshold may read as energized when not.
2. External electrical stress may short this diode.
3. Light may also go dim slowly as a wearout mechanism.

Table 5-3. FMEDA for Conventional PLC Input Circuit

Failure Modes and Effects Analysis

1	2	3	4	5	6	7	8	9	10	11	12	13	14	15	16
Name	Code	Function	Mode	Cause	Effect	Criticality	Failure Rate	Remarks	Detect-ability	Diagnostics	Mode	SD	SU	DO	DU
R1-10K	4555-10	Input threshold	Short		Threshold shift	Safe	0.125		0		1	0	0.13	0	0
			Open	Solder Open	Open circuit	Safe	0.5		1	Lose input pulse	1	0.5	0	0	0
			Drift low			Safe	0.01	None until too low	0		1	0	0.01	0	0
			Drift high			Safe	0.01	None until too high	1	Lose input pulse	1	0.01	0	0	0
R2100K	4555-100	Current limit	Short		Short input	Safe	0.125		1		1	0.13	0	0	0
			Open	Solder Open		Safe	0.5		1	Lose input pulse		0	0	0.5	0
			Drift low			Safe	0.01	None until too low	0		1	0	0.01	0	0
			Drift high			Safe	0.01	None until too high	1	Lose input pulse	1	0.01	0	0	0
D1	4200-7	Voltage drop	Short	Surge	Overvoltage	Safe	2		1	Lose input pulse	1	2	0	0	0
			Open		Open circuit	Safe	5		1	Lose input pulse	1	5	0	0	0
D2	4200-7	Voltage drop	Short	Surge	Overvoltage	Safe	2		1	Lose input Pulse	1	2	0	0	0
			Open		Open circuit	Safe	5		1	Lose input pulse	1	5	0	0	0
OC1	4805-25	Isolate	Led dim	Wear	No light	Safe	28		1	Comp. mismatch	1	28	0	0	0
			Tran. short	Internal short	Read logic 1	Dang.	10		1	Comp. mismatch	0	0	0	10	0
			Tran. open		Read logic 0	Safe	6		1	Comp. mismatch	1	6	0	0	0
OC2	4805-25	Isolate	Led dim	Wear	No light	Safe	28		1	Comp. mismatch	1	28	0	0	0
			Tran. short	Internal short	Read logic 1	Dang.	10		1	Comp. mismatch	0	0	0	10	0
			Tran. open		Read logic 0	Safe	6		1	Comp. mismatch	1	6	0	0	0
OC1/OC2			Cross channel	Short	Same signal	Dang.	0.01		0		0	0	0	0	0.01
R3-100K	4555-100	Filter	Short		Lose filter	Safe	0.125		0		1	0	0.13	0	0
			Open		Input float high	Dang.	0.5		1	Comp. mismatch	0	0	0	0.5	0
R4-10K	4555-10	Voltage divider	Short		Read logic 0	Safe	0.125		1	Comp. mismatch	1	0.13	0	0	0
			Open		Read logic 1	Dang.	0.5		1	Comp. mismatch	0	0	0	0.5	0
R5-100K	4555-100	Filter	Short		Lose filter	Safe	0.125		0		1	0	0.13	0	0
			Open		Input float high	Dang.	0.5		1	Comp. mismatch	0	0	0	0.5	0
R6-10K	4555-10	Voltage divider	Short		Read logic 0	Safe	0.125		1	Comp. mismatch	1	0.13	0	0	0
			Open		Read logic 1	Dang.	0.5		1	Comp. mismatch	0	0	0	0.5	0
C1	4350-32	Filter	Short		Read logic 0	Safe	2		1	Comp. mismatch	1	2	0	0	0
			Open		Lose filter	Safe	0.5		0		1	0	0.5	0	0
C2	4350-32	Filter	Short		Read logic 0	Safe	2		1	Comp. mismatch	1	2	0	0	0
			Open		Lose filter	Safe	0.5		0		1	0	0.5	0	0
							110.8					86.9	1.4	22.5	0.01
			Total failure rate				110.8			Safe coverage	0.9839				
			Total safe failure rate				88.29			Dangerous coverage	0.9996				
			Total dangerous failure rate				22.51								
			Safe detected failure rate				86.895								
			Safe undetected failure rate				1.395								
			Dangerous detected failure rate				22.5								
			Dangerous undetected failure rate				0.01								
						Failures per billion hours									

Table 5-4. FMEDA for High Diagnostic Input Circuit

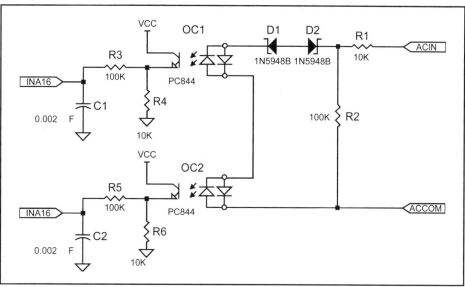

Figure 5-4. High Diagnostic PLC AC Input Circuit.

FMEDA Limitations

The FMEDA technique is most useful on simple circuits and processes where the failure modes of the components are relatively well known. Within the context of electronic components, the technique is less useful on complex VLSI integrated circuits where many possible failure modes may exist. It is generally not considered practical to list all possible failure modes for a microprocessor, for example. However, it is practical to list the failure modes of processor I/O pins where shorts and opens falsely cause highs or lows. Many other failure modes where program execution is interrupted should be listed since several techniques to detect these failures exist.

It is also very useful to list known failure modes of a particular microprocessor based on the microprocessor manufacturer's field failure data. In many cases, only about 20 or so actual failures have occurred. For these known, documented failures, the effects can be stated for a particular microcomputer circuit design. Despite the limitations, the FMEDA has been shown to be effective in finding potential problems in a design.

EXERCISES

5.1 List the steps for an FMEA.

5.2 List the limitations of an FMEA.

5.3 Can results of an FMEDA be verified?

5.4 Should an FMEA or an FMEDA be reviewed?

5.5 Perform an FMEA on a simple process in your plant.

REFERENCES AND BIBLIOGRAPHY

1. MIL-STD-1629, *Failure Mode and Effects Analysis*, Springfield, VA: National Technical Information Service.

2. NPRD, *Nonelectronic Parts Reliability Data*, Rome, NY: Reliability Analysis Center, 1997.

3. FMD-91, *Failure Mode/Mechanism Distributions*, Rome, NY: Reliability Analysis Center, 1991.

4. MIL-HNBK-217F, *Failure Rates for Electronic Components*, Springfield, VA: National Technical Information Service.

5. OREDA-92, *Offshore Reliability Data*, Hovik, Norway: DNV Industry, 1992.

6. Montgomery, T. A., and Marko, K. A., "Quantitative FMEA Automation," *Proceedings of the Annual Reliability and Maintainability Symposium*, New York: IEEE, 1997.

7. Onodera, K., "Effective Techniques of FMEA at Each Life Cycle Stage," *Proceedings of the Annual Reliability and Maintainability Symposium*, New York: IEEE, 1997.

8. Price, C. J., "Effortless Incremental Design FMEA," *Proceedings of the Annual Reliability and Maintainability Symposium*, New York: IEEE, 1996.

9. Price, C. J., Pugh, D. R., Wilson, M. S., and Snooke, N., "The Flame System: Automating Electrical Failure Mode and Effects Analysis," *Proceedings of the Annual Reliability and Maintainability Symposium*, New York: IEEE, 1995.

10. *Guidelines for Process Equipment Reliability Data, with Data Tables*, New York: American Institute of Chemical Engineers, 1989.

6
Fault Tree Analysis

Introduction

A fault tree analysis (FTA) is a top down approach to the identification of system problems. It was developed by H. A. Watson of Bell Laboratories in 1961-62 as part of the Minuteman missile program (Ref. 1). Fault trees were created to help identify design problems in complex systems. Fault trees are very complementary to the FMEA in that they require a deductive approach to problem-finding. The fault tree method is adept at finding combinations of failures that may cause problems and helps the analyst focus on one failure type at a time. A fault tree helps identify which parts of a system are related to a particular failure.

The end result of an FTA is a diagram that graphically shows combinations of events that can cause system failure, thus pinpointing weaknesses in a system. A fault tree, when correctly done, can be a valuable engineering document describing how the system is supposed to operate under various fault conditions. This provides necessary documentation for more detailed reliability and safety analysis.

While it is true that fault trees are used primarily during engineering design to help identify potential design weaknesses, fault trees can also be of great value when investigating causes of failures or accidents. All the trigger events and contributing events can be documented in a graphical format showing the overall relationship between events and the resultant accident.

Fault Tree Method

The process starts by identifying a problem. The fault tree review team must then study system operation and develop a good understanding of how the system works. Knowing how the system is supposed to work, the review team then tries to find all possible ways in which a particular problem could occur. For each item identified, the process continues until trigger events or basic faults are identified on the chart. The process is diagrammed in Figure 6-1.

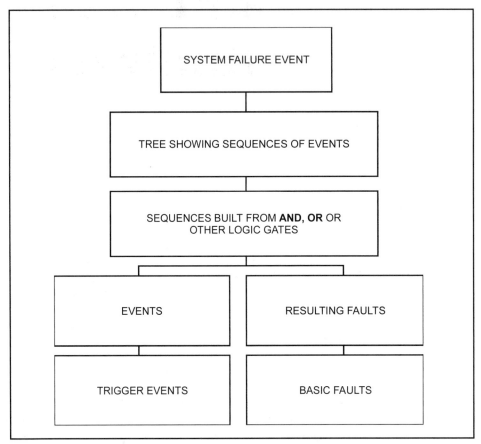

Figure 6-1. Fault Tree Process.

The diagram is constructed from the top. We identify a problem: fire starts. Within a normal atmosphere, we know that two additional factors are required to start a fire: ignition source and combustible material. Working down the tree, we identify sources of combustible material and the basic faults involved. We also identify trigger events that may provide an ignition source. An example fault tree is shown in Figure 6-2.

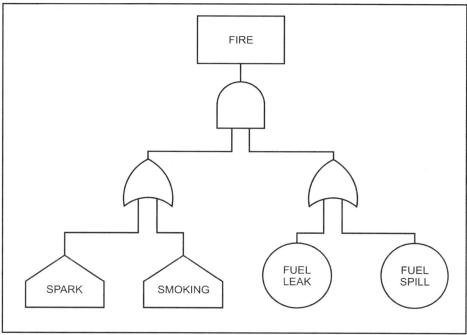

Figure 6-2. Example Fault Tree.

Fault Tree Symbols

The symbols used in a fault tree have special meanings (Figure 6-3). The most common symbol is the "event" or "resulting fault," represented by a rectangle. An event or resulting fault is typically a result of some combination of events as documented by the logic gates: AND/OR. A diamond is used to signify an incomplete event, indicating that perhaps other events could also be involved but are not of interest to the analysis. The hexagon symbol is called an inhibit gate. The inhibit gate is functionally similar to a two-input AND gate except that it indicates an event is not necessarily a direct cause. An example of the inhibit gate is shown in Figure 6-4. There is a chance that the operator may hit the wrong button; this is indicated by the use of the inhibit gate. Figure 6-4 is functionally equivalent to Figure 6-5.

At the bottom of the fault tree are "basic faults" and "trigger events." These are normally considered to be the root cause elements of any failure. The basic fault is represented by a circle. A trigger event is represented by the house symbol.

104   Fault Tree Analysis

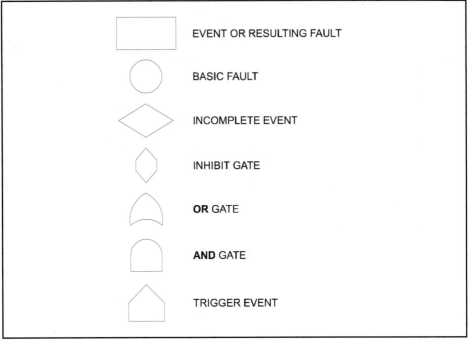

Figure 6-3. Fault Tree Symbols.

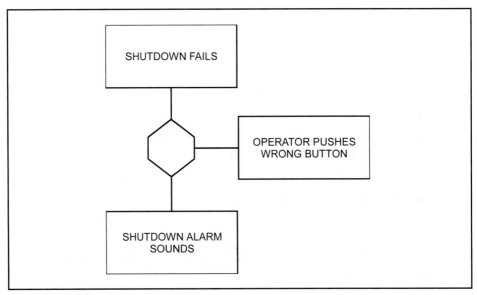

Figure 6-4. Inhibit Gate.

Fault Tree Analysis

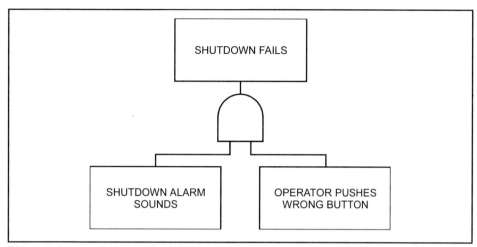

Figure 6-5. Functional Equivalent of Inhibit Gate.

Additional Symbols

Additional symbols are defined in some references. Though these symbols are not common, they may appear in some diagrams (Figure 6-6). The conditional event symbol is used with the inhibit gate to show a probabilistic event. Sometimes an oval is also used as this symbol. The transfer-in and transfer-out symbols are used to connect multiple drawing sheets. A double diamond is sometimes used to indicate that an incomplete event really should be investigated further. The priority AND gate is used to show that inputs must be received in a particular sequence. These symbols are not used as frequently as those in Figure 6-3.

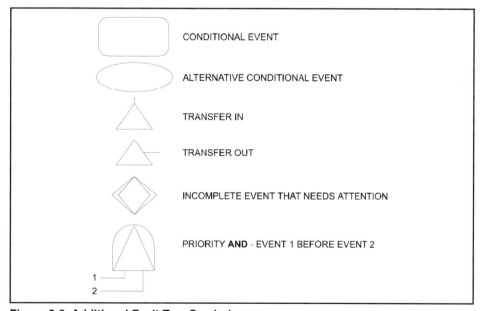

Figure 6-6. Additional Fault Tree Symbols.

Qualitative Fault Tree Analysis

In many cases, a fault tree is used to document system design. It is used as an inspection tool when doing system design review. Many potential weaknesses in system design are discovered. Consider the fault tree in Figure 6-7, where a power system is reviewed. The system has three independent sources of power: the commercial utility, a local generator, and a battery system. The AND gate at the top of the drawing indicates that all three sources must fail in order to lose power.

There are many ways in which utility power can fail, but the main concern is the protective breaker. Therefore, the drawing shows the incomplete event symbol OR utility breaker blown. Generator failure occurs when the generator is off-line and the operator fails to restart. The battery system fails when the batteries are discharged and the charger fails.

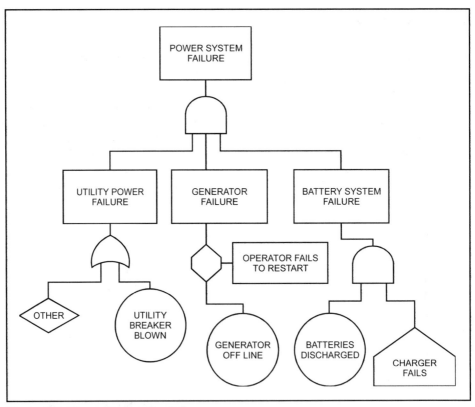

Figure 6-7. Power System Fault Tree.

Quantitative Fault Tree Analysis

Fault trees can be used as a quantitative analysis tool. Probabilities are assigned to basic faults and trigger events. The rules of probability are used to combine these numbers.

AND Gates

With an AND gate, all inputs must be present for the gate to be active. For example, with a two-input AND gate, events A and B both must occur for the gate be active. Referring to failure events, both failures must occur for the gate to be failed. If these events are independent (see Appendix C for a review of probability theory, including definitions of independent events and mutually exclusive events), then the probability of getting failure event A and failure event B is given by:

$$P(A \cap B) = P(A) \times P(B) \qquad (6\text{-}1)$$

Example 6-1

Problem: A power system has a fault tree shown in Figure 6-8. Unavailability of the battery system is estimated to be 0.01 (probability of failure at any moment in time). Unavailability of commercial power is estimated to be 0.0001. What is the unavailability of the power system?

Solution: The fault tree can be applied quantitatively. Since both power sources must fail for the system to fail, Equation 6-1 can be used.

$$\text{Unavailability for the system} = 0.01 \times 0.0001$$
$$= 0.000001$$

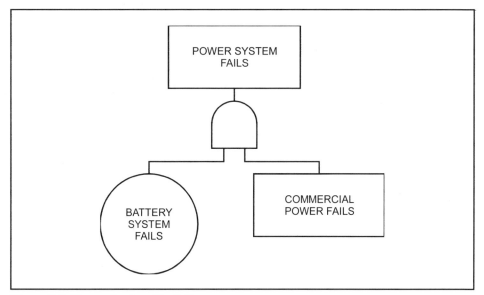

Figure 6-8. Power System Fault Tree.

> **Example 6-2**
>
> **Problem:** Consider the inhibit gate of Figure 6-4. There is a 5% chance that the alarm will sound. If the alarm sounds, there is a 1% chance that the operator pushes the wrong button. What is the chance that the shutdown fails?
>
> **Solution:** Since the inhibit gate is functionally the same as an AND gate, Equation 6-1 applies.
>
> $$\text{Chance of shutdown failure} = 0.01 \times 0.05$$
> $$= 0.0005$$

OR Gates

With an OR gate, any active input indicates the gate will be active. Quantitative evaluation requires summation of the probabilities. For mutually exclusive events:

$$P(A \cup B) = P(A) + P(B) \tag{6-2}$$

For nonmutually exclusive events (normally the case):

$$P(A \cup B) = P(A) + P(B) - P(A \cap B) \tag{6-3}$$

> **Example 6-3**
>
> **Problem:** Two pressure transmitters are used to sense pressure in a reactor. The two analog signals are wired into a safety PLC, and a comparison function is used to detect if the transmitters ever differ by more than 2%. If they do, the sensor subsystem is assumed to have failed. (This is equivalent to a 1oo2 analog voting scheme.) The fault tree for this configuration is shown in Figure 6-9. If unavailability of a pressure transmitter is 0.01, what is the unavailability of the subsystem?
>
> **Solution:** The failure of one pressure transmitter does not preclude failure of the second; therefore, these two events are not mutually exclusive. Equation 6-3 applies:
>
> $$P(\text{subsystem failure}) = 0.01 + 0.01 - (0.0001)$$
> $$= 0.0199$$

Fault Tree Analysis

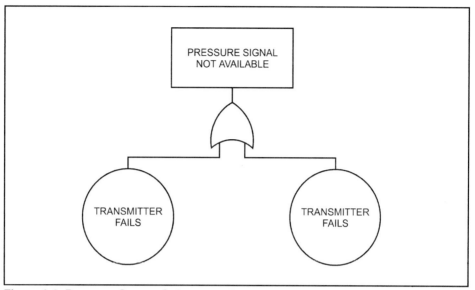

Figure 6-9. Pressure Sensor Subsystem Fault Tree.

Example 6-4

Problem: Three thermocouples are used to sense temperature in a reactor. The three signals are wired into a safety PLC, and a comparison function is used to detect if the signals ever differ by more than 5%. If they do, the sensor subsystem is assumed to have failed. (This is equivalent to a 1oo3 analog voting scheme.) The fault tree for this configuration is shown in Figure 6-10. If unavailability of a thermocouple is 0.005, what is the unavailability of the subsystem?

Solution: The failure of one thermocouple does not preclude failure of the second; therefore, these three events are not mutually exclusive. An expanded version of Equation 6-3 (see Appendix C) applies:

$$P(\text{subsystem failure}) = P(A) + P(B) + P(C) - P(A \text{ and } B)$$
$$- P(A \text{ and } C) - P(B \text{ and } C) + P(A \text{ and } B \text{ and } C)$$

$$P(\text{subsystem failure}) = 0.005 + 0.005 + 0.005 - (0.005 \times 0.005)$$
$$- (0.005 \times 0.005) - (0.005 \times 0.005) + (0.005 \times 0.005 \times 0.005)$$
$$= 0.014925$$

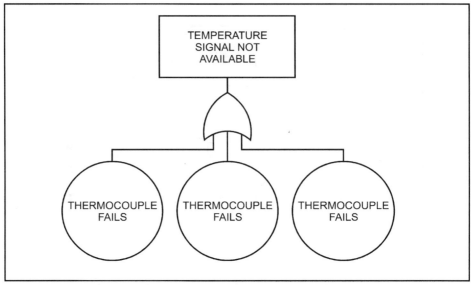

Figure 6-10. Temperature Sensor Subsystem Fault Tree.

Approximation Techniques

Often, in order to speed up and simplify the calculation, the faults and events in a fault tree are assumed to be mutually exclusive and independent. Under this assumption, probabilities for the OR gates are added. Probabilities for the AND gates are multiplied. This approximation technique can provide rough answers when probabilities are low, as is frequently the case with failure probabilities. The approach is generally conservative when working with failure probabilities because the method gives an answer that is larger than the accurate answer. For failure probability, this may be sufficient.

Example 6-5

Problem: Three thermocouples are used to sense temperature in a reactor as in Example 6-4. Use an approximation technique to estimate the probability of subsystem failure.

Solution: If mutually exclusive events are assumed, an expanded version of Equation 6-2 applies:

$$P(\text{subsystem failure}) = P(A) + P(B) + P(C)$$
$$= 0.005 + 0.005 + 0.005$$
$$= 0.015$$

This answer is more conservative than the answer for Example 6-4, which was 0.014925.

Fault Tree Analysis

> **Example 6-6**
>
> **Problem:** The fault tree of Figure 6-11 adds quantitative probability of failure numbers to the power system example. What is the probability of power system failure?
>
> **Solution:** Working from the bottom up, the failure probability for battery system failure = $0.01 \times 0.1 = 0.001$. The failure probability for a generator failure = $0.2 \times 0.01 = 0.002$. The failure probability for utility power failure = 0.001. The system failure probability = 0.000000002.

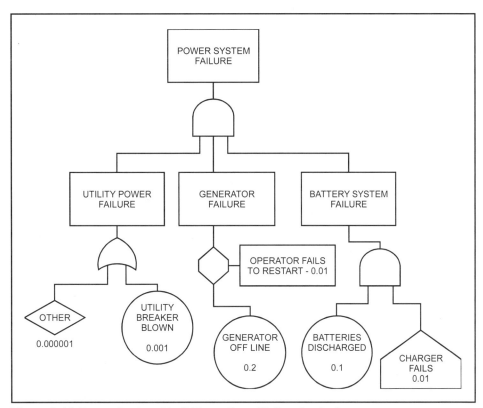

Figure 6-11. Power System Fault Tree - Quantitative Analysis.

Using a Fault Tree as Documentation

A fault tree can be used to document system operation during the design phase. This is the most common usage. When a system is analyzed and design problems are revealed, the design should be changed and the fault tree should be updated. Ideally, at the end of the design phase no serious errors exist and the fault tree looks good. In some cases, probabilities are assigned to failure events and system failure probabilities are calculated.

Fault trees also are used to document system failure events. They can represent a rather complex set of circumstances and describe the situation(s) that led to the failure in a way much clearer than words.

> **Example 6-7**
>
> **Problem:** On September 17, 1991, air traffic control systems in New York failed (Ref. 2). Eventually, more than 85,000 passengers were delayed. The failure was caused by a number of events. Air traffic control system communications were lost when telephone transmission equipment lost its power. Although transmission equipment power could be supplied from three "independent" sources, all three were unavailable. The outside power lines were disconnected as part of a load-shedding agreement with the power company because it was a hot day. The internal diesel generator was automatically taken off-line by a protection relay that was accidentally set with a low trip point. The trip-point testing required by procedures was not performed. The emergency batteries did supply power for 6 hr until they were exhausted.
>
> When batteries are supplying power, an alarm sounds to allow maintenance personnel to respond to the power problem before the batteries are depleted. In this case, three alarms were intended. None of the alarms was acknowledged. Alarms on one floor of the building went unacknowledged because that floor was normally unmanned and the maintenance inspection walkthrough that should occur every time power is switched did not happen because the normal maintenance staff was attending a training session. No substitute staff had been assigned.
>
> Two independent alarms on another floor also went unacknowledged. In one case, the normally de-energized audible alarms had a cut wire that was not detected. The normally de-energized visual alarm had a burned-out light bulb. Alarms in another area went unacknowledged because this area was unmanned.
>
> Document the failure with a fault tree.
>
> **Solution:** The fault tree in Figure 6-12 is an excellent graphical way to show the situation. The top level object is "communications failure." That can be caused by a number of factors that are of no concern in this drawing, so the incomplete event symbol is used. The primary failure event for this example is "telephone lines fail." Note that this was a problem only because the backup communication lines were not yet in service.
>
> The telephone lines were lost because the switching equipment power was lost. The power failed only because all three sources failed. The various reasons for the power source failures are shown as events in the fault tree. Note the use of the inhibit gate for the "shutdown relay mis-set" event. There is a chance this event will happen based on a number of factors; hence the inhibit gate instead of an AND gate. Many contributing factors are indicated via the AND gates. All these things had to fail before the system would fail.

Fault Tree Analysis

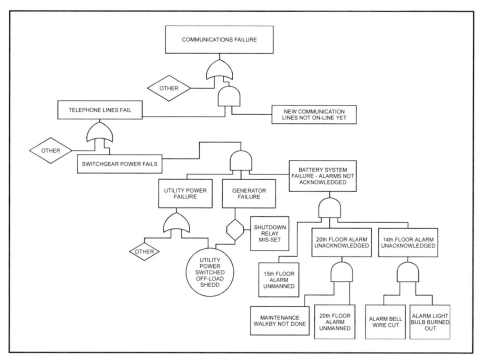

Figure 6-12. Communications Failure - Air Traffic Control.

Alternative Gate Input Method

As fault trees grow in size, the gate input connections can get complicated. Figure 6-13 shows an alternative method that is frequently used to connect inputs to fault tree symbols. A horizontal line with a single connection to input of a gate is used. Connections are drawn to the horizontal line to represent as many conditions as required. This method makes a fault tree easier to draw and even more practical.

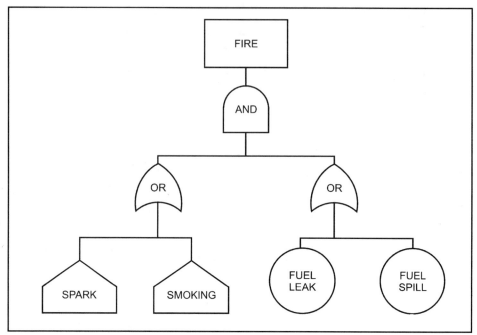

Figure 6-13. Alternative Gate Input Method.

EXERCISES

6.1 The thermocouples of Example 6-4 have an unavailability of 0.05. What is the unavailability of the sensor subsystem? Use the full, accurate method.

6.2 What is the answer to Exercise 6-1 using the approximate method? What is the error?

6.3 The power system of Figure 6-11 has probabilities of failure as follows:

> Utility breaker blown = 0.1
> Utility other = 0.02
> Generator off-line = 0.1
> Operator fails to restart = 0.02
> Batteries discharged = 0.2
> Charger failed = 0.02

What is the probability of system failure?

6.4 What are the advantages and limitations of fault tree analysis?

6.5 Draw a fault tree to document a failure in your plant.

REFERENCES

1. Henley, E. J., and Kumamoto, H., *Probabilistic Risk Assessment — Reliability Engineering, Design and Analysis*, New York: IEEE, 1992.

2. "Faults and Failures," *IEEE Spectrum*, February 1992.

3. Brombacher, A. C., "Fault Tree Analysis: Why and When," *Advances in Instrumentation and Control — Anaheim*, Research Triangle Park, NC: ISA—The Instrumentation, Systems, and Automation Society, 1994.

4. Gulati, R., and Dugan, J. B., "A Modular Approach for Analyzing Static and Dynamic Fault Trees," *Proceedings of the Annual Reliability and Maintainability Symposium*, New York: IEEE, 1997.

5. Kocza, G., and Bossche, A., "Automatic Fault Tree Synthesis and Real-Time Tree Trimming, Based on Computer Models," *Proceedings of the Annual Reliability and Maintainability Symposium*, New York: IEEE, 1997.

7
Network Modeling

Reliability Networks

Many modules and systems used in industrial control applications can be modeled through the use of simple networks. These networks show components in series, in parallel, or in combination configurations.

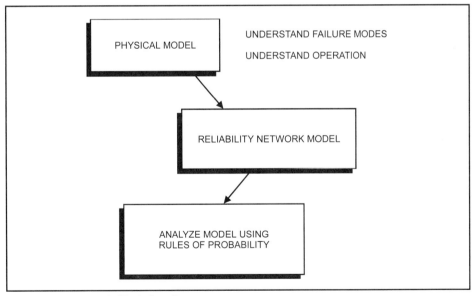

Figure 7-1. Network Modeling Process.

The first step in the process of reliability network modeling (Figure 7-1) is to convert from a physical system into a network model. This step is often the hardest and certainly the most critical. A good qualitative understanding of system operation during normal conditions and during failure conditions must exist. A reliability network model is drawn with boxes that represent the modules or the components that constitute the system. Lines are drawn between the boxes to indicate operational dependency. The network model may connect very differently from the physical model.

A reliability network may be viewed as showing the "success paths." If the viewer can find a path from left to right through the reliability network, those components are sufficient to allow the system to operate. Consider the control system in Figure 7-2. Three sensors are present, all of which are wired to each of two controllers. The controller implements a "voting" algorithm in order to tolerate the failure of one sensor. Either controller is capable of operating the valve.

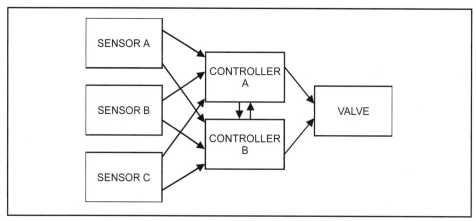

Figure 7-2. Physical Model of Control System.

The reliability network model for this system is illustrated in Figure 7-3. This model shows several success paths. One such path through the network model consists of sensor A, sensor B, controller A, and valve. If only these four components operate, the system operates.

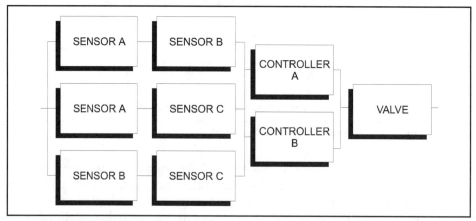

Figure 7-3. Reliability Network Model.

Given a network, the rules of probability are used to evaluate the network. Normally, work is done with network element probabilities and independent components are assumed. Sometimes it is easier to use probability of failure, and sometimes it is easier to use probability of success. The general term "probability of success" may mean reliability or it may mean availability. If working with nonrepairable systems, reliability (probability of success over the time interval of zero to t) is used. In repairable systems, availability (probability of success at time t) is used as the measure of probability of success.

Series Systems

A "series" system (Figure 7-4) is defined as any system in which all components must work for the system to work. Taking the pessimistic perspective, a series system fails if any component fails. A prime example of a series system would be the string of Christmas tree lights with which the author struggles each year. The lights are wired in such a way that when one bulb fails, all the lights go out! A series system offers no fault tolerance; there is no redundancy.

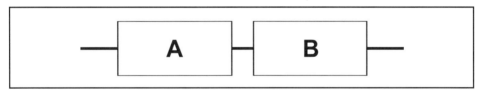

Figure 7-4. Series System.

Consider the two-component series system shown in Figure 7-4. The system has two components: A and B. Therefore,

R_A = Probability of success for component A

F_A = Probability of failure for component A

R_B = Probability of success for component B

F_B = Probability of failure for component B

R_S = Probability of success for the system

F_S = Probability of failure for the system

For the two-component system (assuming independent failures):

$$R_S = R_A \times R_B \tag{7-1}$$

For an n-component system:

$$R_S = \prod_{i=1}^{n} R_i \tag{7-2}$$

These equations are a direct result of one of the rules of probability which states that

$$P(A \cap B) = P(A) \times P(B) \text{ for independent events}$$

In a series network it is generally easier to work with success probabilities. Substituting $1 - F(A) = R(A)$ and $1 - F(B) = R(B)$ into Equation 7-1 gives

$$R_S = 1 - F(A) - F(B) + F(A) \times F(B) \tag{7-3}$$

This form is generally harder to use.

If a series system is composed of nonrepairable components with constant failure rates (exponential pdf), it is possible to substitute

$$R_A = e^{-\lambda_A t}$$

$$R_B = e^{-\lambda_B t}$$

into Equation 7-1, which gives

$$R_S = e^{-\lambda_A t} \times e^{-\lambda_B t}$$

This equals

$$R_S = e^{-(\lambda_A + \lambda_B)t} \tag{7-4}$$

Thus, failure rates for components in a series system can be added to obtain the failure rate for the system:

$$\lambda_S = \lambda_A + \lambda_B \tag{7-5}$$

The general form of the equation is

$$\lambda_S = \sum_{i=1}^{n} \lambda_i \tag{7-6}$$

EXAMPLE 7-1

Problem: A system consists of a sensor, a controller module, and a valve. The system fails if any one of the components fails. The components are not repairable. The two components and the module all have an exponential pdf with failure rates (failures per hour) as follows:

$$\lambda_{SENSOR} = 0.0001, \lambda_{CONTROLLER} = 0.00001, \text{ and } \lambda_{VALVE} = 0.0005$$

What is the system reliability for a one-month period?

Solution: Since the system will fail if any one of the elements fail, the reliability network (Figure 7-5) consists of a series system. Using Equation 7-6, the failure rates are added:

$$\lambda_{TOTAL} = 0.0001 + 0.0005 + 0.00001$$
$$= 0.00061$$

A one-month time period consists of 730 hr (8,760 hr in a year divided by 12). Therefore, system reliability is

$$R(730) = e^{-0.00061 \times 730}$$
$$= 0.6406$$

This system has a 64% chance of operating successfully in the next month.

Figure 7-5. Example 7-1 Series System.

EXAMPLE 7-2

Problem: A system consists of three sensors, a valve, a controller module, and a power source. The system fails if any component or module fails. All elements are repairable. The power source has an availability of 0.9; the controller module has an availability of 0.95; the sensors have an availability of 0.7; and the valve has an availability of 0.8. What is the system availability?

Solution: A series reliability network is applicable (Figure 7-6). Availability ratings for all the elements are given. Since availability is the probability of success at time t, Equation 7-2 can be used to obtain the overall availability (probability of success) for the system:

$$A_S = 0.95 \times 0.9 \times 0.7 \times 0.7 \times 0.7 \times 0.8$$
$$= 0.2346$$

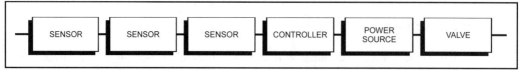

Figure 7-6. Example 7-2 Series System.

EXAMPLE 7-3

Problem: A control computer module is built from:

Parts List	Failure Rate
17 Integrated circuits	10 FITS each
1 Microprocessor	260 FITS
2 Memory circuits (1M)	200 FITS each

NOTE: 1 FIT = 1 failure per billion hours. What is the module failure rate? What is the module reliability for a five-year period?

Solution: Assume a series reliability network and components with a constant failure rate. Using Equation 7-6, the total failure rate is

$$\lambda_{TOTAL} = (17 \times 10) + 260 + (2 \times 200)$$
$$= 830 \text{ FITS}$$

The FIT is in units of failures per billion (10^9) hours.

One year contains 8,760 hr. Substituting,

$$R(5 \text{ years}) = e^{-0.00000083 \times 8760 \times 5}$$
$$= 0.9643$$

EXAMPLE 7-4

Problem: The memory in the control computer of Example 7-3 is increased to 8 megabytes. This requires 16 memory circuits instead of two. What is the module failure rate? What is the module reliability for a five-year period?

Solution: The additional failure rate of the extra memory must be added to the failure rate of the other components.

$$\lambda_{TOTAL} = (17 \times 10) + 260 + (16 \times 200)$$
$$= 3630 \text{ FITS}$$

The FIT is in units of failures per billion (10^9) hours. One year contains 8,760 hr. Substituting,

$$R(5 \text{ years}) = e^{-0.00000363 \times 8760 \times 5}$$
$$= 0.853$$

Network Modeling

These examples show an important phenomenon occurring in series networks. Since each component has a probability of success less than unity, the probability of system success is always less than that of any component. As more components are added, the probability of system success drops.

MTTF for a Series System

Mean time to failure (MTTF) has been previously defined as

$$MTTF = \int_0^\infty R(t)dt$$

For a series system:

$$MTTF_{SYSTEM} = \int_0^\infty R_{SYSTEM}(t)dt$$

$$= \int_0^\infty [R_1(t) \times R_2(t) \times \ldots \times R_n(t)]dt$$

For a series system with constant failure rates (exponential pdf):

$$MTTF_{SYSTEM} = \int_0^\infty e^{-\left(\sum_{i=1}^n \lambda_i\right)t} dt$$

Evaluating this gives

$$MTTF_{SYSTEM} = \frac{1}{\sum_{i=1}^n \lambda_i} \quad (7\text{-}7)$$

$$= \frac{1}{\lambda_1 + \lambda_2 + \ldots + \lambda_n}$$

Therefore, for a series system of exponentially distributed components where

$$\lambda_{TOTAL} = \lambda_1 + \lambda_2 + \ldots + \lambda_n$$

$$MTTF_{SERIES\ SYSTEM} = \frac{1}{\lambda_{TOTAL}} \quad (7\text{-}8)$$

Parallel Systems

A "parallel" system is defined as a system that is successful if any one of the components is successful. From the pessimistic perspective, the parallel system fails only if all its components fail. A parallel system offers fault tolerance that is accomplished through redundancy.

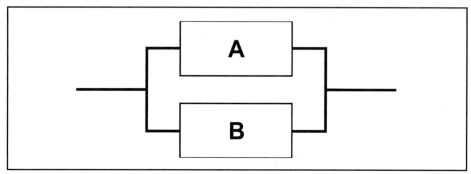

Figure 7-7. Parallel System.

Consider a two-component parallel system as shown in Figure 7-7. Two components are present, A and B. Assume that the component failures are independent. The system operates if either A or B operates.

R_A = Probability of success for component A

F_A = Probability of failure for component A

R_B = Probability of success for component B

F_B = Probability of failure for component B

R_S = Probability of success for the system

F_S = Probability of failure for the system

Since the system works if either A or B works, the probability summation rule can be used to obtain a formula for system success:

$$R_S = R_A + R_B - R_A \times R_B \qquad (7\text{-}9)$$

Note that the system fails only if both A and B fail. The formula for system failure is

$$F_S = F_A \times F_B \qquad (7\text{-}10)$$

For an n-component system:

$$F_S = \prod_{i=1}^{n} F_i \tag{7-11}$$

These equations result from the fact that all components must fail for a parallel system to fail. Working with probabilities of failure, the product rule for probabilities of independent components is used.

To obtain probability of success for an n-component system, the rule of complementary events is used:

$$R_S = 1 - F_S = 1 - \prod_{i=1}^{n} F_i \tag{7-12}$$

EXAMPLE 7-5

Problem: A system consists of three pressure sensors. The system operates if any one of the sensors operates (Figure 7-8). Each sensor has a probability of success of 0.8. What is the probability of system success?

Solution: Using Equation 7-12:

$$R_S = 1 - (1 - 0.8)^3$$
$$= 1 - 0.008$$
$$= 0.992$$

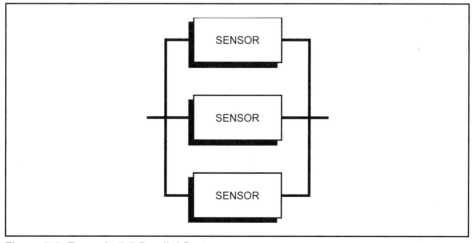

Figure 7-8. Example 7-5 Parallel System.

EXAMPLE 7-6

Problem: A reactor cooling jacket is fed from two independent cooling systems. Each cooling system consists of a water tank, a pump, and a power source with component reliabilities of 0.95, 0.7, and 0.85, respectively. A cooling system will fail if any of its components fail. The jacket will be cooled if either cooling system works. What is the probability that the jacket cannot be cooled?

Solution: This is a combination series-parallel system (Figure 7-9). First, obtain the reliability of a cooling system:

$$R_{\text{COOLING SYSTEM}} = 0.95 \times 0.7 \times 0.85$$
$$= 0.56525$$

The reactor jacket will be cooled if either cooling system works. Therefore,

$$R_{\text{COOLED}} = 1 - (1 - 0.56525)^2$$
$$= 0.811$$

The probability that the reactor will not be cooled is

$$F_{\text{COOLED}} = 1 - 0.811$$
$$= 0.189$$

In this case, remember that a quicker way to calculate the needed answer is to use Equation 7-6:

$$F_{\text{COOLED}} = (1 - 0.56525)^2$$
$$= 0.189$$

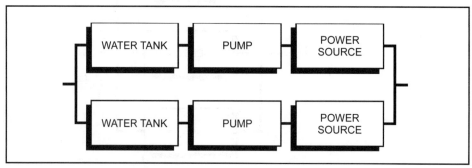

Figure 7-9. Example 7-6 System.

EXAMPLE 7-7

Problem: A second sensor and a second valve can be added in parallel with the sensor and with the valve for the system of Example 7-1. The resulting reliability network model is shown in Figure 7-10. What is the probability of system success for the same one-month period?

Solution: First, calculate the reliability of each sensor and each valve. For a 730-hr period:

$$R_{SENSOR} = e^{-0.0001 \times 730}$$
$$= 0.9296$$

and

$$R_{VALVE} = e^{-0.0005 \times 730}$$
$$= 0.6942$$

Using these probabilities, calculate the reliability of a parallel combination of sensors:

$$R_{P\ SENSOR} = 1 - (1 - 0.9296)^2$$
$$= 0.9950$$

Calculate the reliability of a parallel combination of valves:

$$R_{P\ VALVES} = 1 - (1 - 0.6942)^2$$
$$= 0.9065$$

Calculate the reliability of the controller for the 730-hr period.

$$R_{CONTROLLER} = e^{-0.00001 \times 730}$$
$$= 0.9927$$

Finally, evaluate the series combination of three components: the parallel sensors, the controller, and the parallel valves.

$$R_{SYSTEM} = 0.9950 \times 0.9927 \times 0.9065$$
$$= 0.8954$$

The chances of successful operation for the next month are now 89%. This is a substantial improvement over the results of Example 7-1, in which a one-month reliability of 0.6406 was calculated.

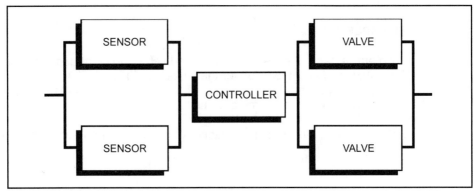

Figure 7-10. Example 7-7 System.

MTTF of a Parallel System

Using the defining equation of MTTF and substituting Equation 7-9:

$$MTTF_{SYSTEM} = \int_0^\infty [R_A + R_B - R_A \times R_B]dt$$

for a two-component parallel system. If these two components have constant failure rates (exponential pdf), then

$$MTTF_{SYSTEM} = \int_0^\infty [e^{-\lambda_A \times t} + e^{-\lambda_B \times t} - e^{-(\lambda_A + \lambda_B) \times t}]dt$$

This evaluates to

$$\left[-\frac{1}{\lambda_A} e^{-\lambda_A \times t} - \frac{1}{\lambda_B} e^{-\lambda_B \times t} + \frac{1}{\lambda_A + \lambda_B} e^{-(\lambda_A + \lambda_B) \times t} \right]_0^\infty$$

Evaluating the exponentials:

$$MTTF_{PARALLEL\ SYSTEM} = \frac{1}{\lambda_A} + \frac{1}{\lambda_B} - \frac{1}{\lambda_A + \lambda_B} \qquad (7\text{-}13)$$

Notice that MTTF does not equal $1/\lambda$. A parallel system of exponentially distributed components no longer has an exponential pdf.

$$MTTF_{PARALLEL\ SYSTEM} \neq \frac{1}{\lambda_{TOTAL}}$$

The equation "MTTF = $1/\lambda$" does not apply to any system that has parallel components. This includes any system that has redundancy. Triple modular redundant systems, dual systems, and partially redundant systems are all in this category.

k-out-of-*n* Systems

Many systems, especially "load-sharing" systems, are designed with extra elements. In systems where the extra elements can share the load without complicated switching mechanisms, this scheme can be an effective form of redundancy.

Consider an example of a temperature-sensing system with five thermocouples. The physical model is shown in Figure 7-11. The system, which requires only four of the five thermocouples in order to operate, is known as a "four-out-of-five" (4oo5) system. If two or more thermocouples fail, the system will fail.

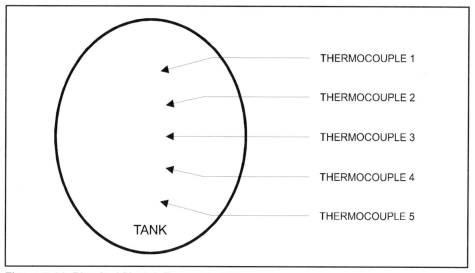

Figure 7-11. Physical Model, Four-out-of-Five.

Any combination of the four out of five thermocouples will allow system operation. The number of those combinations is given by

$$\binom{5}{4} = \frac{5!}{4!(5-4)!}$$

$$= 5$$

Notice that two numbers stacked within parentheses is the notation for combinations of numbers.

Given that there are five combinations of four elements, we can build a series/parallel reliability network. The network in Figure 7-12 has five parallel paths. Each path has four elements. An examination of the network shows that each path is a different combination of four elements (out of a possible five).

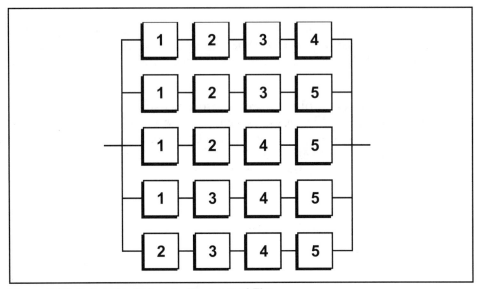

Figure 7-12. Reliability Network, Four-out-of-Five.

It is easy to calculate the probability of success for one path. Using Equation 7-2, the probability of success for the top path is

$$R_{PATH1} = R_1 \times R_2 \times R_3 \times R_4$$

The probability of success for the remaining paths is:

$$R_{PATH2} = R_1 \times R_2 \times R_3 \times R_5$$

$$R_{PATH3} = R_1 \times R_2 \times R_4 \times R_5$$

$$R_{PATH4} = R_1 \times R_3 \times R_4 \times R_5$$

$$R_{PATH5} = R_2 \times R_3 \times R_4 \times R_5$$

The probability of success for the system is the union of these four path probabilities. These numbers cannot merely be added. The four paths are *not* mutually exclusive. To obtain the union, the path probabilities must be added and then the combinations of two intersections must be subtracted; the combinations of three intersections must be added, then the combinations of four intersections subtracted; and the combinations of five intersections must be added. This equation would be quite long!

Network Modeling

Fortunately, if all elements are the same, they can be drastically simplified. Note that the intersection of path 1 and path 2 is

$$R_{PATH1} \times R_{PATH2} = R_1 \times R_2 \times R_3 \times R_4 \times R_5$$

$$= R_I$$

The same result occurs for all intersections, including intersections of two at a time, three at a time, four at a time, and even five at a time. Taking advantage of this fact, a reasonable equation can be written for the union of the five paths.

$$R_{SYSTEM} = R_{PATH1} + R_{PATH2} + R_{PATH3} + R_{PATH4} + R_{PATH5}$$

$$-\binom{5}{2}R_I + \binom{5}{3}R_I - \binom{5}{4}R_I + \binom{5}{5}R_I$$

Expanding the combinations, this equals

$$R_{SYSTEM} = R_{PATH1} + R_{PATH2} + R_{PATH3} + R_{PATH4} + R_{PATH5}$$

$$-10R_I + 10R_I - 5R_I + 1R_I$$

Adding the terms produces

$$R_{SYSTEM} = R_{PATH1} + R_{PATH2} + R_{PATH3} + R_{PATH4} + R_{PATH5} - 4R_I$$

If all the elements in the model are the same, further simplification is possible. In the thermocouple system example, all the thermocouples are the same. Therefore:

$$R_1 = R_2 = R_3 = R_4 = R_5 = R$$

The equation simplifies to

$$R_{SYSTEM} = 5R^4 - 4R^5 \qquad (7\text{-}14)$$

> **EXAMPLE 7-8**
>
> **Problem:** The system of Figure 7-11 uses thermocouples that have a one-year reliability of 0.9. What is the system reliability for a one-year interval?
>
> **Solution:** This four-out-of-five system has a system reliability given by Equation 7-14. Using a value of 0.9:
>
> $$R_{SYSTEM} = 5(0.9)^4 - 4(0.9)^5$$
> $$= 3.2805 - 2.362$$
> $$= 0.9185$$

The general form of the equation can be derived by answering the question, "When is the system successful?" The system is successful when:

1. All five thermocouples are good, or
2. Four thermocouples are good and one is bad.

What is the probability of getting one combination of four good and one bad? This can be written as

$$R \times R \times R \times R \times (1 - R)$$

In the general sense, this is written as

$$R^k \times (1 - R)^{n-k}$$

for k out of n elements. Since more than one combination of k out of n elements allows system success, the number of combinations is multiplied by the probability of each combination. This is written as

$$\binom{n}{k} R^k (1 - R)^{n-k} \qquad (7\text{-}15)$$

This is the probability of realizing any combination of k out of n elements. The system is successful if k *or more* elements are successful, which is mathematically described by

$$R_{SYSTEM} = \sum_{i=k}^{n} \binom{n}{i} R^i (1 - R)^{n-i} \qquad (7\text{-}16)$$

For the four-out-of-five thermocouple problem, n equals five and k equals four. Substituting these values into Equation 7-16:

$$R_{SYSTEM} = 5[R^4(1-R)^1] + 1[R^5(1-R)^0]$$

Simplifying yields

$$R_{SYSTEM} = 5R^4 - 4R^5 \tag{7-17}$$

Of course, this is the same as Equation 7-14.

EXAMPLE 7-9

Problem: Three control computers are wired to a voting logic circuit. A system diagram is shown in Figure 7-13. The outputs of the computers are compared. The output of the voting logic circuit equals the majority of its inputs. If one computer fails, the other two successful units will outvote the bad unit. The system remains successful. If a second computer fails, the two bad units will outvote the single good unit and the system will fail. All computers are repairable. Each computer has an availability of 0.95. The voting logic circuit has an availability of 0.98. What is the system availability?

Solution: This system requires that two out of three (2oo3) control computers be successful. The voting logic circuit must also be successful. The reliability network is shown in Figure 7-14. We will first solve the left side of the reliability network.

Using Equation 7-16 (substituting numbers),

$$\begin{aligned} A_{2oo3} &= 3[(0.95)^2(1-0.95)^1] + [(0.95)^3(1-0.95)^0] \\ &= 3(0.95)^2 - 2(0.95)^3 \\ &= 0.99275 \end{aligned}$$

Since the voting logic is in series,

$$A_{SYSTEM} = 0.99275 \times 0.98 = 0.9729$$

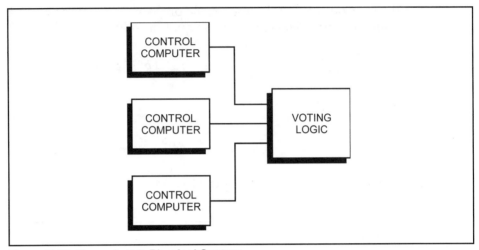

Figure 7-13. Example 7-9, Physical System.

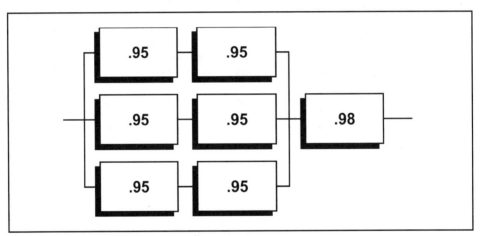

Figure 7-14. Example 7-9, Reliability Model.

Quantitative Network Evaluation

Some reliability networks are more than simple series/parallel combinations. These network topologies require different solution techniques. One of most straightforward methods is called the "event space method."

Event Space Method

The event space method involves the creation of all combinations of possible component (or module) failures. It is necessary to identify which combinations represent system success and which combinations represent system failure. The probability of system success is obtained by adding the combinational probabilities. Those probabilities can be summed because each combination is mutually exclusive. If a system has n components, each with one failure mode, there will be 2^n combinations. It is obvious that the method will grow impractical as the number of components increases. For smaller networks, the method can be effective because it is very good at showing the overall picture.

The control system shown in Figure 7-15 has three sensors, two controllers, and two valves. The two controllers each have two inputs. Each controller is wired to two of the sensors and can operate successfully if one of the two sensors to which it is wired is successful. The sensors are identical and have an availability of 0.8; the controllers have an availability of 0.95; and the valves have an availability of 0.7.

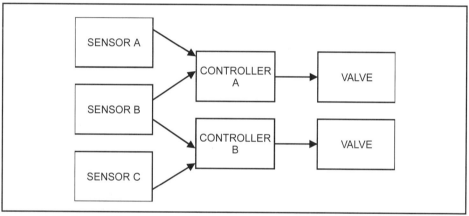

Figure 7-15. Control System.

Each controller and its associated valve can be treated as a series system with an availability of

$$A_{CONT/VALVE} = 0.95 \times 0.7 = 0.665$$

The simplified system is shown in Figure 7-16.

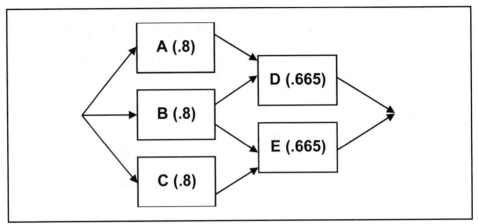

Figure 7-16. Simplified System.

The probability of success for this system can be obtained using the event space method. With five elements in the network, 32 ($2^5 = 32$) combinations of elements can be expected. The combinations are listed in groups according to the number of failed elements. Group 0 has one combination: all elements successful. It is listed as item 1 in Table 7-1.

	Item	Combination
Group 0	1	A B C D E
Group 1	2	A* B C D E
	3	A B* C D E
	4	A B C* D E
	5	A B C D* E
	6	A B C D E*

Table 7-1. Events Listing: Group 0, Group 1

Next, all combinations of one failure are listed. With five elements, five combinations are expected. These are listed as items 2 to 6 in Table 7-1. A failed element is marked with an asterisk. In the next step, all combinations of two failures are listed. The equation for combinations (Equation C-16 in Appendix C) is used to determine that there are 10 combinations of five items taken two at a time.

$$\binom{5}{2} = \frac{5!}{2!(5-2)!} = 10$$

These combinations are listed as items 7 to 16 in Table 7-2. Notice that combinations of A, B, and C were developed. Group 3 consists of all combinations of three failures. Again, from Equation C-16, 10 combinations of five items, taken three at a time, are expected.

$$\binom{5}{3} = \frac{5!}{3!(5-3)!} = 10$$

Network Modeling

	Item	Combination
Group 2	7	A* B* C D E
	8	A* B C* D E
	9	A* B C D* E
	10	A* B C D E*
	11	A B* C* D E
	12	A B* C D* E
	13	A B* C D E*
	14	A B C* D* E
	15	A B C* D E*
	16	A B C D* E*

Table 7-2. Event Listing: Group 2

These combinations are listed as items 17 to 26 in Table 7-3.

	Item	Combination
Group 3	17	A* B* C* D E
	18	A* B* C D* E
	19	A* B* C D E*
	20	A B* C* D* E
	21	A B* C* D E*
	22	A* B C* D* E
	23	A B C* D* E*
	24	A* B C D* E*
	25	A B* C D* E*
	26	A* B C* D E*

Table 7-3. Event Listing: Group 3

Group 4 consists of all combinations of four failures. Five combinations are expected. These are listed along with the single combination from Group 5 (all failures) as items 27 to 32 in Table 7-4.

	Item	Combination
Group 4	27	A B* C* D* E*
	28	A* B C* D* E*
	29	A* B* C D* E*
	30	A* B* C* D E*
	31	A* B* C* D* E
Group 5	32	A* B* C* D* E*

Table 7-4. Event Listing: Group 4, Group 5

After listing all combinations of failed elements, the combinations that will cause system failure are identified. Of course, the combination in Group 0 represents system success since no elements have failed. Continuing, Figure 7-16 is examined. A path across the reliability network exists for any combination of one failed element. It can be concluded, then, that all Group 1 combinations represent system success. Group 2 must be examined carefully, however. Combination 7 has elements A and B failed. Looking again at Figure 7-16, a path still exists across the network using elements C and E; therefore, the system is still successful with combination 7. As the other combinations of two failures are examined, no system failures can be found until combination 16. That combination has elements D and E failed. There is no path across the network when these elements fail. This is illustrated in Figure 7-17.

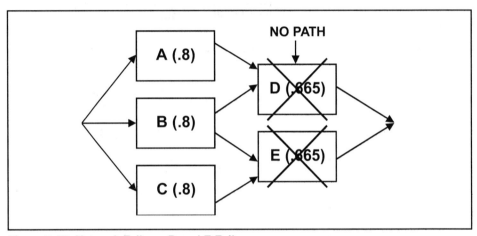

Figure 7-17. Network Failure: D and E Fail.

Examination of Group 3 shows a system failure with combination 17: Elements A, B, and C are failed. With all sensors failed, the system cannot work. A close examination of the remainder of Group 3 shows that only four combinations will allow system success: items 18, 21, 22, and 26. These combinations have one sensor and a connecting controller/valve operating. All combinations of Group 4 and Group 5 cause system failure. The system requires at least one sensor and at least one controller/valve working, which does not happen in any combinations of those groups. Table 7-5 compiles the list of all combinations and shows which combinations are successful.

	Item	Combination	System
Group 0	1	A B C D E	Success
Group 1	2	A* B C D E	Success
	3	A B* C D E	Success
	4	A B C* D E	Success
	5	A B C D* E	Success
	6	A B C D E*	Success
Group 2	7	A* B* C D E	Success
	8	A* B C* D E	Success
	9	A* B C D* E	Success
	10	A* B C D E*	Success
	11	A B* C* D E	Success
	12	A B* C D* E	Success
	13	A B* C D E*	Success
	14	A B C* D* E	Success
	15	A B C* D E*	Success
	16	A B C D* E*	Failure
Group 3	17	A* B* C* D E	Failure
	18	A* B* C D* E	Success
	19	A* B* C D E*	Failure
	20	A B* C* D* E	Failure
	21	A B* C* D E*	Success
	22	A* B C* D* E	Success
	23	A B C* D* E*	Failure
	24	A* B C D* E*	Failure
	25	A B* C D* E*	Failure
	26	A* B C* D E*	Success
Group 4	27	A B* C* D* E*	Failure
	28	A* B C* D* E*	Failure
	29	A* B* C D* E*	Failure
	30	A* B* C* D E*	Failure
	31	A* B* C* D* E	Failure
Group 5	32	A* B* C* D* E*	Failure

Table 7-5. Event Combination List

To calculate probability of system success, the probabilities of the successful combinations are added. This can be done because each combination is mutually exclusive. The probability of each combination is obtained by multiplying the probabilities of element success and element failure. For example, the probability of combination 1, where all elements are successful, is

$$A_{COMB1} = 0.8 \times 0.8 \times 0.8 \times 0.665 \times 0.665 = 0.226419$$

The probability of combination 2, in which element A has failed, is

$$A_{COMB2} = 0.2 \times 0.8 \times 0.8 \times 0.665 \times 0.665 = 0.056605$$

Note that the probability of failure for element A is multiplied by the probability of success for other elements when calculating the combination 2 probability. When calculating combination probabilities, the element probabilities of success and failure are used. The probability of combination 3 and of combination 4 is the same as for combination 2. The probability of combination 5 is

$$A_{COMB5} = 0.8 \times 0.8 \times 0.8 \times 0.335 \times 0.665 = 0.114061$$

A complete listing of all element probabilities and the combination probabilities is presented in Table 7-6. If the combination probabilities are added, the result should equal one. This is a good checking mechanism.

Adding the successful combination probabilities gives an answer of 0.866415. The failure combination probabilities total 0.133585. These two numbers also sum to one. The event space method appears quite tedious, but personal computer spreadsheet programs can be quickly set to solve the problems automatically. This tool makes the method quite acceptable, considering the qualitative overview that the method provides.

Item	Element 1	Element 2	Element 3	Element 4	Element 5	Combination
1	0.8	0.8	0.8	0.665	0.665	0.226419
2	0.2	0.8	0.8	0.665	0.665	0.056605
3	0.8	0.2	0.8	0.665	0.665	0.056605
4	0.8	0.8	0.2	0.665	0.665	0.056605
5	0.8	0.8	0.8	0.335	0.665	0.114061
6	0.8	0.8	0.8	0.665	0.335	0.114061
7	0.2	0.2	0.8	0.665	0.665	0.014151
8	0.2	0.8	0.2	0.665	0.665	0.014151
9	0.2	0.8	0.8	0.335	0.665	0.028515
10	0.2	0.8	0.8	0.665	0.335	0.028515
11	0.8	0.2	0.2	0.665	0.665	0.014151
12	0.8	0.2	0.8	0.335	0.665	0.028515
13	0.8	0.2	0.8	0.665	0.335	0.028515
14	0.8	0.8	0.2	0.335	0.665	0.028515
15	0.8	0.8	0.2	0.665	0.335	0.028515
16	0.8	0.8	0.8	0.335	0.335	0.057459
17	0.2	0.2	0.2	0.665	0.665	0.003538
18	0.2	0.2	0.8	0.335	0.665	0.007129
19	0.2	0.2	0.8	0.665	0.335	0.007129
20	0.8	0.2	0.2	0.335	0.665	0.007129
21	0.8	0.2	0.2	0.665	0.335	0.007129
22	0.2	0.8	0.2	0.335	0.665	0.007129
23	0.8	0.8	0.2	0.335	0.335	0.014365
24	0.2	0.8	0.8	0.335	0.335	0.014365
25	0.8	0.2	0.8	0.335	0.335	0.014365
26	0.2	0.8	0.2	0.665	0.335	0.007129
27	0.8	0.2	0.2	0.335	0.335	0.003591
28	0.2	0.8	0.2	0.335	0.335	0.003591
29	0.2	0.2	0.8	0.335	0.335	0.003591
30	0.2	0.2	0.2	0.665	0.335	0.001782
31	0.2	0.2	0.2	0.335	0.665	0.001782
32	0.2	0.2	0.2	0.335	0.335	0.000898
					Total	1

Table 7-6. Probability Combination Listing

Approximation of Failure Probability

Reliability calculations are done for a number of reasons. At the design stage of a project, reliability comparisons are made between alternative designs. For this purpose, an approximation that can be done quickly may be a better use of engineering time. As long as the comparative numbers are not close, the approximation is sufficient to make a decision. At other times, reliability calculations are done to determine whether the proposed system meets a specification. Such a specification might state, "The system shall have an unavailability of no greater than 10%." In such situations an approximation may serve if the approximation always provides worst-case answers (probability of failure numbers should always come out high; probability of success numbers should always come out low).

Bounding approximations using reliability networks can be done with little relative effort. The approximation techniques result in worse-case answers. If these answers can serve the intended purpose, engineering productivity can be improved. Consider the system of Figure 7-15. We can roughly approximate system failure if we look at Figure 7-16. The system fails if blocks A and B and C fail or if D and E fail or if A and B and E fail or if B and C and D fail. This is approximated by

$$F_{SYSTEM} \approx F_{ABC} + F_{DE} + F_{ABE} + F_{BCD} \tag{7-18}$$

Substituting the numerical values produces

$$F_{SYSTEM} \approx 0.008 + 0.112225 + 0.0134 + 0.0134$$
$$= 0.147025$$

This is known as the "upper bound" on failure probability. Compare this to the previous result, 0.133585. The approximate answer is clearly more pessimistic. If this result indicates that the system will meet specification, the additional work of getting a precise answer is not necessary.

Approximation of Success Probability

In a manner similar to the failure approximation techniques, the probability of system success can be approximated. The "upper bound" of success probability is obtained using only the probabilities of success for each successful set of elements in the network. For the example:

$$R_{SYSTEM} = R_{AD} + R_{BD} + R_{BE} + R_{CE} \tag{7-19}$$

Substituting numerical values gives a result of 2.128. Since a probability cannot exceed one, this "upper bound" result is not very useful. The "lower bound" on probability of success is obtained by subtracting the "two at a time" intersections. For the example:

$$R_{SYSTEM} = R_{AD} + R_{BD} + R_{BE} + R_{CE}$$

$$- (R_{AD} \cap R_{BD}) - (R_{AD} \cap R_{BE}) \quad (7\text{-}20)$$

$$- (R_{AD} \cap R_{CE}) - (R_{BD} \cap R_{BE}) - (R_{BD} \cap R_{CE}) - (R_{BE} \cap R_{CE})$$

Substituting the numerical values gives

$$R_{SYSTEM} = 0.532 + 0.532 + 0.532 + 0.532 - 0.4256 - 0.0283024$$

$$- 0.0283024 - 0.35378 - 0.283024 - 0.4256$$

$$= 0.07395$$

Again, this is not a very useful result. The only thing determined from the success approximation approach is that there is a probability of success somewhere between 0.07395 and 1. The method works much better when element probabilities of success are low. In our example, the numbers are near one.

In general, if element success probabilities are low, use a success approximation model; if element success probabilities are high, use a failure approximation model. The approximation techniques will then be useful.

EXERCISES

7.1 A system consists of a power source, a pump, and a valve. The system operates only if all three components operate. The power source has an availability of 0.8; the pump has an availability of 0.7; and the valve has an availability of 0.75. Draw a reliability network of the system. What is the system availability?

7.2 A second pump and a second valve are added in parallel with the system of Exercise 7.1. This is shown in Figure 7-18. What is the system availability?

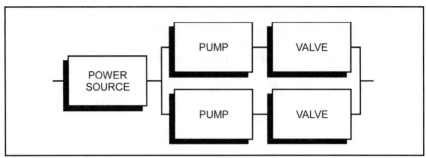

Figure 7-18. Exercise 7-2 System.

7.3 A control system consists of four sensors, an input module, a controller, an output module, and two valves. If any component or module fails, the system fails. All items are repairable. The component/module availabilities are: sensor, 0.8; input module, 0.95; controller, 0.9; output module, 0.95; and valve, 0.75. What is the system availability?

7.4 An executive from corporate headquarters suggests that your control system (the system from Exercise 7.3) could be improved by adding a redundant controller. If a second controller is put in parallel with the first, how much does system availability improve?

7.5 A nonrepairable controller module has a constant failure rate of 500 FITS. What is the MTTF for the controller?

7.6 Two nonrepairable controller modules with a constant failure rate of 500 FITS are used in parallel. What is the system MTTF?

7.7 You wish to approximate the probability of failure for a reliability network. All network elements have a probability of success in the range of 0.95 to 0.998. Which model method should you use?

7.8 You have a reliability network with six elements, each of which has one failure mode. How many combinations must be listed in an event space evaluation?

7.9 You have a reliability network with four elements, each of which has two failure modes. How many combinations must be listed in an event space evaluation?

BIBLIOGRAPHY

1. Billinton, R., and Allan, R. N., *Reliability Evaluation of Engineering Systems: Concepts and Techniques*, New York: Plenum Press, 1983.

2. Dhillon, B. S., *Reliability Engineering in Systems Design and Operation*, New York: Van Nostrand Reinhold, 1983.

3. Henley, E. J., and Kumamoto, H., *Probabilistic Risk Assessment: Reliability Engineering, Design, and Analysis*, Piscataway, NJ: IEEE Press, 1992.

4. Shooman, M. L., *Probabilistic Reliability: An Engineering Approach*, 2nd Ed., Malabar, FL: Robert E. Krieger Publishing, 1990.

8
Markov Modeling

Repairable Systems

Repairable systems are typical in an industrial environment. Systems should be installed where they can be repaired, usually by replacing modules. Such systems offer many advantages in terms of system availability and safety. Several different fault-tolerant system configurations of repairable modules have been created, which can run nonstop with very low system failure probability for many years.

Repairs take time. Simple reliability network modeling methods do not directly account for repair time. Probability methods can provide approximate solutions that account for repair time, but errors that are due to approximation techniques remain low only for low failure rates and short repair times. A different method is needed to model the failure performance of a repairable system over a wide range of failure rates and repair times. The method must account for realistic repair times, various system configurations, and realistic system features, including diagnostic testing. The technique must apply to systems that are fully repairable and systems that are partially repairable.

Markov Models

Markov modeling, a reliability and safety modeling technique that uses state diagrams, fulfills these goals. The Markov modeling technique uses only two simple symbols (Figure 8-1). It provides a complete set of evaluation tools when compared with many other reliability and safety evaluation techniques (Refs. 1 and 2).

The Markov model in Figure 8-2 demonstrates how the symbols are used. Circles (states) show combinations of successfully operating components and failed components. Possible component failures and repairs are

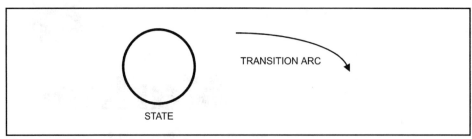

Figure 8-1. Markov Model Symbols.

shown with transition arcs, arrows that go from one state to another. A number of different combinations of failed and successful components are possible. Some represent system success states, while others represent system failure states. It should be noted that multiple failure modes can be shown on one drawing.

A Markov model can show on a single drawing the entire operation of a fault-tolerant control system. If the model is created completely, it will show full system success states. It will also show degraded states where the system is still operating successfully but vulnerable to further failures. The drawing will also show all failure modes.

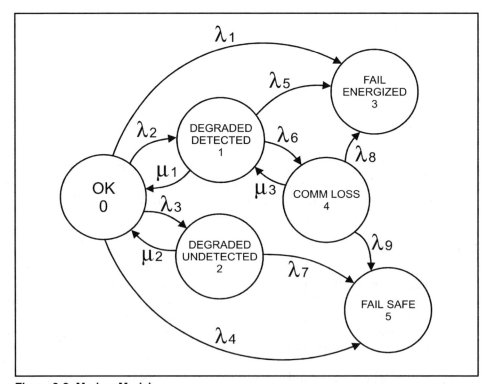

Figure 8-2. Markov Model.

Markov Model Metrics

A number of different reliability and safety measurements can be generated from a Markov model. The probability of continuous successful operation over a time interval (as stated in the definition of reliability) can be calculated. In addition, the probability of system success at time t, known as availability, can be calculated. Other measures of system success, including MTTF, can be obtained by using certain solution techniques on the Markov model.

For fully repairable systems, availability is obtained either as a function of time or as a steady-state value. System reliability as a function of time is obtained by ignoring repair arcs from failure states to success states and by using similar solution techniques. MTTF or MTTR values can be obtained using linear algebra techniques.

A full suite of safety metrics can also be calculated from a Markov model. Since the technique is oriented toward providing state probabilities as a function of time, the probability of failure on demand (PFD) can be obtained by adding probabilities from states where the system will not respond to a demand (fail-danger states). This value is provided as a function of time as part of the solution process. MTTFS, MTTFD (Chapter 4), and other mean failure time measurements can be precisely calculated using Markov model solution techniques. Overall, Markov modeling is probably the most flexible modeling technique available for control system evaluation.

Solving Markov Models

Andrei Andreyevich Markov (1856-1922), a Russian mathematician, studied probability while teaching at St. Petersburg University in the late 1800s. He defined the "Markov process," in which the future variable is determined by the present variable but is independent of predecessors (Ref. 3). Markov emphasized sequences where the variable takes on particular discrete values; these sequences are known as Markov chains. That work has been extensively developed over the years. These methods apply nicely to the failure/repair process because combinations of failures create discrete system states. In addition, the failure/repair process moves between discrete states only as a result of current state and current failure.

The Markov model building technique involves definition of all mutually exclusive success/failure states in a system. These are represented by labeled circles. The system can transition from one state to another whenever a failure or a repair occurs. Transitions between states are shown with arrows (transition arcs) and are labeled with the appropriate failure or repair probabilities (often approximated using failure/repair rates). This model is used to describe the behavior of the system with time. If time is modeled in discrete increments (for example, once per hour), simulations can be run using the probabilities shown in the models.

Calculations can be made showing the probability of being in each state for each time interval. Since some states represent system success, the probabilities of these states are added to obtain either system reliability or system availability as a function of time. Many other reliability metrics are also obtained using various techniques.

A Markov model for a single nonrepairable component with one failure mode is shown in Figure 8-3. Two states are present. In state 0, the component is operating successfully. In state 1, the component has failed. One transition that represents component failure shows movement from state 0 to state 1. That arrow is labeled with the lowercase Greek letter lambda (λ), which represents the instantaneous failure rate of the component.

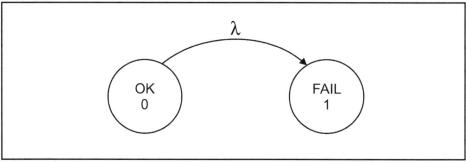

Figure 8-3. Markov Model, Single Nonrepairable Component.

A single repairable component with one failure mode has a Markov model as shown in Figure 8-4. The two states are the same as previously described for the nonrepairable component. Two transitions are present. The upper transition represents a failure — movement from state 0 to state 1. The lower transition represents a repair — movement from failure state 1 to success state 0. The repair rate is represented by the lowercase Greek letter mu (μ).

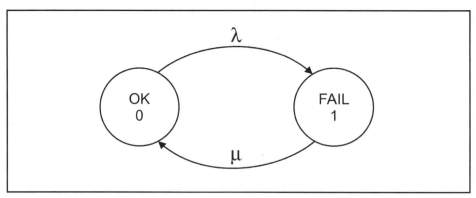

Figure 8-4. Markov Model, Single Repairable Component.

Markov models can represent nonrepairable, partially repairable, or fully repairable systems. Multiple failure modes can be modeled using as many failure states as required. Failures detected by on-line diagnostics can be distinguished from those that are undetected by defining separate states. Common-cause failures are also easily shown. Figure 8-2 shows the Markov model for a dual control system. It is partially repairable with two failure states. Failures that are detected by computerized diagnostics (state 1) are distinguished from those that are not (state 2). The models can grow as complicated as necessary to serve the needed level of modeling accuracy.

The Markov approach to reliability modeling of control systems is not only flexible enough to account for the realities of the industrial environment, but is also a systematic approach to modeling which can reveal unexpected failure states. The construction of the Markov model can be a valuable qualitative reliability tool.

Discrete-Time Markov Modeling

Markov modeling techniques have been applied to a wide range of problems in many disciplines. The techniques apply to any process in which the probability of moving from one state (condition) to another depends only on the current state and not on the previous states. A process meeting this criterion is known as a "memory-less" system. The process is modeled by defining the mutually exclusive states and the transitions between states. A transition is drawn between states whenever there is a probability of going from one state to another during a time interval. If probabilities do not change with time, the models can be solved using linear algebra or calculus to obtain formulas for average state probabilities, state probabilities as a function of time, or average times between states, as appropriate. If probabilities do change with time, computer simulations can be run that show state probabilities as a function of time. The other reliability and safety metrics can be calculated from that data.

A Markov model can be solved for time-dependent conditions. Time can be viewed two different ways: discrete or continuous. Discrete-time models change once every "time increment." The time increment depends on the model. It may be once an hour, 10 times an hour, once a day, once a week, or some other suitable time increment. In continuous-time models, the same concepts are used. As in calculus, the time increment is reduced to the limit approaching zero (the limit as Δt approaches zero). In reliability and safety modeling, the discrete-time approach works well.

Time-Dependent Probabilities

Consider an industrial forging process. A forging machine stamps out large pieces once every 10 min (six times an hour). Records show that one time out of 100, the machine fails. The average repair time is 20 min.

A discrete-time Markov model would do an excellent job of modeling this process. A good selection for the time interval would be 10 min. Two states are required; state 0, defined as success, and state 1, defined as failure.

The system starts in state 0, success. From state 0, the system will either stay in state 0 or move to state 1 in each time interval. There is a one in 100 (0.01) probability that the system will move state 0 to state 1. For each time interval, the system must either move to the new state or stay in the present state with a probability of one. The probability that the system will stay in state 0, therefore, is 99 out of 100 (1 − 0.01 = 0.99). Once the system has failed, it will either stay in state 1 (has not been repaired) or move to state 0 (has been repaired). The probability of moving from state 1 to state 0 in any time interval is 0.5 (10-min interval/20-min repair time). The system will stay in state 1 with a probability of 0.5 (1 − 0.5). The Markov model for this process is shown in Figure 8-5.

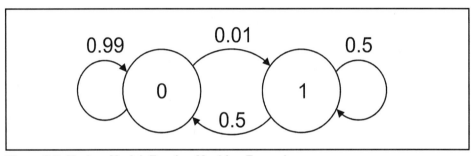

Figure 8-5. Markov Model, Forging Machine Example.

Transition Matrix

The model can be represented by showing its probabilities in matrix form. An $n \times n$ matrix is drawn (n equals the number of states) showing all probabilities. This matrix is known as the stochastic transition probability matrix, and is often called the "transition matrix," nicknamed **P**. The transition matrix, **P**, for the forging machine is written as follows:

$$\mathbf{P} = \begin{bmatrix} 0.99 & 0.01 \\ 0.5 & 0.5 \end{bmatrix} \quad (8\text{-}1)$$

Each row and each column represents one of the states. In Equation 8-1, row 0 and column 0 represent state 0, while row 1 and column 1 represent state 1. If more states existed, they would be represented by additional rows and columns. The numerical entry in a given row and column is the probability of moving from the state represented by the row to the state represented by the column. For example, in Equation 8-1, the number in row 0, column 1 (0.01) represents the probability of moving from state 0 to state 1 during the next time interval. The number in row 0, column 0 (0.99) represents the probability of moving from state 0 to state 0 (i.e., of remaining in state 0) during the next time interval. The other entries have similar interpretations. A transition matrix contains all the necessary information about a Markov model. It is used as the starting point for further calculation methods.

Steady-State Availability

The behavior of the forging system can be seen by following the tree diagram in Figure 8-6. Starting at state 0, the system moves to state 1 or stays in state 0 during the first time interval. Behavior during subsequent time intervals is shown as the tree diagram branches out. The step probabilities are marked above each arrow in the diagram.

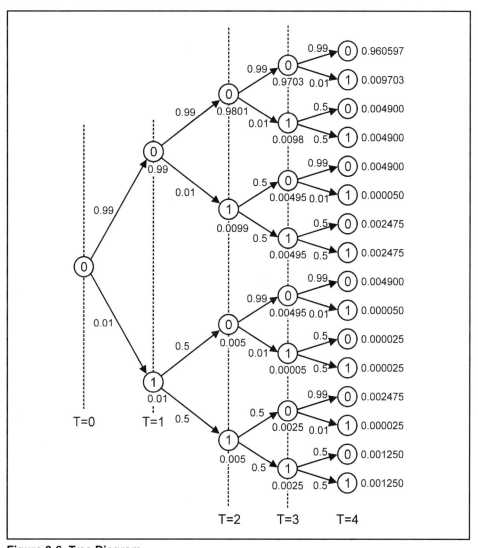

Figure 8-6. Tree Diagram.

One of the most commonly asked questions about a system like this is, "How much system downtime should we expect?" A system is "down" when it is in the failure state. With a Markov model, we can translate this question into, "What is the average probability of being in state 1?" The probability of being in state 1 can be calculated using the tree diagram.

Using rules of probability, multiply the probabilities from each step to obtain the individual probabilities for each path. Consider the uppermost path. The system starts in state 0. After one time interval, the system remains in state 0. It also remains in state 0 after time intervals two, three, and four. The probability of following this path is calculated in Equation 8-2.

$$\mathbf{P}_{(upper\ path)} = 0.99 \times 0.99 \times 0.99 \times 0.99 \qquad (8\text{-}2)$$
$$= 0.960597$$

This procedure can be followed to find path probabilities for each time interval. At time interval one, two paths exist. The upper path has one step with a probability of 0.99. The lower path has one step with a probability of 0.01. To find total probabilities of being in each state after the time interval, add all paths to a given state. For time interval one, there is only one path to each state so no addition is necessary. The probability of being in state 0 equals 0.99 and the probability of being in state 1 equals 0.01. Notice that the probability of being in either state 0 or state 1 is one (0.99 + 0.01 = 1). At any time interval, the state probabilities should sum to one. This is a good checking mechanism.

After the second time interval, the probability of being in state 0 equals the sum of two paths as calculated in Equation 8-3.

$$\mathbf{P}\ (state\ 0\ at\ t = 2) = 0.9801 + 0.005 = 0.9851 \qquad (8\text{-}3)$$

The probability of being in state 1 also equals the sum of two paths. The same method is used repeatedly to obtain the path probabilities for time intervals three and four as shown in Table 8-1. Note that the values change less and less with each time interval. A plot of the two probabilities is shown in Figure 8-7. The values are heading toward a steady state. This behavior is characteristic of fully repairable systems. Following the tree diagram further, note that the steady-state probability of being in state 1 is 0.01961. This is the answer to the question about downtime. We should expect the system to be "down" 1.961% of the time on average. For such fully repairable systems, "downtime" is known as unavailability.

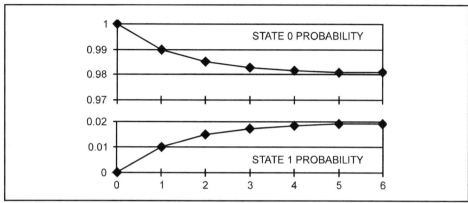

Figure 8-7. State Probabilities per Time Interval.

Time Interval	State 0	State 1
0	1.0	0.0
1	0.99	0.01
2	0.9851	0.0149
3	0.9827	0.0173
4	0.9815	0.0185
5	0.9809	0.0191
6	0.9806	0.0194

Table 8-1. State Probabilities

The tree diagram approach is not practical for Markov models of realistic size. However, there is another method for calculating steady-state probabilities. Remember that the transition matrix, **P**, is a matrix showing probabilities for moving from any one state to another state in one time interval. This matrix can be multiplied by itself to get transition probabilities for multiple time intervals.

When **P** is squared, the result is another $n \times n$ matrix that gives probabilities of going between states in two steps.

$$\mathbf{P}^2 = \begin{bmatrix} 0.99 & 0.01 \\ 0.5 & 0.5 \end{bmatrix} \begin{bmatrix} 0.99 & 0.01 \\ 0.5 & 0.5 \end{bmatrix}$$

$$= \begin{bmatrix} 0.9851 & 0.0149 \\ 0.745 & 0.255 \end{bmatrix}$$

If this matrix is multiplied by **P** again, a matrix of three-step probabilities is obtained.

$$\mathbf{P}^3 = \mathbf{P} \times \mathbf{P}^2 = \begin{bmatrix} 0.99 & 0.01 \\ 0.5 & 0.5 \end{bmatrix} \begin{bmatrix} 0.9851 & 0.0149 \\ 0.7450 & 0.2550 \end{bmatrix}$$

$$= \begin{bmatrix} 0.9827 & 0.0173 \\ 0.8650 & 0.1350 \end{bmatrix}$$

This process can be continued as long as desired to obtain the n-step probability transition matrix. For example:

$$\mathbf{P}^4 = \mathbf{P} \times \mathbf{P}^3 = \begin{bmatrix} 0.99 & 0.01 \\ 0.5 & 0.5 \end{bmatrix} \begin{bmatrix} 0.9827 & 0.0173 \\ 0.8650 & 0.1350 \end{bmatrix}$$

$$= \begin{bmatrix} 0.9815 & 0.0185 \\ 0.9239 & 0.0761 \end{bmatrix}$$

$$\mathbf{P}^5 = \mathbf{P} \times \mathbf{P}^4 = \begin{bmatrix} 0.99 & 0.01 \\ 0.5 & 0.5 \end{bmatrix} \begin{bmatrix} 0.9815 & 0.0185 \\ 0.9239 & 0.0761 \end{bmatrix}$$

$$= \begin{bmatrix} 0.9809 & 0.0191 \\ 0.9527 & 0.0473 \end{bmatrix}$$

$$\mathbf{P}^6 = \mathbf{P} \times \mathbf{P}^5 = \begin{bmatrix} 0.99 & 0.01 \\ 0.5 & 0.5 \end{bmatrix} \begin{bmatrix} 0.9809 & 0.0191 \\ 0.9527 & 0.0473 \end{bmatrix}$$

$$= \begin{bmatrix} 0.9807 & 0.0193 \\ 0.9668 & 0.0332 \end{bmatrix}$$

After multiplying these further, note that the result changes less and less with each step:

$$\mathbf{P}^{18} = \mathbf{P} \times \mathbf{P}^{17} = \begin{bmatrix} 0.99 & 0.01 \\ 0.5 & 0.5 \end{bmatrix} \begin{bmatrix} 0.98039 & 0.01961 \\ 0.98039 & 0.01961 \end{bmatrix}$$

$$= \begin{bmatrix} 0.98039 & 0.01961 \\ 0.98039 & 0.01961 \end{bmatrix}$$

A point is reached where $\mathbf{P}^{n+1} = \mathbf{P}^n$. The numbers will not change further. This matrix, labeled $\mathbf{P}^L$, is known as the "limiting state probability" matrix. Note that the top and bottom rows of the limiting state probability matrix are the same numbers. The probability of going to state 0 in n steps is the same regardless of starting state.

However, the starting state does affect the time-dependent probabilities in early time steps, as shown in Figure 8-7. Starting-state probabilities are specified with a row matrix (a $1 \times n$ matrix), $\mathbf{S}$. This row matrix is a list of numbers that indicate the probability that the system will be in each state. $\mathbf{S}^0$ is the starting probability list (time interval 0). For example, if a system always starts in one particular state, $\mathbf{S}^0$ will contain a single one and a quantity of zeros. The forging machine example always starts in state 0. The starting probability $\mathbf{S}$ would be

$$\mathbf{S}^0 = [1 \quad 0] \tag{8-4}$$

The $\mathbf{S}^n$ matrix for any particular time interval is obtained by multiplying $\mathbf{S}^{n-1}$ times $\mathbf{P}$ or $\mathbf{S}^0$ times $\mathbf{P}^{n-1}$. For example:

$$\mathbf{S}^1 = \mathbf{S}^0 \times \mathbf{P} = \begin{bmatrix} 1 & 0 \end{bmatrix} \begin{bmatrix} 0.99 & 0.01 \\ 0.5 & 0.5 \end{bmatrix}$$

$$= \begin{bmatrix} 0.99 & 0.01 \end{bmatrix}$$

$$\mathbf{S}^2 = \mathbf{S}^0 \times \mathbf{P}^2 = \begin{bmatrix} 1 & 0 \end{bmatrix} \begin{bmatrix} 0.9851 & 0.0149 \\ 0.7450 & 0.2550 \end{bmatrix}$$

$$= \begin{bmatrix} 0.9851 & 0.0149 \end{bmatrix}$$

$$\mathbf{S}^2 = \mathbf{S}^1 \times \mathbf{P} = \begin{bmatrix} 0.99 & 0.01 \end{bmatrix} \begin{bmatrix} 0.99 & 0.01 \\ 0.5 & 0.5 \end{bmatrix}$$

$$= \begin{bmatrix} 0.9851 & 0.0149 \end{bmatrix}$$

$$\mathbf{S}^3 = \mathbf{S}^2 \times \mathbf{P} = \begin{bmatrix} 0.9851 & 0.0149 \end{bmatrix} \begin{bmatrix} 0.99 & 0.01 \\ 0.5 & 0.5 \end{bmatrix}$$

$$= \begin{bmatrix} 0.9827 & 0.0173 \end{bmatrix}$$

This process can be continued as necessary.

$$S^4 = S^3 \times P = \begin{bmatrix} 0.9827 & 0.0173 \end{bmatrix} \begin{bmatrix} 0.99 & 0.01 \\ 0.5 & 0.5 \end{bmatrix}$$

$$= \begin{bmatrix} 0.9815 & 0.0185 \end{bmatrix}$$

$$S^5 = S^4 \times P = \begin{bmatrix} 0.9815 & 0.0185 \end{bmatrix} \begin{bmatrix} 0.99 & 0.01 \\ 0.5 & 0.5 \end{bmatrix}$$

$$= \begin{bmatrix} 0.9809 & 0.0191 \end{bmatrix}$$

$$S^6 = S^5 \times P = \begin{bmatrix} 0.9809 & 0.0191 \end{bmatrix} \begin{bmatrix} 0.99 & 0.01 \\ 0.5 & 0.5 \end{bmatrix}$$

$$= \begin{bmatrix} 0.9807 & 0.0193 \end{bmatrix}$$

As with P, the numbers change less and less each time. Eventually, there is no significant change:

$$S^{18} = S^{17} \times P = \begin{bmatrix} 0.98039 & 0.01961 \end{bmatrix} \begin{bmatrix} 0.99 & 0.01 \\ 0.5 & 0.5 \end{bmatrix}$$

$$= \begin{bmatrix} 0.98039 & 0.01961 \end{bmatrix}$$

Again, the "limiting state probability" row, labeled S^L, has been reached. The limiting state probability matrix can be created by merely replicating the rows as often as necessary, taking advantage of the fact that all rows in the limiting state probability matrix are the same.

Although the matrix multiplication method is quicker than the tree method, either can be time consuming for a realistic Markov model with many states. Fortunately, another direct method is used to calculate limiting state probabilities.

Markov Modeling

The limiting state probabilities have been reached when $\mathbf{S}^{n+1}$ multiplied by $\mathbf{P}$ equals $\mathbf{S}^n$. This fact allows an algebraic relationship to solve the problem directly. Limiting state probabilities exist when

$$\begin{bmatrix} S_1^L & S_2^L \end{bmatrix} = \begin{bmatrix} S_1^L & S_2^L \end{bmatrix} \begin{bmatrix} 0.99 & 0.01 \\ 0.5 & 0.5 \end{bmatrix} \tag{8-5}$$

If the matrix is multiplied, we obtain the algebraic relations:

$$0.99 S_1^L + 0.5 S_2^L = S_1^L \tag{8-6}$$

and

$$0.01 S_1^L + 0.5 S_2^L = S_2^L \tag{8-7}$$

With algebraic manipulation it is apparent that both equations contain the same information:

$$0.01 S_1^L = 0.5 S_2^L$$

The problem has two variables and only one equation; thus, it would appear that no further progress can be made. However, an earlier rule can help. The probabilities in a row should always add to one. This gives the additional equation:

$$S_1^L + S_2^L = 1 \tag{8-8}$$

Substituting

$$S_2^L = \frac{0.01}{0.5} S_1^L$$

into the second equation:

$$S_1^L = \frac{0.01}{0.5} S_1^L = 1$$

Adding the first terms:

$$1.02 S_1^L = 1$$

Finally:

$$S_1^L = \frac{1}{1.02} = 0.98039$$

$$S_2^L = 1 - 0.98039 = 0.01961$$

Of the three different methods presented for determining limiting state probabilities, the direct algebraic method offers the quickest solution. The method is applicable to all fully repairable systems. These systems have Markov models that allow eventual movement from any state to any other state. Markov models of this type are technically called "regular" Markov models.

Markov models are used to solve many problems outside the scope of reliability analysis. Consider a package delivery process. A company has three locations in a city: corporate headquarters, central engineering, and a manufacturing plant site. Each location has a package dispatch center where a courier goes to pick up packages to be delivered. Each morning 100 couriers begin work at the corporate headquarters dispatch center. When a package has been delivered to the person at the designated site, a courier goes to the message center at that site to pick up the next message to be delivered. On average, four packages per hour are delivered. At the end of each day, the couriers who happen to be at each site are transported back to corporate headquarters.

Historical log records of the couriers indicate that packages from headquarters go to the plant site 25% of the time. Packages from headquarters go to engineering 50% of the time and to other areas within headquarters 25% of the time. Packages from the plant site go to headquarters 50% of the time and to engineering 50% of the time. Packages from engineering go to headquarters 20% of the time and to the plant site 60% of the time. Remaining packages go to other areas within engineering.

A Markov model of this process is developed by noting that a courier can be in one of three states: headquarters, engineering, and the plant site. Three circles are drawn. Directed arcs are drawn showing possible package movement. These arcs are labeled with the associated probabilities. The resulting Markov model is shown in Figure 8-8.

Markov Modeling

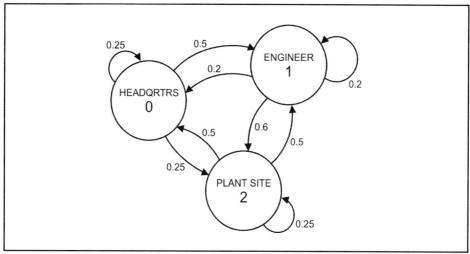

Figure 8-8. Delivery Process.

In the model, packages move from state to state four times an hour in discrete time steps. The various possibilities for the next step depend only on the current state, not on previous states. Thus, this system is a memoryless system. Examining the model, we verify that we could eventually get from any state to any other state. Therefore, this is a "regular" Markov model, and limiting state probabilities exist.

Since the company must provide transportation at the end of the day from each site back to corporate headquarters, it would be valuable to know how many couriers are expected at each site.

The number of couriers ending at each site could be calculated by knowing the probability that one courier will end up at each site and multiplying this number by the total number of couriers.

The probability that any one courier starting at headquarters would be in any of the three sites after one quarter hour is calculated by multiplying a starting row vector by **P**. The process is repeated to find the probable location of a courier as a function of time. After 32 steps (four steps per hour times 8 hr), the ending site probabilities for a courier are known. Computer simulations have indicated that if the transition matrix has large numbers (>0.1), limiting state probabilities are quickly reached (10 to 20 steps). In this case, the probabilities are large. Therefore, it is very likely that limiting state probabilities are reached within 32 steps. Since site probabilities as a function of time are not needed to solve the problem, limiting state probabilities can be directly calculated. **P** for the delivery process is:

$$\mathbf{P} = \begin{bmatrix} 0.25 & 0.5 & 0.25 \\ 0.2 & 0.2 & 0.6 \\ 0.5 & 0.5 & 0 \end{bmatrix}$$

The limiting state probabilities are calculated by multiplying:

$$\begin{bmatrix} S_0^L & S_1^L & S_2^L \end{bmatrix} \begin{bmatrix} 0.25 & 0.5 & 0.25 \\ 0.2 & 0.2 & 0.6 \\ 0.5 & 0.5 & 0 \end{bmatrix} = \begin{bmatrix} S_0^L & S_1^L & S_2^L \end{bmatrix}$$

Multiplying the row matrix by **P**, three algebraic relationships are obtained:

$$S_0^L = 0.25 S_0^L + 0.2 S_1^L + 0.5 S_2^L$$

$$S_1^L = 0.5 S_0^L + 0.2 S_1^L + 0.5 S_2^L$$

$$S_2^L = 0.25 S_0^L + 0.6 S_1^L$$

Discarding one of these equations (the equation set provides redundant information) and using the fact that all probabilities sum to one:

$$S_0^L + S_1^L + S_2^L = 1$$

allows a solution for limiting state probabilities. With a little algebraic manipulation:

$$S_1^L = 1.25 S_2^L$$

Substituting,

$$S_0^L = S_2^L$$

Substituting again,

$$S_2^L + 1.25 S_2^L + S_2^L = 1$$

Summing,

$$3.25 S_2^L = 1$$

Finishing the problem, the results are:

$$S_2^L = 0.307692$$

$$S_0^L = 0.307692$$

$$S_1^L = 0.384615$$

Knowing the limiting state probabilities for one courier, the probable distribution of couriers at the end of the day can be calculated:

$$\text{Number at headquarters} = 100 S_0^L = 30.77$$

$$\text{Number at engineering} = 100 S_1^L = 38.46$$

$$\text{Number at plant site} = 100 S_2^L = 30.77$$

If the process had no variability, transportation would be provided for 39 persons from engineering and 31 persons from the plant site. However, due to possible variability in the process, it would be prudent to add extra transportation capacity. Transportation should be provided for 45 persons each day from engineering and 35 persons each day from the plant site.

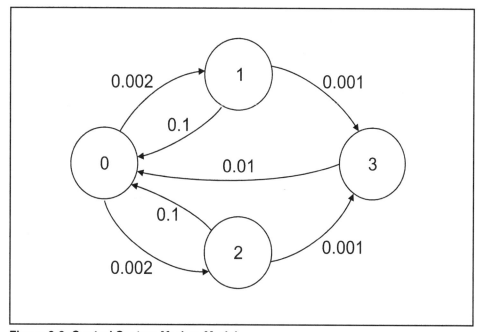

Figure 8-9. Control System Markov Model.

EXAMPLE 8-1

Problem: A control system has the Markov model shown in Figure 8-9. The process is shut down when the controller is in state 3. Therefore, state 3 is classified as system failure. States 0, 1, and 2 are classified as system success. Failure probabilities and repair probabilities are shown on the Markov model on a per hour basis. What is the expected system downtime?

Solution: The transition matrix is

$$P = \begin{bmatrix} 0.996 & 0.002 & 0.002 & 0 \\ 0.1 & 0.899 & 0 & 0.001 \\ 0.1 & 0 & 0.899 & 0.001 \\ 0.01 & 0 & 0 & 0.99 \end{bmatrix}$$

Solving for limiting state probabilities, the results are

$$S_0^L = 0.958254$$

$$S_1^L = 0.018975$$

$$S_2^L = 0.018975$$

$$S_3^L = 0.003794$$

Since the system is down in state 3, the average steady-state downtime is 0.3794%.

Availability/Unavailability

Once the limiting state probabilities of a Markov model are known, it is a simple matter to calculate steady-state availability or unavailability. Each state in the fully repairable Markov model represents either system success or system failure. To calculate availability, identify the system success states and sum their probabilities. The failure state probabilities sum to provide unavailability.

One success state, state 0, is present in the forging process example (Figure 8-4). Thus, steady-state availability for the forging process is 0.98039, or 98.039%. One failure state exists, state 1; therefore, steady-state unavailability is 0.01961, or 1.96%.

Markov Modeling

EXAMPLE 8-2

Problem: Calculate the steady-state availability of the control system from Example 8-1.

Solution: The control system is successful in states 0, 1, and 2. Therefore, add the limiting state probability of the success states.

$$A(s) = S_0^L + S_1^L + S_2^L \tag{8-9}$$

$$= 0.958255 + 0.018975 + 0.018975$$

$$= 0.996205$$

EXAMPLE 8-3

Problem: A tank has two identical drain valves. The tank will drain properly if only one valve works. Each valve has a failure probability of 0.001 per hour. There is a small chance (0.0001 per hour) that both valves will fail simultaneously due to a common stress. A valve repair takes an average time of 20 hr. The system is fully repairable. What is the availability of the drain valve system?

Solution: The drain valve system is fully repairable. Therefore, we can calculate limiting state probabilities. Success state probabilities can be added to obtain steady-state availability.

The Markov model is developed first. State 0 is defined as the state where both valves are operating. One valve can fail. This would take the system to a state where one valve operates and the other does not. This is defined as state 1. The system will move from state 0 to state 1 if either valve fails. Therefore, the directed arc from state 0 to state 1 is labeled with the number 0.002 (0.001 from the first valve + 0.001 from the second valve). Another arc exits from state 0. This is the result of common stress that fails both valves (called a "common-cause" failure). This arc goes from state 0 to state 2.

From state 1, the remaining valve could fail. This action would take the system to a state where both valves have failed. This state is labeled state 2. One valve can be repaired in an average time of 20 hr. The discrete time step is defined as 1 hr. Therefore, the repair probability is 0.05 (1 divided by 20). The completed Markov model is illustrated in Figure 8-10.

EXAMPLE 8-3 (continued)

The transition matrix is

$$P = \begin{bmatrix} 0.9979 & 0.002 & 0.0001 \\ 0.05 & 0.949 & 0.001 \\ 0 & 0.05 & 0.95 \end{bmatrix}$$

To calculate the limiting state probabilities, set up the row matrix multiplication:

$$\begin{bmatrix} s_0^L & s_1^L & s_2^L \end{bmatrix} \begin{bmatrix} 0.9979 & 0.002 & 0.0001 \\ 0.05 & 0.949 & 0.001 \\ 0 & 0.05 & 0.95 \end{bmatrix}$$

$$= \begin{bmatrix} s_0^L & s_1^L & s_2^L \end{bmatrix}$$

Performing the multiplication yields three equations:

$$0.9979 s_0^L + 0.05 s_1^L = s_0^L$$

$$0.002 s_0^L + 0.949 s_1^L + 0.05 s_2^L = s_1^L$$

$$0.0001 s_0^L + 0.001 s_1^L + 0.95 s_2^L = s_2^L$$

As in past examples, only two of these equations can be used. Again, the additional equation results from the fact that the state probabilities sum to one.

$$s_0^L + s_1^L + s_2^L = 1$$

Using these equations, limiting state probabilities are obtained:

$$s_0^L = 0.957085$$

$$s_1^L = 0.040197$$

$$s_2^L = 0.002718$$

(Stop! This is an excellent opportunity for the reader to verify understanding by cranking through the algebra and hopefully obtaining the same answers!) The system is successful in state 0 and state 1; therefore:

$$A(s) = 0.957085 + 0.040197 = 0.997282$$

Markov Modeling

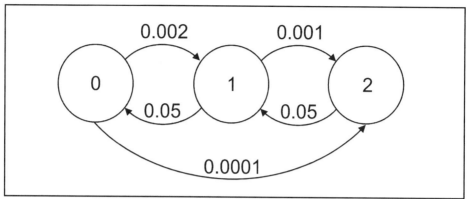

Figure 8-10. Dual Drain System Markov Model.

Time-Dependent Availability

When a discrete-time Markov model begins at $t = 0$ in a particular starting state, the availability will vary with time. Figure 8-7 shows this behavior for the forging system example. If the system is fully repairable, the availability and unavailability will eventually reach a steady-state value. Until steady-state is reached, the availability and unavailability values may change in significant ways.

We have already developed two different methods of calculating time-dependent availability: the tree diagram/path probability method and the **S/P** multiplication method. The **S/P** method is vastly more practical, especially since many calculators and all personal computer spreadsheet programs are capable of matrix multiplication.

To calculate time-dependent availability, follow the previously developed procedure of multiplying the state probability state row, **S**, by **P**. At each time step add the probabilities of the success states to obtain availability. Add the probabilities of the failure states to obtain unavailability.

EXAMPLE 8-4

Problem: Calculate the time-dependent availability for the system described in Example 8-1. The system always starts in state 0.

Solution: The state probabilities after step 1 (1 hr) are

$$S^1 = \begin{bmatrix} 1 & 0 & 0 & 0 \end{bmatrix} \begin{bmatrix} 0.996 & 0.002 & 0.002 & 0 \\ 0.1 & 0.899 & 0 & 0.001 \\ 0.1 & 0 & 0.899 & 0.001 \\ 0.01 & 0 & 0 & 0.99 \end{bmatrix} = \begin{bmatrix} 0.996 & 0.002 & 0.002 & 0 \end{bmatrix}$$

EXAMPLE 8-4 (continued)

The availability at $t = 1$ hr is the sum of states 0, 1, and 2. This equals 1.0. Continuing this process, we obtain availability and unavailability for each time increment.

For step two, the state probabilities are

$$s^2 = \begin{bmatrix} 0.996 & 0.002 & 0.002 & 0 \end{bmatrix} \begin{bmatrix} 0.996 & 0.002 & 0.002 & 0 \\ 0.1 & 0.899 & 0 & 0.001 \\ 0.1 & 0 & 0.899 & 0.001 \\ 0.01 & 0 & 0 & 0.99 \end{bmatrix}$$

$$= \begin{bmatrix} 0.992416 & 0.00379 & 0.00379 & 0.000004 \end{bmatrix}$$

Availability equals the sum of the probabilities for states 0, 1, and 2. This equals 0.999996. Continuing the process, availability is plotted in Figure 8-11 and unavailability in Figure 8-12.

It can be seen that availability decreases from a value of 1.0 toward the steady-state value (0.996205 from Example 8-2) as time passes. The time increments for this model are 1 hr. The numbers on the time line shown in Figure 8-11 and Figure 8-12 are hours.

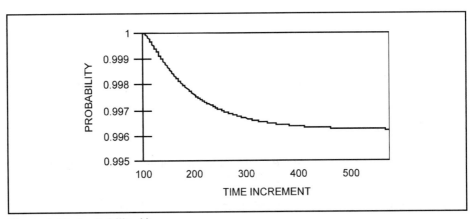

Figure 8-11. Availability (*t*).

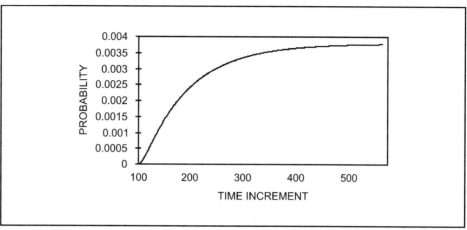

Figure 8-12. Unavailability (*t*).

EXAMPLE 8-5

Problem: Calculate the time-dependent availability of the dual-drain system from Example 8-3, assuming the system starts in state 0.

Solution: Using a discrete Markov model with a time increment of 1 hr:

$$S^1 = S^0 \times P = \begin{bmatrix} 1 & 0 & 0 \end{bmatrix} \begin{bmatrix} 0.9979 & 0.002 & 0.0001 \\ 0.05 & 0.949 & 0.001 \\ 0 & 0.05 & 0.95 \end{bmatrix}$$

$$= \begin{bmatrix} 0.9979 & 0.002 & 0.0001 \end{bmatrix}$$

Adding the probabilities from states 0 and 1 gives the result:

$$A(1 \text{ hr}) = 0.9979 + 0.002$$
$$= 0.9999$$

Continuing the process, the availability for each hour is obtained. (The calculations can be done very quickly by setting up a spreadsheet.) The results are plotted in Figure 8-13. The plot settles relatively quickly toward the steady-state value of 0.997282, as calculated in Example 8-3.

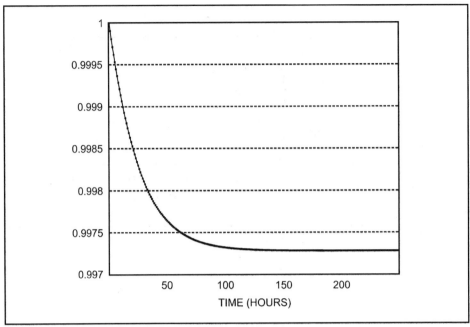

Figure 8-13. Availability (*t*).

Absorbing States

When modeling failure behavior of control systems, the system often fails in such a way that it cannot be repaired. An obvious example of this is an explosion or other permanent destruction. In other situations, failure behavior of a system is modeled for a specific time period when certain failures cannot be repaired. Or it might be the time needed for an industrial process to run a campaign of product batches. That time period might be the time between scheduled plant shutdowns. At shutdown, the control system is inspected and overhauled as necessary.

A Markov model of such system behavior would show one or more failure states from which there is no exit. State 1 in Figure 8-3 is an example. Such states are known as "absorbing states." They are typically used whenever a failure occurs from which there is no practical repair during the time period of interest.

Systems that are modeled with absorbing states have a steady-state availability of zero. As time periods tick by, the model will eventually reach an absorbing failure state and stay there. As a measure of system success, steady-state availability is not useful to systems modeled with absorbing states.

Time-Dependent Reliability

Earlier, reliability was formally defined as the probability of system success over a time interval. The definition of reliability does not allow for system failure and repair. This fits in perfectly with systems that cannot be fully repaired at the system level.

Repairs can be made, though. A system that has a discrete-time Markov model with more than one success state may move between those states without violating the definition of reliability. Component or module failures and subsequent repairs that cause movement only between system success states do not cause system failure. When a component or module failure causes the system to move from a success state to an absorbing failure state, system failure occurs without repair.

Reliability can be calculated for any time interval using methods similar to those used to calculate availability as a function of time. The initial probability state row is multiplied by **P** to obtain a probability state row for each time step. The probabilities from all successful states are added to obtain reliability as a function of time. The probabilities from all failure states are added to obtain unreliability.

EXAMPLE 8-6

Problem: A partially repairable control system has the Markov model shown in Figure 8-14. The process is shut down when the controller is in state 3. Once the process is shut down, it cannot be restarted without a complete overhaul of the process equipment. There is a scheduled shutdown once a month; therefore, we are interested in a plot of reliability for a one-month time period. Failure probabilities and repair probabilities are shown on the Markov model on a per hour basis. The system always starts in state 0.

Solution: Using a discrete-time Markov model with a 1-hr time increment, reliability is calculated by multiplying each successive **S** by **P**. The transition matrix for this model is

$$P = \begin{bmatrix} 0.996 & 0.002 & 0.002 & 0 \\ 0.1 & 0.899 & 0 & 0.001 \\ 0.1 & 0 & 0.899 & 0.001 \\ 0 & 0 & 0 & 1 \end{bmatrix}$$

EXAMPLE 8-6 (continued)

The state probabilities for the first hour are given by:

$$S^1 = S^0 \times P = \begin{bmatrix} 1 & 0 & 0 & 0 \end{bmatrix} \begin{bmatrix} 0.996 & 0.002 & 0.002 & 0 \\ 0.1 & 0.899 & 0 & 0.001 \\ 0.1 & 0 & 0.899 & 0.001 \\ 0 & 0 & 0 & 1 \end{bmatrix}$$

$$= \begin{bmatrix} 0.996 & 0.002 & 0.002 & 0 \end{bmatrix}$$

Adding the probabilities from states 0, 1, and 2, the reliability for the first hour equals 1.0. Continuing the process, the reliability values for 750 hr are calculated. These are plotted in Figure 8-15.

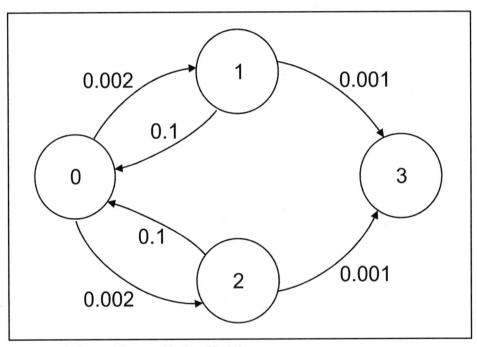

Figure 8-14. Control System Markov Model.

Markov Modeling

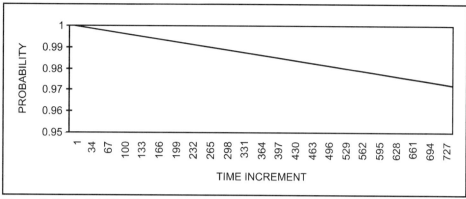

Figure 8-15. Reliability (*t*).

Reliability for Fully Repairable Systems

These methods can be used to obtain reliability for a fully repairable system. Although steady-state availability and time-dependent availability are the typical measures of system success, system reliability information can also be useful. To obtain reliability, all repair arcs from system failure states to system success states are deleted from the Markov model. A new transition matrix is obtained.

EXAMPLE 8-7

Problem: Calculate and plot the reliability for the dual-drain system of Example 8-3.

Solution: There is one repair arc from a system failure state to a system success state. That arc is from state 2 to state 1. When that arc is eliminated, the new transition matrix is

$$\mathbf{P} = \begin{bmatrix} 0.9979 & 0.002 & 0.0001 \\ 0.05 & 0.949 & 0.001 \\ 0 & 0 & 1 \end{bmatrix}$$

Compare this equation to the first equation of Example 8-3. The bottom row is different. Can you explain why?

Multiplying this **P** by the state rows for each time interval generates new probability state rows. The sum of probabilities for successful states provides the reliability for each time interval. These values are plotted in Figure 8-16.

Comparison of Figures 8-13 and 8-16 clearly shows how availability compares to reliability. The availability value drops in a decreasing manner to the steady-state value of 0.997282. The reliability continues to drop. Eventually, reliability would go to zero, as nothing lasts forever. For fully repairable systems, availability generally is greater than reliability.

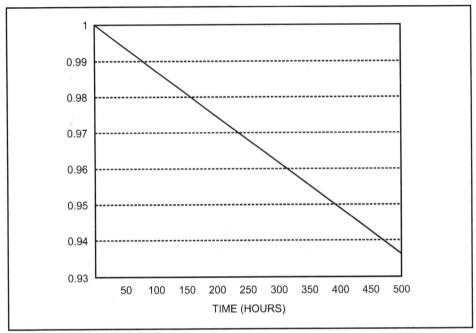

Figure 8-16. Reliability (*t*).

Mean Time to Failure

A single measure of success is required for systems modeled with absorbing states. The mean time to failure (MTTF) often is used for this purpose. In terms of a discrete-time Markov model, the time to failure is represented by the average number of time increments between system startup and system failure over many independent starts.

If many computer simulations of a discrete-time Markov model were run, counting the time increments between start and failure (absorption), we could average the time increment numbers obtained. That average number represents the MTTF for the system being modeled.

It is not necessary to run computer simulations and average the results in order to obtain the MTTF. The MTTF for a Markov model can be calculated from the transition matrix. The method is well documented (Ref. 4). The first step is to create a truncated matrix that contains only the transient states of the system. This is done by crossing out the rows and the columns of the "absorbing" states. Using the control system defined in Example 8-6, the truncated matrix, called the **Q** matrix, is

$$\mathbf{Q} = \begin{bmatrix} 0.996 & 0.002 & 0.002 \\ 0.1 & 0.899 & 0 \\ 0.1 & 0 & 0.899 \end{bmatrix} \qquad (8\text{-}10)$$

The **Q** matrix is subtracted from the identity matrix, known as **I** (Ref. 5).

$$\mathbf{I} - \mathbf{Q} = \begin{bmatrix} 1 & 0 & 0 \\ 0 & 1 & 0 \\ 0 & 0 & 1 \end{bmatrix} - \begin{bmatrix} 0.996 & 0.002 & 0.002 \\ 0.1 & 0.899 & 0 \\ 0.1 & 0 & 0.899 \end{bmatrix}$$

$$= \begin{bmatrix} 0.004 & -0.002 & -0.002 \\ -0.1 & 0.101 & 0 \\ -0.1 & 0 & 0.101 \end{bmatrix} \qquad (8\text{-}11)$$

Another matrix, called the **N** matrix, is obtained by inverting the (**I** − **Q**) matrix. Matrix inversion can be done analytically for small matrices, but this is impractical for large, realistic matrices that represent more complicated systems. A shortcut is available, however. Many spreadsheet programs in common use (Refs. 6 and 7) can numerically invert a matrix, making quick work of previously time-consuming MTTF calculations.

A numerical matrix inversion of the (**I** − **Q**) matrix using a spreadsheet provides the following:

$$[\mathbf{I} - \mathbf{Q}]^{-1} = \mathbf{N} = \begin{bmatrix} 25{,}250 & 500 & 500 \\ 25{,}000 & 504.95 & 495.05 \\ 25{,}000 & 495.05 & 504.95 \end{bmatrix} \qquad (8\text{-}12)$$

The **N** matrix provides the expected number of time increments that the system dwells in each system success state (a transient state) as a function of starting state. In our example, the top row states the number of time increments per transient state if we start in state 0. The middle row gives the number of time increments if we start in state 1. The bottom row states the number of time increments if we start in state 2. If a system always starts in state 0, we can add the numbers from the top row to get the total number of time increments in all system success states. When this is multiplied by the time increment, we obtain the MTTF when the system starts in state 0. In our example, this number equals 26,250 hr, since we used a time increment of 1 hr. If we started this system in state 1, we would expect the system to fail after 26,000 hr on average. If we start the system in state 2, we would also expect 26,000 time increments to pass until absorption (26,000 hr until failure).

> **EXAMPLE 8-8**
>
> **Problem:** Calculate the MTTF of the dual-drain valve system described in Example 8-3. The system always starts in state 0.
>
> **Solution:** Given that the system always starts in state 0, we can calculate the MTTF of the system by modeling the failure state as absorbing. The **P** matrix is given in Example 8-7. The truncated matrix **Q** is
>
> $$\mathbf{Q} = \begin{bmatrix} 0.9979 & 0.002 \\ 0.05 & 0.949 \end{bmatrix}$$
>
> Subtracting the **Q** matrix from the identity matrix yields:
>
> $$\mathbf{I} - \mathbf{Q} = \begin{bmatrix} 0.0021 & -0.002 \\ -0.05 & 0.051 \end{bmatrix}$$
>
> Inverting the (**I** − **Q**) matrix provides the **N** matrix:
>
> $$\mathbf{N} = \begin{bmatrix} 7183.1 & 281.7 \\ 7042.2 & 295.8 \end{bmatrix}$$
>
> Then, adding the state 0 row gives the result:
>
> $$MTTF = 7183.1 + 281.7 = 7464.8 \text{ hr}$$

Limiting State Probability without Full Repair

In Markov models with absorbing states, we have learned that the model eventually goes to an absorbing state and stays there. A three-state Markov model with one absorbing state (numbered last) will eventually have a probability state row that equals [0 0 1].

The eventual probability state row, $\mathbf{S}^L$, for a Markov model with one absorbing state is easy to predict. However, some Markov models have more than one absorbing failure state.

Consider a model that consists of four states, two of which are absorbing. We know that the limiting state probability row will have two zeros and that the other two numbers will sum to one. But other than that, we cannot yet calculate the actual limiting state probability row for multiple absorbing states.

Control systems are modeled with multiple failure states whenever more than one failure mode is important. Controllers can fail with outputs energized, outputs de-energized, or outputs frozen. In many situations (protection systems), it is important to distinguish these modes. There are many other examples of multiple failure states in the realm of control systems modeling.

The probability of absorption into a particular failure state can be calculated from the transition matrix, **P**. If the states are numbered such that the absorbing states are either the first assigned numbers or the last assigned numbers, **P** can be broken into four submatrices. Two of these submatrices are combined to provide the information needed. Figure 8-17 shows a Markov model with two absorbing failure states: state 2 and state 3. The system is successful in states 0 and 1. The transition matrix, **P**, is

$$\mathbf{P} = \begin{bmatrix} 0.973 & 0.02 & 0.002 & 0.005 \\ 0.05 & 0.933 & 0.002 & 0.015 \\ 0 & 0 & 1 & 0 \\ 0 & 0 & 0 & 1 \end{bmatrix}$$

P is divided into four submatrices as shown in Figure 8-18. The upper left submatrix is the **Q** matrix used when calculating MTTF. It gives probabilities of moving between transient states (normally, success states). The lower left submatrix is a null matrix that contains only zeros. The lower right submatrix is an **I** matrix. The upper right submatrix is called the **R** matrix. It relates the transient (success) states to the absorbing (failure) states.

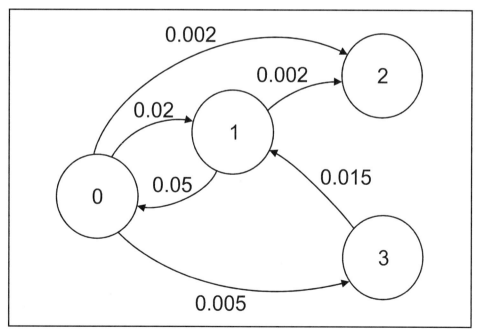

Figure 8-17. Multiple Failure States.

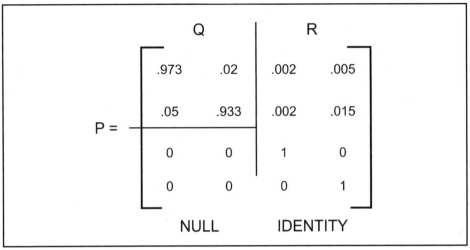

Figure 8-18. Submatrices.

In order for a system to fail into a particular absorbing state, it must be in a transient state that has an arc to the particular absorbing state. For every time increment that the model is in such a state, there is a probability that the model will transition to the absorbing state. These probabilities are given by the **R** submatrix. We now need to determine how many time increments the model spends in each transient state. If we multiply the number of time increments spent in each transient state by the probability of absorption per time increment, we obtain the total probability of absorption.

Fortunately, we have already determined that the **N** matrix provides the number of time increments spent in each transient state, given a particular starting state. Thus, we can multiply the **N** matrix by the **R** submatrix to obtain a matrix of absorption probabilities, given a starting state.

EXAMPLE 8-9

Problem: Calculate limiting state probabilities for the system modeled by Figure 8-17 if the system always starts in state 0.

Solution: Figure 8-18 shows that the **Q** matrix is:

$$\mathbf{Q} = \begin{bmatrix} 0.973 & 0.02 \\ 0.05 & 0.933 \end{bmatrix}$$

Subtracting this from the identity matrix yields:

$$\mathbf{I} - \mathbf{Q} = \begin{bmatrix} 0.027 & -0.02 \\ -0.05 & 0.067 \end{bmatrix}$$

EXAMPLE 8-9 (continued)

Inverting the (**I** − **Q**) matrix provides the **N** matrix:

$$\mathbf{N} = \begin{bmatrix} 82.82 & 24.72 \\ 61.8 & 33.37 \end{bmatrix}$$

Figure 8-18 also shows that the **R** matrix is

$$\mathbf{R} = \begin{bmatrix} 0.002 & 0.005 \\ 0.002 & 0.015 \end{bmatrix}$$

Multiplying the **N** matrix by the **R** matrix results in:

$$\mathbf{NR} = \begin{bmatrix} 0.215080 & 0.784919 \\ 0.190358 & 0.809641 \end{bmatrix}$$

Given that the system always starts in state 0, the probability of absorption in state 2 is given by $\mathbf{NR}_{02}$ (row 0, column 2) and equals 0.215080. The probability of absorption in state 3 is given by $\mathbf{NR}_{03}$ and equals 0.784919. The limiting state probability row equals:

$$\mathbf{S}^L = \begin{bmatrix} 0 & 0 & 0.215080 & 0.784919 \end{bmatrix}$$

Note: In the case of a fully repairable system, the limiting state probability is the same regardless of starting state. In the case of models with absorbing states, the starting state will affect the results.

The limiting state probability row can be verified by performing **S**/**P** multiplication to obtain state probabilities as a function of time. Starting with

$$\mathbf{S}^0 = \begin{bmatrix} 1 & 0 & 0 & 0 \end{bmatrix}$$

we multiply the subsequent **S** rows by **P**. The results are plotted in Figures 8-19 to 8-22. Figure 8-19 shows the probability of being in state 0 as a function of 1-hr time increments. The plot starts at a value of one and drops to zero as expected. Figure 8-20 plots state 1 probability for the same time period. Since it is a transient state, the value eventually drops to zero as expected. Figure 8-21 plots the probability as a function of time for state 2. The value increases toward the limiting state value. The same result is obtained for state 3, as plotted in Figure 8-22.

System reliability is the sum of probabilities for states 0 and 1. Figure 8-23 shows reliability as a function of time. Careful comparison with Figures 8-19 and 8-20 indicates that Figure 8-23 is the sum of those two plots.

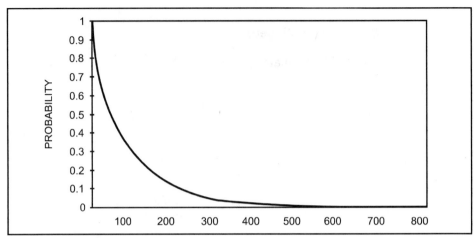
Figure 8-19. State 0 Probability.

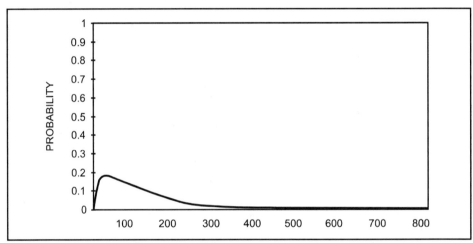
Figure 8-20. State 1 Probability.

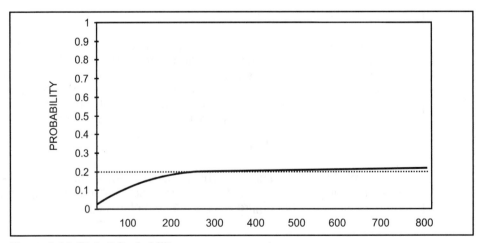
Figure 8-21. State 2 Probability.

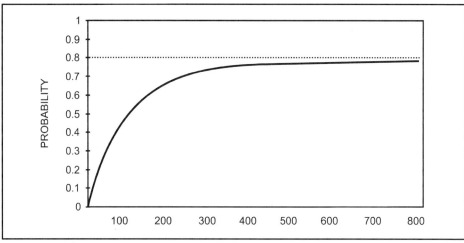
Figure 8-22. State 3 Probability.

Time Increments

The time increments in a discrete Markov model should be as small as is practical. Larger time increments save calculation steps, especially for long time periods, but add error to the model. Time-dependent reliability was plotted using 1-hr increments in Figure 8-23 for the system modeled by Figure 8-17. The calculation was repeated using 1/2-hr increments. The transition matrix is recomputed for 1/2-hr time increments:

$$\mathbf{P} = \begin{bmatrix} 0.9865 & 0.01 & 0.001 & 0.0025 \\ 0.025 & 0.9665 & 0.001 & 0.0075 \\ 0 & 0 & 1 & 0 \\ 0 & 0 & 0 & 1 \end{bmatrix}$$

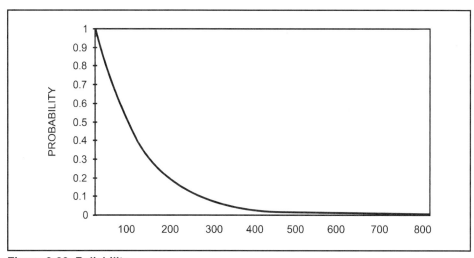
Figure 8-23. Reliability.

The results are shown in Table 8-2. The values calculated for the 1-hr time increment are shown in Table 8-3.

Hours	Time Increment	Increment Number	Reliability
100	0.5	200	0.395849
200	0.5	400	0.151746
300	0.5	600	0.057960
400	0.5	800	0.022178

Table 8-2. Reliability Values (Increment = 1/2 hr)

Hours	Time Increment	Increment Number	Reliability
100	1	100	0.394933
200	1	200	0.150771
300	1	300	0.057558
400	1	400	0.021973
500	1	500	0.008388

Table 8-3. Reliability Values (Increment = 1 hr)

In order to save calculation steps (and computer memory), we can increase the time increment to 2 hr. The **P** for a 2-hr increment is

$$\mathbf{P} = \begin{bmatrix} 0.946 & 0.04 & 0.004 & 0.01 \\ 0.1 & 0.866 & 0.004 & 0.03 \\ 0 & 0 & 1 & 0 \\ 0 & 0 & 0 & 1 \end{bmatrix}$$

The reliability values obtained using a 2-hr time increment are listed in Table 8-4. Table 8-5 lists values obtained for a 5-hr time increment.

Hours	Time Increment	Increment Number	Reliability
100	2	50	0.393090
200	2	100	0.149366
300	2	150	0.056755
400	2	200	0.021565
500	2	250	0.008194

Table 8-4. Reliability Values (Increment = 2 hr)

Markov Modeling

Hours	Time Increment	Increment Number	Reliability
100	5	20	0.387466
200	5	40	0.145120
300	5	60	0.054358
400	5	80	0.020357
500	5	100	0.005147

Table 8-5. Reliability Values (Increment = 5 hr)

A comparison of the 400-hr values shows the errors. Using the 1/2-hr increment as a base value, Table 8-6 shows the differences. The error is due to the fundamental difference between failure probability and failure rate. Discrete Markov models were developed with constant *probability* values. We learned in Chapter 4 that devices with constant failure rates have exponentially decreasing failure probability. We also learned that failure probability is approximated by multiplying failure rate by the time interval. As this product gets larger, the errors increase.

Time Increment	400-hr Value	%Difference
0.5	0.022178	0
1	0.021973	0.924
2	0.021565	2.764
5	0.020357	8.210

Table 8-6. Errors due to Time Increment

Always use the smallest possible time increment. The trade-offs in such a decision usually relate to computation time and memory limitations of the machine performing the calculation. These trade-offs are moving rapidly in favor of lower time increments as computer speed and memory rapidly increase. However, in large Markov models that use hundreds of states, the trade-offs still exist. Fortunately, the errors go in the direction of pessimism. The 400-hr reliability value for the 5-hr time increment is lower than the equivalent value for the 1-hr time increment. This is generally in the "safe" direction.

Another approach can be taken to save computation time and computer memory when doing discrete Markov calculations. Instead of doubling the time increment, use $\mathbf{P}^2$ [Ref. 8]. $\mathbf{P}^2$ is the two-step transition matrix. Values are obtained for each two steps, and accuracy is not sacrificed. The approach can be carried further. Instead of using 5-hr increments in the above example, $\mathbf{P}^5$ could be used. Data values would be obtained with the same resolution, and accuracy would be maintained.

Time-Dependent Failure Rate

Many real components do not exhibit a constant failure rate. As seen in Chapter 4, the failure rates of some components vary with time in a manner that appears like a "bathtub" when plotted. Such changes in failure rate can have a drastic effect on reliability or availability plots, especially in the wearout region.

A time-dependent failure rate can be accommodated in a discrete-time Markov model by changing the **P** as required each time interval. In one approach, **P** is a full three-dimensional matrix where a new **P** is used for each time interval. In other approaches, each variable element in **P** is characterized by a piecewise linear equation. To account for wearout, two segments are required: a constant segment for the first portion of time and a linearly increasing segment thereafter.

The ability to accommodate time-dependent failure rates is one of the great strengths of the discrete-time Markov model approach. As the need for realistic reliability modeling becomes stronger, such abilities will grow in importance.

EXERCISES

8.1 A system contains three modules. Each module can be either operating successfully or failed. How many possible states may exist in the Markov model for this system?

8.2 A fully repairable system has the Markov model shown in Figure 8-24, where the arcs are labeled with probabilities in units of failure probability per hour. The system is successful in states 0 and 1. Calculate the limiting state probability row. What is the steady-state availability?

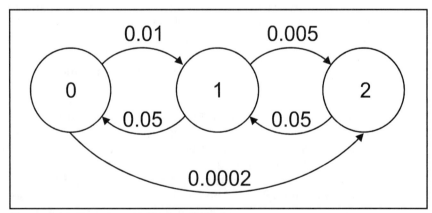

Figure 8-24. Exercise 8.2 Markov Model.

Markov Modeling

8.3 Using the system modeled in Exercise 8.2, calculate availability as a function of time, using 1-hr time increments. How many hours are required for the system to reach a steady-state value (within six digits)?

8.4 A system has the Markov model shown in Figure 8-25, where the arcs are labeled with probabilities in units of failure probability per hour. The system is successful in states 0 and 1. Calculate the limiting state probability row. What is the steady-state availability?

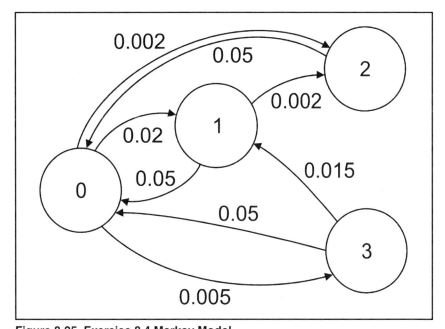

Figure 8-25. Exercise 8.4 Markov Model.

8.5 What is the MTTF for the system modeled in Exercise 8.4, assuming that the system always starts in state 0?

8.6 What is the MTTF for the system modeled by Figure 8-3?

REFERENCES AND BIBLIOGRAPHY

1. Rouvroye, J. L., Goble, W. M., and Brombacher, A. C., "A Comparison Study of Qualitative and Quantitative Analysis Techniques for the Assessment of Safety in Industry," *Proceedings of the Probabilistic Safety Assessment and Management Conference*, June 1996.

2. Rouvroye, J. L., Stavrianidis, P., Spiker, R. Th. E., Nieuwenhuizen, J. K., and Brombacher, A. C., "Uncertainty in Safety: New Techniques for the Assessment and Optimisation of Safety in the Process Industry," *Proceedings Winter Annual Meeting ASME*, New York: ASME, 1995.

3. *The Academic American Encyclopedia (Electronic Version)*, Danbury, CT: Grolier Electronic Publishing, 1991.

4. Maki, D. P., and Thompson, M., *Mathematical Models and Applications*, Chapter 3, Englewood Cliffs, NJ: Prentice-Hall, 1973.

5. Kreyszig, E., *Advanced Engineering Mathematics*, Chapter 4.4, New York: John Wiley & Sons, 1967.

6. *123 Reference Manual, Release 2*, Cambridge, MA: Lotus Development Corp., 1987.

7. *Excel Reference Manual, Version 3.0*, Redmond, WA: Microsoft, 1991.

8. Bukowski, J. V., Suggestion made during private meeting, 1992.

9. Bukowski, J. V., and Goble, W. M., "PC Computation of MTTF: Fail-Safe vs. Fail-Danger," *Proceedings of the 29th Annual IEEE Boston Reliability Conference*, New York: IEEE, 1991.

10. Papoulis, A., *Probability, Random Variables, and Stochastic Processes*, Chapter 12-4, New York: McGraw-Hill, 1984.

11. Fleming, R. E., Josselyn, J. V., Dolny, L. J., and DeHoff, R. L., "Complex System RMA and T, Using Markov Models," *1985 Proceedings of Annual Reliability and Maintainability Conference*, New York: IEEE, 1985.

12. Josselyn, J. V., Fleming, R. E., Frenster, J. A., and DeHoff, R. L., "Application of Markov Models for RMA Assessment," *1986 Proceedings of Annual Reliability and Maintainability Conference*, New York: IEEE, 1986.

13. Billinton, R., and Allan, R. N., *Reliability Evaluation of Engineering Systems: Concepts and Techniques*, New York: Plenum Press, 1983.

9
Diagnostics

Improving Safety and MTTF

The reliability and safety of automatic control systems depends on a number of factors. One of the most important factors in both redundant and nonredundant architectures is the ability of the system to detect component failures: on-line diagnostics. Good diagnostics improve both safety and availability (Refs. 1 and 2). The importance of diagnostic capability was pointed out clearly in a 1969 paper (Ref. 3) as part of work done on fault-tolerant computing for the U.S. space program, where the term "diagnostic coverage factor, C" was defined. Diagnostic coverage factor is the probability (a number between 0 and 1) that a failure will be detected given that it occurs.

All repairable control systems (redundant and nonredundant) have improved availability and safety when diagnostics are added. Depending on exact architecture, other benefits include:

- The amount of time the system operates in a dangerous mode can be reduced.

- The amount of time spent in a degraded (not fully operational) mode is reduced.

- The diagnostics directly affect system operation to improve both safety and availability.

Diagnostic Coverage Effects

The significance of diagnostic capability in a redundant system can be seen by looking at an ideal dual architecture using a Markov model evaluation (Figure 9-1). The model assumes perfect coverage, no common-cause failure, no common components, perfect repair, and no switching mechanism. State 0 represents successful operation of both units. If one unit fails, the system moves to state 1. Since two controller modules are successfully operating in state 0, the failure probability is $2 \times \lambda$. From state 1, a repair will return the system to state 0. A failure from state 1 will move the system to state 2. In state 2 the system has failed. The repair arcs are both marked with a repair rate, μ. This model is known as the "single repairman" model.

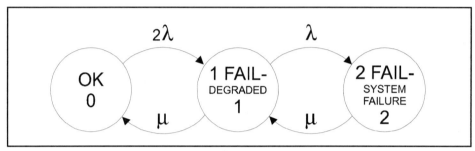

Figure 9-1. Markov Model of Ideal Dual Controller.

The system MTTF can be calculated using matrix algebra. Since MTTF is a measure that does not consider system failure and subsequent repair, the model is modified by eliminating the repair arc from state 2 to state 1. The **P** matrix for the modified model is

$$\mathbf{P} = \begin{bmatrix} 1 - 2\lambda & 2\lambda & 0 \\ \mu & 1 - (\lambda + \mu) & \lambda \\ 0 & 0 & 1 \end{bmatrix}$$

Using this new **P** matrix, solve for MTTF by following the steps detailed in Chapter 8 (see Appendix B for matrix math). Assume that the system starts in state 0. After inverting the $(\mathbf{I} - \mathbf{Q})$ matrix and adding the top row:

$$MTTF = \frac{3\lambda + \mu}{2\lambda^2} \qquad (9\text{-}1)$$

The previous analysis made a number of idealistic assumptions that do not reflect a real control system implementation. One of these assumptions is that all failures are detected and are repaired. In many simple analysis problems, perfect coverage (100%) is implicitly assumed. If perfect coverage is not assumed, failures that are covered must be distinguished from those that are not. This distinction must be made because coverage affects repair time.

Diagnostics

If a failure is detected by the automatic diagnostics and annunciated, it can be repaired immediately. If a failure is not detected by diagnostics, it may remain completely hidden. Consider the short-circuit failure of an output transistor on a controller module. If the transistor is in a circuit configuration that normally conducts, there is no difference between normal operation and short-circuit failure. These types of failures will likely remain undiscovered until a plant shutdown or, even worse, until a hazardous event occurs. In order to discover these latent failures sooner, some installations do a periodic maintenance test. Inspections are made by qualified technicians, and each piece of equipment is manually tested for proper operation. It is assumed that the manual inspection will discover any problems.

Two specific repair rates result from this situation. The on-line repair rate is

$$\mu_O = \frac{1}{T_R} \tag{9-2}$$

The variable T_R refers to average repair time. The on-line repair rate applies to all failures that are covered (detected by on-line diagnostics).

The repair rate for failures detected during periodic maintenance is much lower. The time required both to detect and to repair the problem must be included in an analysis. In a situation in which scheduled periodic maintenance testing is done, a failure may occur right before the test, right after the test, or at any time in between. If we assume that the chance of failure is equally likely at all times, then the average time to detect the problem is $T_I/2$, where T_I is the periodic inspection time. The total time to detect and repair the problem is

$$\frac{T_I}{2} + T_R$$

Therefore, the repair rate due to periodic inspection is

$$\mu_P = \frac{1}{\frac{T_I}{2} + T_R} \tag{9-3}$$

EXAMPLE 9-1

Problem: Once a year a process is shut down and the control system is completely tested for proper operation. Any failures are discovered during the inspection and are repaired. If the average repair time is 8 hr, what are the on-line and periodic repair rates?

Solution: Using Equation 9-2, a simple calculation shows the on-line repair rate to be 0.125 repairs per hour.

$$\mu_O = \frac{1}{8}$$
$$= 0.125$$

Using Equation 9-3, the periodic repair rate is calculated by substituting $T_I = 8{,}760$ hr (one year) and $T_R = 8$ hr.

$$\mu_P = \frac{1}{(8760/2) + 8}$$
$$= 0.0002279$$

In Figure 9-2 a new Markov model is presented that accounts for the difference between detected failures and undetected failures. From state 0, a detected failure will take the system to state 1. Repairs are made from state 1 to state 0 at the on-line repair rate. An undetected failure will take the system to state 2. From state 2, repairs are made at the periodic repair rate. It is assumed that all latent failures are discovered during the periodic inspection. When the system is in state 1 or in state 2, one controller is operating. From these states another failure will cause system failure. Repair rates are not included from the system failure state because the modeling objective is to calculate MTTF starting from state 0.

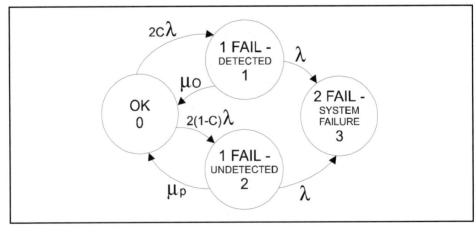

Figure 9-2. Markov Model of a Dual Controller with Coverage.

Diagnostics

An analysis of the model using the matrix inversion method presented in Chapter 8 (see Appendix B for matrix math) provides an equation for MTTF:

$$MTTF_{COV} = \frac{3\lambda^2 + (3 - 2C)\lambda\mu_O + (1 + 2C)\lambda\mu_P + \mu_O\mu_P}{2\lambda^3 + 2C\lambda^2\mu_P + 2(1 - C)\lambda^2\mu_O} \quad (9\text{-}4)$$

The reader will want to spend an algebraic evening duplicating the results. Figure 9-3 shows how MTTF varies as a function of coverage using these values: $\lambda = 0.0005$ failures per hour, $T_R = 8$ hr, and $T_I =$ one year. MTTF drops very sharply as coverage is reduced. When coverage drops from 1.0 to 0.94, MTTF drops by an order of magnitude. While most of the drop occurs in the first two steps, MTTF continues to drop rapidly until coverage is reduced below 0.95.

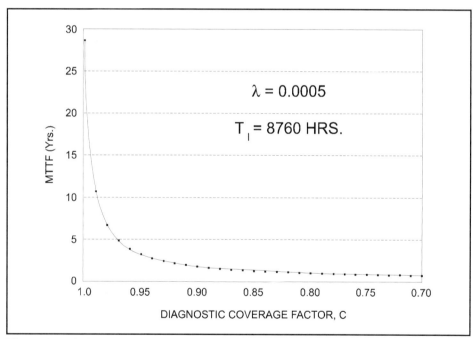

Figure 9-3. MTTF versus Diagnostic Coverage with Yearly Inspections.

The effect of coverage can be lowered by decreasing the time between manual inspections. If the inspection period is reduced to one month (730 hr), the slope of the curve is less severe. Figure 9-4 shows the same plot with an inspection period of one month.

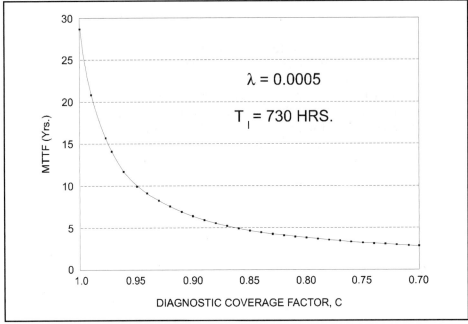

Figure 9-4. MTTF versus Diagnostic Coverage with Monthly Inspections.

Figure 9-5 shows a plot of MTTF versus diagnostic coverage when λ is lowered to 0.00005 failures per hour and the inspection period is one year. The shape of the curve is the same as Figure 9-3. The actual MTTF numbers are much larger. This illustrates a classic trade-off between higher coverage and higher failure rate. Controller modules with higher failure rates require higher coverage factors to achieve the same system MTTF. Fortunately, the trend in control system technology is a decreasing failure rate and higher coverage factor.

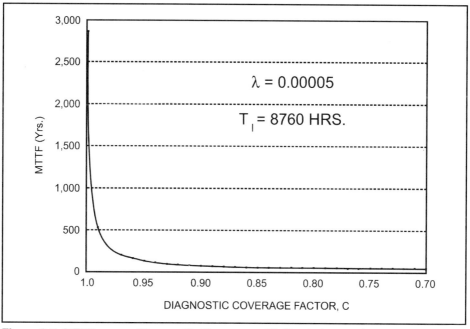

Figure 9-5. MTTF versus Diagnostic Coverage with Yearly Inspections.

EXAMPLE 9-2

Problem: A system consists of dual controller units. Each controller unit is built from a computer module, an input module, an output module, and a power supply module. The failure rates are

Computer module	12,000 FITS
Input module	2,400 FITS
Output module	3,600 FITS
Power supply module	5,000 FITS

All failure rates are assumed to be constant. Each controller unit has a coverage factor of 0.90. No periodic inspections are done. The average repair time is 8 hr.

> **EXAMPLE 9-2 (continued)**
>
> A second input module could be added to each controller unit. That module would be used to measure power supply voltages, signal voltages, and output voltages. Such inputs are used with diagnostic programs to increase the coverage factor to 0.94. How will the second input module with its diagnostics affect the system MTTF?
>
> **Solution:** The MTTF for the system without the second input module is calculated using Equation 9-4. We assume that each controller unit fails if any module in the controller unit fails. For this series system, we add the module failure rates to obtain the controller unit failure rate of 23,000 FITS. The system on-line repair rate is 0.125 repairs per hour. The system periodic repair rate is zero. Substituting these values into Equation 9-4:
>
> $$MTTF = \frac{3(0.000023)^2 + [3 - 2(0.9)](0.000023)(0.125)}{2(0.000023)^3 + 2(1 - 0.9)(0.000023)^2(0.125)}$$
>
> $$= 260{,}510 \text{ hr}$$
>
> When we add the second input module, the controller unit failure rate increases to 25,400 FITS. The coverage increases to 0.94. Substituting the new values into Equation 9-4:
>
> $$MTTF = \frac{3(0.0000254)^2 + [3 - 2(0.94)](0.0000254)(0.125)}{2(0.0000254)^3 + 2(1 - 0.94)(0.0000254)^2(0.125)}$$
>
> $$= 366{,}413 \text{ hr}$$
>
> We conclude that the extra input module will result in a significant increase in system MTTF.

Imperfect Switching

Another assumption made in the ideal model of Figure 9-1 was that no switching mechanism is present. Many practical implementations of redundancy require a switching mechanism to select the appropriate module output. The new system is shown in Figure 9-6. It has an output selector switch that chooses which module output to route to the system output.

The output selector switch depends on good diagnostics. This is an example of a form of fault tolerance where actual system operation depends on diagnostic output. Assume that each module generates a logic-one signal if it detects a failure. The switch logic looks at both signals. If both signals are logic zero, the switch may select either output, as both are deemed correct. If one signal is logic zero, that output will be selected. If both signals are logic one, the controller modules have both

Diagnostics

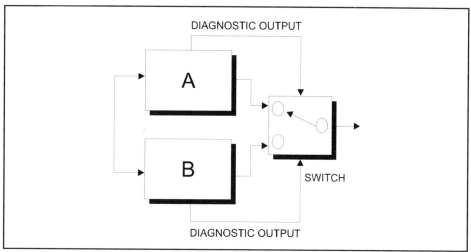

Figure 9-6. Dual System with Switch.

failed. The switch should act in the safe manner. (In many systems the switch should not connect either output and simply de-energize.)

A Markov model of the system is shown in Figure 9-7. As long as a module failure is detected, the diagnostic output from a module will properly control the switch. The system moves from state 0 to state 1. If the failure is not detected by the diagnostics, one of two things will happen. If the switch is selecting the module in which the undetected failure has occurred, the system will fail immediately. The Markov model has an arc from state 0 to state 3 to show this action. Note that the model includes a failure rate for the switch. That failure also causes immediate system failure.

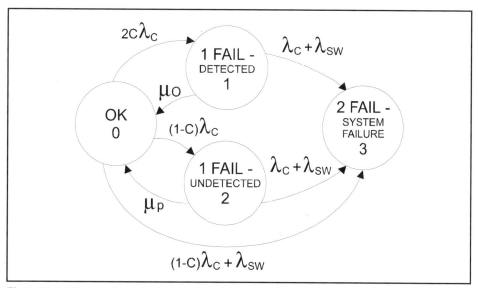

Figure 9-7. Markov Model of a Dual System with Switch.

If an undetected failure occurs in one module and the switch is selecting the other module, the system moves to state 2 where an undetected failure is present. It is reasonable to assume that the system has an equal chance of being in either switch position when the undetected failure occurs. Therefore, we split the undetected failure rate evenly between the two alternatives. A plot of MTTF versus coverage for this system is shown in Figure 9-8. The effect of diagnostic coverage is again critical, because one-half of the uncovered failures cause immediate failure.

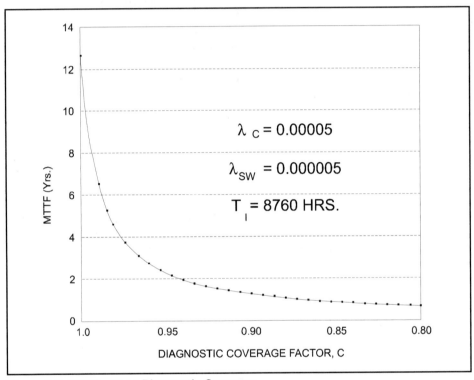

Figure 9-8. MTTF versus Diagnostic Coverage.

Diagnostic Coverage Effects on Other PLC Architectures

Safety, as measured by probability of failure on demand (PFD) or risk reduction factor (RRF), is improved even in a nonredundant architecture. Consider a normally energized safety protection application. If the conventional 1oo1 PLC (one out of one, Figure 9-9) architecture fails with outputs de-energized, the process is inadvertently shut down. This is called a false trip. Most processes do not require on-line diagnostics to detect when the process has shut down; a false trip is usually quite apparent. If, however, the 1oo1 fails with its output energized, the PLC cannot respond to a dangerous condition. Unfortunately, the process keeps operating with no safety protection and no indication that something is wrong. Diagnostics can detect these dangerous failures and

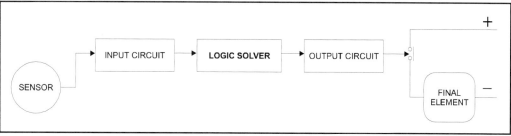

Figure 9-9. 1oo1 PLC Architecture.

allow the system to be quickly repaired. There is a significant improvement in RRF when a system has good diagnostics (and plant maintenance policies that ensure reasonably quick repair).

Diagnostics reduce time in a degraded mode for architectures with redundancy. The 1oo2 architecture (Figure 9-10) was designed to tolerate dangerous failures (in a normally energized protection application). Two PLC units are wired in series. If one unit fails with its outputs energized, the other unit can still de-energize the load and provide a safety protection function. Diagnostics improve the safety of this architecture as well. If one unit fails dangerously, the system is degraded and a second dangerous failure will fail the system dangerously. Diagnostics capable of detecting a dangerous failure will allow repairs to be made more quickly, thus minimizing the amount of time this system operates in a degraded mode. Figure 9-11 shows a plot of RRF versus dangerous failure diagnostic capability; RRF is highest when the dangerous diagnostic coverage is 100%.

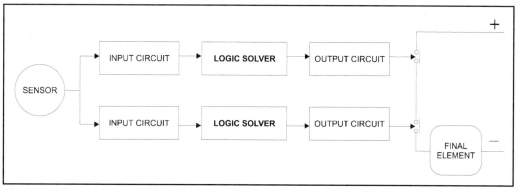

Figure 9-10. 1oo2 PLC Architecture.

198   Diagnostics

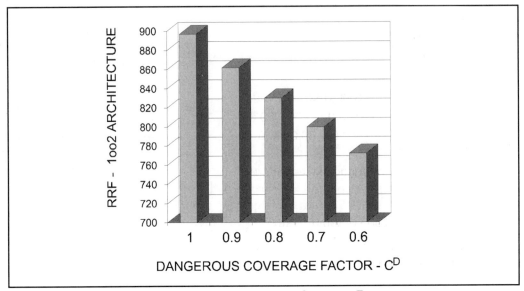

Figure 9-11. 1oo2 RRF versus Dangerous Diagnostic Coverage Factor.

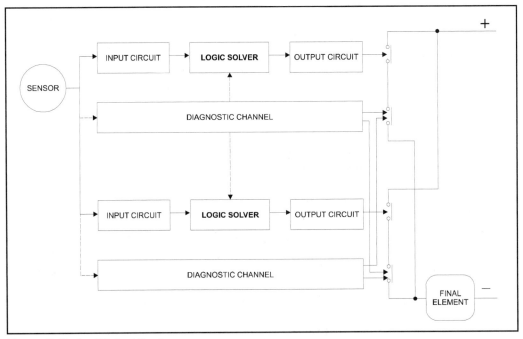

Figure 9-12. 1oo2D Architecture.

The 1oo2D architecture (Figure 9-12) was designed to provide high RRF and maximum uptime (high availability). It is implemented with four channels of electronics, typically in two physical sets. Each set of electronics includes one PLC channel with the input circuitry, a logic solver, output circuitry, and another channel of diagnostic circuitry. When two sets of electronics are combined, a four-channel architecture with diverse safety channels is created. Safety is achieved when the diagnostics detect the failure and de-energize the output of the bad unit. The system continues to operate successfully since the second unit keeps the load energized. Figure 9-13 shows a plot of RRF (a measure of safety, where higher is better) versus dangerous detected failure probability; RRF drops as diagnostic capability goes down.

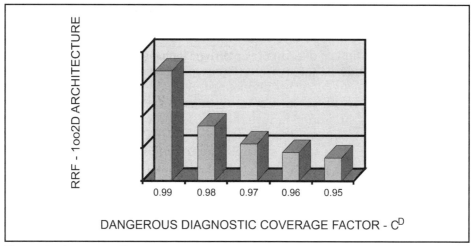

Figure 9-13. 1oo2D RRF versus Dangerous Diagnostic Coverage Factor.

Figure 9-14 shows how MTTFS (a measure of the false trip rate, where higher is better) changes with diagnostics. When the diagnostics detect a failure, the system can be repaired relatively quickly. The amount of time a system spends in a degraded mode is minimized. A false trip is more likely in the degraded mode.

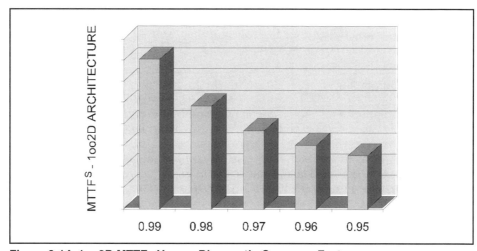

Figure 9-14. 1oo2D MTTFs Versus Diagnostic Coverage Factor.

In all fault-tolerant architectures, diagnostics can be used to improve safety and/or availability. The ability of a control system to detect component failures is important and must be measured to properly evaluate system reliability and safety metrics.

Measuring Diagnostic Coverage

Since diagnostics are one of the major factors in safe and reliable control systems, the ability to measure and evaluate those diagnostics is important. This is done using a FMEDA (failure modes, effects, and diagnostic analysis) (Chapter 5) and verified with fault injection testing.

The definition of diagnostic coverage factor includes the probability of failure. Failures that are more likely to occur count more than failures than are unlikely to occur. As an example, consider a system of three component/failure modes. Component 1 is detected by the on-line diagnostics and has a failure rate of 0.0998 failures per hour. The other two components are not detected by the on-line diagnostics and have a failure rate of 0.0001 failures per hour. The total failure rate is 0.1 failures per hour. The diagnostic coverage factor is calculated by adding the failure rates of detected components divided by the total failure rate of all components. In this example, the coverage factor equals 0.998, or 99.8%. The coverage factor does not equal 0.33, as might be assumed. Coverage is not calculated by counting the number of detected failures and dividing by the total number of failures.

Failure Mode Diagnostics

When the architectures are reviewed, it becomes apparent that the ability to detect dangerous failures is the important characteristic in safety protection applications. Therefore, in safety PLCs it is generally necessary to distinguish the coverage factor for safe failures from the coverage factor for dangerous failures. The superscript S is used for the safe coverage factor, C^S. The superscript D is used for the dangerous coverage factor, C^D. This is a key parameter for safety measurements such as PFD and RRF.

Example 9-3

Problem: An electronic circuit has the FMEDA chart listed in Table 9-1. What is the overall coverage factor?

Solution: Twelve of the 22 possible component/failure modes are detectable. The total detected failure rate is 80.03 FITS (failures per billion hours). The total failure rate is 93.56 FITS. The overall coverage factor is 80.03/93.56 = 0.855.

Diagnostics

		Failures/Billion Hours			
1	4	7	8	10	12
Name	Mode	Criticality	λ	Detected	Mode
R1-1K	Short	Safe	0.5	0	1
	Open	Safe	1	1	1
	Drift low	Safe	0.01	0	1
	Drift high	Safe	0.01	1	1
C1-0.18	Short	Safe	4	1	1
	Open	Safe	2	0	1
R2-200K	Short	Dangerous	0.5	0	0
	Open	Safe	1	1	1
	Drift low	Dangerous	0.01	0	0
	Drift high	Safe	0.01	1	1
R3-10K	Short	Safe	0.5	0	1
	Open	Dangerous	1	1	0
	Drift low	Safe	0.01	0	1
	Drift high	Safe	0.01	1	0
D1	Short	Safe	4	1	1
	Open	Dangerous	4	0	0
D2	Short	Dangerous	4	1	0
	Open	Dangerous	4	0	0
OC1	Diode short	Safe	45	1	1
	Diode open	Safe	2	0	1
	Tran. short	Dangerous	10	1	0
	Tran. open	Safe	10	1	1

Table 9-1. FMEDA Chart

Example 9-4

Problem: For Example 9-3, what is the dangerous coverage factor? What is the safe coverage factor?

Solution: The total safe failure rate in Table 9-1 is 70.04 FITS. The safe detected failure rate is 65.02; therefore, the safe coverage factor $C^S = 0.928$. The total dangerous failure rate is 23.5 FITS. The dangerous detected failure rate is 15 FITS. The dangerous coverage factor is 0.64. This circuit is based on a conventional PLC input circuit with added diagnostics. Many would judge such conventional PLC circuits to be insufficient for safety applications.

Measurement Limitations

Although the FMEDA technique can provide good diagnostic coverage factor numbers when done accurately, the main limitation is that the FMEDA creator and its reviewers must know about all possible failure modes of the components used in the circuit/module/unit. When unknown failure modes occur, they may or may not be detected by on-line diagnostics as designed. Fortunately, the failure rates of those failure modes are likely to be very small (otherwise, someone would know about the failure mode). This is especially true on components that are widely used in many applications. The possibility of unknown/undetected failure modes is higher for new components. When the possibility of unknown failure modes is considered likely, this can be indicated on the FMEDA as shown in Table 9-2. This is the FMEDA of Table 5-4 with an unknown failure mode added. Note that the dangerous diagnostic coverage dropped from 99.96% to 99.73%.

Diagnostic Techniques

As control computer capability increases, we expect better diagnostic coverage. The computer "HAL" from the movie *2001* said, "I've just picked up a fault in the AE35 unit. It's going to 100% failure within 72 hours." While our newest machines are not yet quite at this level, many new techniques are constantly being used to improve diagnostic coverage.

A moderate level of coverage can easily be achieved. Even early microcomputer systems used simple watchdog timer circuits. These circuits consist of retriggerable timing devices. As long as the system triggers the timer within the timeout period, the timer is retriggered and does not time out. If something in the system fails in such a way that the timer is not retriggered, the timer does time out and indicates the failure.

These simple mechanisms provide some coverage. It is estimated that the coverage from such devices is between 50 and 90% (Ref. 4). High coverage factors ($C > 95\%$) are difficult to achieve. Controllers must be designed from the ground up, with self-diagnostics as a goal. Electronic hardware must be carefully designed to monitor correct operation of a module. Software must properly interact with the hardware. In the past, control computers have been estimated at 93 and 95% coverage [Refs. 5 and 6]. To achieve even these levels of coverage, a number of diagnostic techniques have been developed. They can be classified in two ways: a comparison to a predetermined reference and a comparison to a known good operating unit.

1	2	3	4	5	6	7	8	9	10	11	12	13	14	15	16
Name	Code	Function	Mode	Cause	Effect	Criticality	λ	Remarks	Det.	Diagnostics	Mode	SD	SU	DD	DU
R1-10K	4555-10	Input threshold	Short		Threshold shift	Safe	0.125		0		1	0	0.13	0	0
			Open	Solder open	Open circuit	Safe	0.5		1	Lose input pulse	1	0.5	0	0	0
			Drift low			Safe	0.01	None until too low	0		1	0	0.01	0	0
			Drift high			Safe	0.01	None until too high	1	Lose input pulse	1	0.01	0	0	0
R2100K	4555-100	Current limit	Short		Short input	Safe	0.125		1		1	0.13	0	0	0
			Open	Solder open		Safe	0.5		1	Lose input pulse		0	0	0.5	0
			Drift low			Safe	0.01	None until too low	0		1	0	0.01	0	0
			Drift high			Safe	0.01	None until too high	1	Lose input pulse	1	0.01	0	0	0
D1	4200-7	Voltage drop	Short	Surge	Overvoltage	Safe	2		1	Lose input pulse	1	2	0	0	0
			Open		Open circuit	Safe	5		1	Lose input pulse	1	5	0	0	0
D2	4200-7	Voltage drop	Short	Surge	Overvoltage	Safe	2		1	Lose input pulse	1	2	0	0	0
			Open		Open circuit	Safe	5		1	Lose input pulse	1	5	0	0	0
OC1	4805-25	Isolate	Led dim.	Wear	No light	Safe	28		1	Comp. mismatch	1	28	0	0	0
			Tran. short	Internal short	Read logic 1	Dang.	10		1	Comp. mismatch	0	0	0	10	0
			Tran. open		Read logic 0	Safe	6		1	Comp. mismatch	1	6	0	0	0
OC2	4805-25	Isolate	Led dim.	Wear	No light	Safe	28		1	Comp. mismatch	1	28	0	0	0
			Tran. short	Internal short	Read logic 1	Dang.	10		1	Comp. mismatch	0	0	0	10	0
			Tran. open		Read logic 0	Safe	6		1	Comp. mismatch	1	6	0	0	0
OC1/OC2			Cross channel	Short	Same signal	Dang.	0.01		0		0	0	0	0	0.01
R3-100K	4555-100	Filter	Short		Lose filter	Safe	0.125		0		1	0	0.13	0	0
			Open		Input float high	Dang.	0.5		1	Comp. mismatch	0	0	0	0.5	0
R4-10K	4555-10	Voltage divider	Short		Read logic 0	Safe	0.125		1	Comp. mismatch	1	0.13	0	0	0
			Open		Read logic 1	Dang.	0.5		1	Comp. mismatch	0	0	0	0.5	0
R5-100K	4555-100	Filter	Short		Lose filter	Safe	0.125		0		1	0	0.13	0	0
			Open		Input float high	Dang.	0.5		1	Comp. mismatch	0	0	0	0.5	0
R6-10K	4555-10	Voltage divider	Short		Read logic 0	Safe	0.125		1	Comp. mismatch	1	0.13	0	0	0
			Open		Read logic 1	Dang.	0.5		1	Comp. mismatch	0	0	0	0.5	0
C1	4350-32	Filter	Short		Read logic 0	Safe	2		1	Comp. mismatch	1	2	0	0	0
			Open		Lose filter	Safe	0.5		0		1	0	0.5	0	0
C2	4350-32	Filter	Short		Read logic 0	Safe	2		1	Comp. mismatch	1	2	0	0	0
			Open		Lose filter	Safe	0.5		0		1	0	0.5	0	0
U		Unknown				Safe	0.05		0		1	0	0.05	0	0
U		Unknown				Dang.	0.05		0		0	0	0	0	0.05
							110.9					86.9	1.45	22.5	0.06
			Total Failure Rate				110.9			Safe Coverage	0.9834				
			Total Safe Failure Rate				88.34			Dangerous Coverage	0.9973				
			Total Dangerous Failure Rate				22.56								
			Safe Detected Failure Rate				86.895								
			Safe Undetected Failure Rate				1.445								
			Dangerous Detected Failure Rate				22.5								
			Dangerous Undetected Failure Rate				0.06								
			Failures per billion hours												

Table 9-2. FMEDA Chart

The diagnostic technique where a comparison is made between an operating value and a known reference is called *reference diagnostics*. Reference diagnostics can be done by a single PLC. The coverage factor of reference diagnostics will vary widely depending on the implementation, with most results ranging from 0.0 to 0.999. The comparison of two operational units is called *comparison diagnostics*. This, of course, requires two or more operating units. The coverage factor again depends on implementation but results are generally good, with most ranging from 0.9 to 0.999.

The notation C_1 is used for reference diagnostics since a single unit is involved. The notation C_2 is used for comparison diagnostics since two units are involved. It should be noted that many safety PLCs use a combination of both reference diagnostics and comparison diagnostics for maximum effectiveness. The symbol C, for diagnostic coverage factor, is used to denote a failure detected by either reference diagnostics, comparison diagnostics, or a combination of both.

Reference Diagnostics

A comparison to a predetermined reference is the most common and universally used diagnostic technique. The auto mechanic measures oil pressure, wet and dry compression pressure, and mechanical clearances. Then these results are compared with the manufacturer's specifications in order to determine whether anything is wrong with the car's engine. The TV repairman would retrieve the "Sams-Fotofact" schematic and compare the voltage levels and waveforms to the references on the schematic. The same techniques are used in a control computer.

Certain results are expected during on-line operation. The analog converter should operate within 15 μsec after receiving a convert command. The "data received" signal should acknowledge within two clock cycles after a "store" operation. The calibration adjustment should never exceed 10%. The noise level of the analog signal should always be above 2 bits. The data tables in memory should always start with a zero. These expected results can be stored in the computer program and compared to actual operation. Many diagnostic algorithms use the comparison-to-reference technique. Time period measurement is one of the most commonly used techniques.

Watchdog timers are a simple diagnostic tool. A timer is reset periodically by a properly operating cyclic circuit. Any failure that prevents the circuit from finishing its cycle in the allotted time will be detected by the watchdog. A "windowed watchdog" is a little more sophisticated. The timer looks for pulses within the stated time period and for pulses that are too early (Figure 9-15). The method is especially effective in detecting failures in a computer that executes a periodic program. System software must operate correctly to maintain the time period.

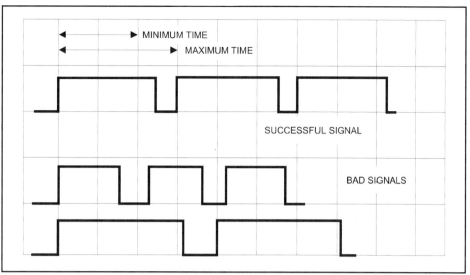

Figure 9-15. Windowed Watchdog Operation.

Analog-to-digital converters are useful for more than process input measurement. Certain voltages and currents within the module indicate failure. Circuits can monitor the voltage of any power source. If a component failure causes the supply voltage to exceed bounds, the failure is detected. Power consumption can be measured. Many failures are indicated by an increase in current.

Computer memory can be tested by storing a checksum for any memory block whenever a memory write occurs. The memory can be reread and the checksum recalculated. Known failure modes of memory are detected if the checksums do not match. Many digital circuits have known sequences of bit patterns. If the sequence of binary numbers is added, a sum results. The same number should be calculated every time the sequence repeats. If the number is different, a failure has occurred.

Input/output circuitry is present on many modules used in industrial controllers. Special techniques are necessary to obtain good coverage in such circuits. Output readback is one of the first techniques used (Figure 9-16). An input circuit is used to read the result of all output circuits. This allows output switch failures to be detected whenever differences are discovered. This works well on dynamic signals, the values of which change frequently. It will only detect failures in one mode for signals that are static.

Output current sensors can be used to detect open and short circuits in output devices. If the current rises above an upper bound, a short failure is detected. Outputs can be pulsed briefly to verify that the output is able to switch. The normal output state is restored automatically as soon as the switch is verified good. The I/O power can be monitored, which allows detection of failed I/O power or possibly a failed cutoff switch.

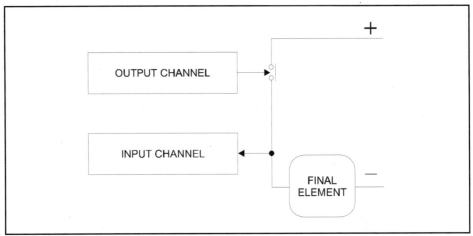

Figure 9-16. Output Readback Diagnostic.

Input circuits have component failures that are hard to detect, especially when the input signals are static. In the case of static discrete input switches that are normally closed, a pulsed output (Figure 9-17) can be used to generate a pulsed signal that is read by the input circuit. The application program looks for a pulsing signal to indicate that the switch is closed. If the pulse stops, either the switch is open or the input circuit has failed. This diagnostic method will detect stuck-at-one and stuck-at-zero input circuit failures.

Analog input signals have better diagnostic potential. In normal situations, the analog signal varies. One good diagnostic is the use of a "stuck signal" detector. If the analog signal gives the exact same reading for several scans in a row, it has probably failed.

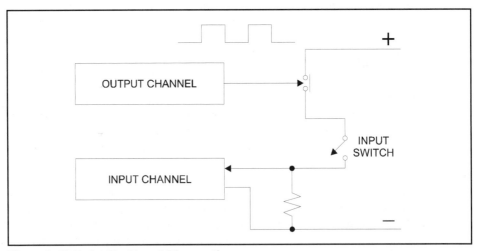

Figure 9-17. Pulse Test Input Diagnostic

Comparison Diagnostics

Comparison to another operating unit is also useful. Results of dynamic calculations can be compared. In a dual configuration, any disagreement may identify a fault. In a triple configuration, a voting circuit is used to identify when one module disagrees with the other two. It is likely that the disagreeable module has failed. This technique can provide a useful diagnostic and fault identification technique for many failures.

Comparison diagnostic techniques depend on comparing some data between two or more PLC units. The concept is simple. If a failure occurs in the circuitry, processor, or memory of one PLC, there will a difference between data tables in that unit when compared to another unit. Comparisons can be made of input scans, calculation results, output readback scans, and other critical data. In the architectures described above, this could be done in the 1oo2 or the 1oo2D. The comparison coverage factor will vary since there are trade-offs between the amount of data compared and the coverage effectiveness.

Field Device Coverage

Diagnostics should be extended to all field I/O devices and associated wiring. Without this extension, the failures in process sensors may dominate the system failure rate and negate the benefits of redundant controllers. A good system safety and reliability analysis must include the field devices: valves, sensors, field transmitters, limit switches, solenoids, and other devices.

Field sensors with microprocessors built into them are known as *smart transmitters*. The addition of such intelligence into a field device adds the ability to embed diagnostics. Environmental stress factors are high in areas where such devices are typically used. This amplifies the need for such diagnostics, since high stress factors can cause higher failure rates. Fortunately, the ability to put a microprocessor into a field device allows diagnostic techniques never before possible. Techniques used in a module are now practical within the field device.

Diagnosing failures in some field devices, even "dumb" field devices, can be done with intelligent input/output circuits in a control system module. Output current sensors can detect field device failures. If the average output current exceeds an upper bound for too long, this indicates a short-circuit failure of the load device or the field wiring. If an output channel is on and a minimum current is not being drawn, this indicates an open-circuit failure of the field device or the associated wiring.

Current sensors on input channels are able to distinguish whether a contact is open or closed, or whether the input wire is open. Special sensors can detect when the I/O power source current does not equal the

I/O power return current. This can be used to detect a failed field device in which a current leak has developed to earth ground.

Comparison diagnostic techniques are popular for use with field sensors, both analog and discrete. For discrete sensors, Boolean comparisons can detect differences in sensors. Be careful, though, that scanning and noise do not cause false diagnostics. The logic of Figure 9-18 has a timer to filter out temporary differences between two discrete inputs labeled A and B. A TOFF timer is used for the normally energized signals. Voting logic is used to compare three discrete inputs labeled A, B, and C in Figure 9-19. The majority signal drives the coil. Additional logic could be added to specifically compare each of the three combinations of two signals. When one signal appears in two comparison mismatches, it is likely to be the bad signal.

An equivalent technique can be used for analog. Analog signals are sent to a median selector. The median output is used as the process signal. In addition, comparisons must be made between analog signals looking for differences greater than a certain magnitude that last longer than a reasonable amount of time. Again, the designer should be careful about scanning differences and make sure the diagnostic limits are not set too narrow.

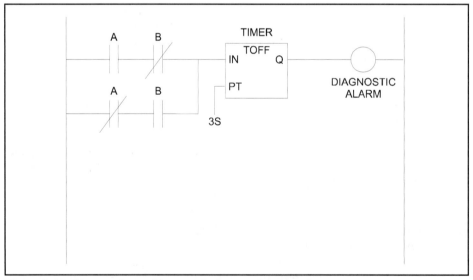

Figure 9-18. Discrete Input Diagnostic Comparison Logic.

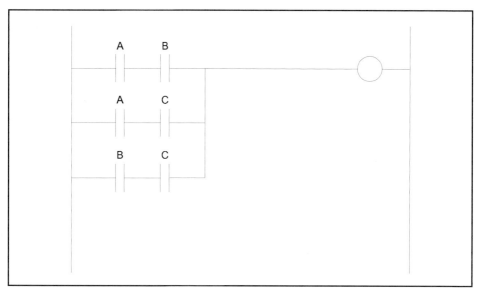

Figure 9-19. 2oo3 Voting Logic for Discrete Signals.

System Coverage

The overall availability and safety of a system can be enhanced with system-level diagnostics. Many techniques can be used in the context of a particular system installation. These techniques can detect failures that are beyond the scope of embedded module diagnostics, and can also supplement embedded module diagnostics to increase overall coverage. When inventing such techniques, remember the two basic approaches: compare to a reference and compare to another unit.

Comparison between units can extend beyond direct comparison. Within a specific control system implementation, process sensors and their related input circuits can be compared whenever system relationships exist. In the case of two sensors that read the same process variable, the comparison can be direct. Boolean variables will agree as long as timing and filtering differences between input circuits are acknowledged. Analog variables will agree within limits. Other relationships exist between process signals. Temperatures relate to pressures. Pump motor current may relate to flow. Once recognized, these relationships can be used to detect failures within the system. The system designer can configure function blocks or ladder logic to calculate as necessary and do the comparison.

Comparison to reference techniques also can work well at the system level. The system designer expects certain things to happen in certain ways. Function blocks can be added to the system configuration to verify that the system is operating as expected. Communication links can be verified with a watchdog timer function. A retriggerable timer is established in one control module. The trigger signal is generated in another control module located across the communication link. If the communication link fails, the timer will detect the failure. Figure 9-20 shows an example configuration. Similar techniques using timers or window timers can detect complex system failures.

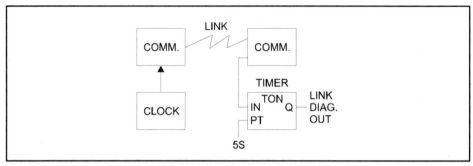

Figure 9-20. Communications Link Diagnostic.

Fault Injection Testing

The theoretical coverage calculation done in a FMEDA can be affirmed by testing. Various test techniques exist to simulate, automatically and manually, component failure modes. On an actual module, the failures can be simulated. For example, a short circuit (or low resistance) can be clipped across a component to simulate a short-circuit failure. Component leads can be cut to simulate open-circuit failures. Other methods have been used to induce component failure, including high-intensity light or physical force (Ref. 7). Remember, however, that not all component failure modes can be physically simulated.

The purpose behind such component failure simulation testing is to verify the corresponding on-line diagnostic test. Some on-line diagnostic tests do not perform as well as expected. Some tests detect failures that were not anticipated. More extensive testing will result in a more accurate coverage factor estimate.

EXERCISES

9.1 Plant maintenance policy states that safety protection systems will be completely inspected once a year. Records indicate that actual repair times for the equipment in the plant are 4 hr. What is the repair rate for failures detected by on-line diagnostics? What is the repair rate for failures detected during the periodic inspection?

9.2 A dual controller system has the Markov model shown in Figure 9-2. Using the repair rates from Exercise 9.1, a controller module failure rate of 0.001 failures/hour, and a coverage of 0.9, what is the system MTTF?

9.3 If the coverage in Exercise 9.2 is increased to 0.97, what is the MTTF?

9.4 Describe methods for diagnosing a burned-out thermocouple.

9.5 A valve has two switches. One switch closes when the valve is in the closed position. When the valve is in the open position, the other switch closes. These switches could fail open. What diagnostic could detect a failed open limit switch?

9.6 What diagnostic could detect a shorted 4-20 mA transmitter?

9.7 A digital-to-analog converter in a smart transmitter could fail in a mode such that the output became frozen. How could such a failure be detected?

9.8 A pressure switch to detect overpressure in a boiler is normally closed. When the pressure becomes high (a dangerous condition), the switch opens. This switch could fail by short circuit and would not indicate the dangerous condition if it occurred. How can this failure be detected?

9.9 The pressure switch in Exercise 9.8 has SPDT contacts. How can these be used to implement a diagnostic?

REFERENCES

1. Smith, S. E., "Fault Coverage in Plant Protection Systems," *ISA Transactions*, Vol. 30, No. 1, 1991.

2. Goble, W. M., and Speader, W. J., "1oo1D — Diagnostics Make Programmable Safety Systems Safer," *Proceedings of the ISA92 Conference and Exhibit*, Research Triangle Park, NC: ISA—The Instrumentation, Systems, and Automation Society, 1992.

3. Bouricius, W. G., Carter, W. C., and Schneider, P. R., "Reliability Modeling Techniques for Self-Repairing Systems," *Proceedings of ACM Annual Conference*, 1969; reprinted in *Tutorial — Fault-Tolerant Computing*, Nelson, V. P., and Carroll, B. N., eds., Washington, DC: IEEE Computer Society Press, 1987.

4. *Programmable Electronic Systems in Safety Related Applications*, Sheffield, U.K.: Health and Safety Executive, 1987.

5. Goble, W. M., "High Availability Systems for Safety and Performance — The Coverage Factor," *Proceedings of the ISA89 Conference and Exhibit*, Research Triangle Park, NC: ISA—The Instrumentation, Systems, and Automation Society, 1989.

6. Harris, D. E., "Built-in Test for Fail-Safe Design," *1986 Proceedings of the Annual Reliability and Maintainability Symposium*, New York: IEEE, 1986.

7. Lasher, R. J., "Integrity Testing of Control Systems," *Control Engineering*, February 1990.

BIBLIOGRAPHY

1. Albert, J., Partridge, M., Fennell, T., and Spillman, R., "Built-in Test Verification Techniques," *1986 Proceedings of the Annual Reliability and Maintainability Symposium*, New York: IEEE, 1986.

2. Amer, H. H., and McCluskey, E. J., "Weighted Coverage in Fault Tolerant Systems," *1987 Proceedings of the Annual Reliability and Maintainability Symposium*, New York: IEEE, 1987.

3. Cusimano, J. A., "PLC Application Techniques in Safety Protection," *Proceedings of the ISA Spring Symposium*, Research Triangle Park, NC: ISA—The Instrumentation, Systems, and Automation Society, 1996.

4. Fisher, T. G.,"Editor's Viewpoint," *ISA Transactions*, Vol. 30, No. 1, 1991.

5. Goble, W. M., "Using PLCs for Safety Applications," *Hydrocarbon Processing*, June 1996.

6. Goble, W. M., Safety of Programmable Electronic Systems — Critical Issues, Diagnostics and Common Cause Strength, *Proceedings of the IChemE Symposium*, Rugby, U.K.: Institution of Chemical Engineers, 1995.

7. Hummel, R. A., "Automatic Fault Injection for Digital Systems," *1988 Proceedings of the Annual Reliability and Maintainability Symposium*, New York: IEEE, 1988.

8. Jager, R., "Integrated Diagnostics — Extension of Testability," *1986 Proceedings of the Annual Reliability and Maintainability Symposium*, New York: IEEE, 1986.

9. Johnson, D. A., "Automatic Fault Insertion," *INTECH*, Research Triangle Park, NC: ISA—The Instrumentation, Systems, and Automation Society, November 1994.

10. Luthra, P., "FMECA: An Integrated Approach," *1991 Proceedings of the Annual Reliability and Maintainability Symposium*, New York: IEEE, 1991.

11. Moreno, F. J., "Built-in Test Design for Cost Effective Fault Isolation," *1992 Proceedings of the Annual Reliability and Maintainability Symposium*, New York: IEEE, 1992.

10 Common Cause

Common-Cause Failures

A common-cause failure is defined as the failure of more than one component, module, unit, or system due to the same stress (cause). A common-cause failure negates the benefits of a fault-tolerant system. Fault-tolerant systems provide two or more modules to prevent system failure when a module failure occurs. If two or more modules fail due to a single stress, a fault-tolerant system will fail. This is not a theoretical problem. Actual study of some field installations has shown that the reliability metrics, such as PFD, MTTF, and so forth, are much worse than reliability models predicted. The "autopsy" reports of failures in these situations indicate that in some cases, more than one redundant device has failed due to a *common-cause stress*.

The door of the control rack cabinet was opened to check on a status display. Just as the maintenance worker finished, a call came on the walkie-talkie. "It's time for lunch." The simple response was, "I'll be there soon." In the cabinet were two processors mounted in the same card rack in a fault-tolerant system. Both were subject to the same electromagnetic field stress and failed. These two processors were part of a safety instrumented system, and a major process unit was immediately shut down. For this failure:

Title	Common-cause failure: system de-energized
Root cause	Radio transmissions from walkie-talkie
Failure type	Physical
Primary stressor	Electrical radiation
Secondary stressors	Cabinet door open

Several loose wire lugs in the bottom of a cabinet created excess resistance in I/O conductors that normally carry several amps. These high-resistance connections were generating heat in the cabinet. Two microprocessor cards in the cabinet were configured redundantly. As the temperature went up, the precise timing needs of the digital signals were no longer met and both microprocessors failed in a short period of time. The system failed and shut down a chemical process. For this failure:

Title	Near-common cause temperature problem
Root cause	Loose wire connections in cabinet
Failure type	Physical
Primary stressor	Temperature
Secondary stressors	Possible electrical noise

An engineer was adding new logic to a dual redundant PLC. When the download command was sent, a "memory full" error message was received just before both units crashed and shut down. For this failure:

Title	Common-cause software failure
Root cause	Adding new logic on-line
Failure type	Functional
Primary stressor	"Memory overstress"
Secondary stressors	Insufficient testing

An equipment cabinet was mounted over the raised floor in a control room. Several rodents found their way into the floor area and into the cabinet. They built a nest above the control rack and began to urinate and defecate into the safety instrumented system (SIS). One carefully aimed stream of liquid hit redundant circuits mounted on a module and caused failure of the entire system. For this failure:

Title	Common-cause failure: SIS
Root cause	Rodent urination
Failure type	Physical
Primary stressor	Liquid urine
Secondary stressors	None

A new engineer notices that the pressure readings in the burner management system (BMS) are a little off. He recalibrates all three transmitters using the wrong procedure and sets all three to the wrong range. If the pressure ever went into shutdown range, none of the three transmitters would send the correct signal.

For this failure:

Title	Common-cause failure: BMS
Root cause	Wrong range on pressure transmitter
Failure type	Functional
Primary stressor	Maintenance error
Secondary stressors	Poor calibration procedures and/or training

Two valves are piped in series to ensure that fuel flow can be shut off when the valves are closed. If one valve sticks open, a second valve will do the job. The system was designed to be energized to trip to avoid power dissipation during the long periods where the valves were not in use. A fire started near the process unit. This was sensed by the SIS and both valve outputs were energized. Unfortunately, the cables for the valves were routed through the same tray, and that tray was over the fire. Both cables had burned and the valves could not close. For this failure:

Title	Common-cause failure: wiring
Root cause	Redundant cables in same physical location
Failure type	Physical/functional
Primary stressor	Fire
Secondary stressors	Design weakness: energize to trip

All the factors that cause failure can be the source of a common-cause failure. They may be internal, such as design errors or manufacturing errors. They may be external, such as maintenance faults, operational errors, or environmental stress. Environmental stressor events include electrical events: power spikes, lightning, and high current levels. Mechanical stress includes shock and vibration. Chemical stress includes corrosive atmospheres, salt air, and humidity. Physical stress includes temperature. Heavy usage — including high data rates — is even a stress, especially to system software.

Design errors are a major source of common-cause failure. Consider the design process. During product development, testing is extensive. Even so, the complexity of control systems prevents complete testing. In many cases the system is not tested in such a way that a design fault is apparent. Then, one day, the system receives inputs that require proper operation of something that was designed incorrectly. System components do not operate as needed, and thus the system does not operate properly. By definition, this is a system failure. If redundant elements are identical, they suffer from the same design errors and may fail in exactly the same way, given the same inputs. Although hardware design errors do occur, the vast majority of design errors are software design errors. These are a major source of common-cause failures. Complexity increases the chances of common-cause design errors.

Operational faults or maintenance faults can be responsible for common-cause failures. During operation of a system, certain actions — including system configuration errors, incorrect shutdown commands, or "force-contact" types of commands — can affect the multiple resources of a redundant system. Maintenance actions, including incorrect upgrade and installation, repair procedure errors, poor calibration, and failure to replace renewable resources (batteries, disks, etc.), can affect multiple components. Often the same procedures are used on all portions of a redundant system. Thus, common-cause failures occur.

Stress-Strength View of Common Cause

Remember that all failures occur when a stress exceeds the associated strength. This concept was presented in Chapter 3, where many different types of stressor events were described. Figure 10-1 shows the stress-strength view of common-cause failures. The strength probability distribution is presented as a dotted line. Three values of strength are shown from within the valid range representing three modules from a fault-tolerant system. A value of stress is shown that is larger than all three strength values. All three units will fail under these conditions.

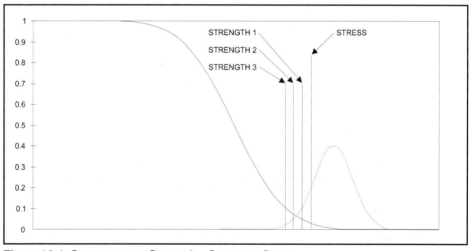

Figure 10-1. Stress versus Strength - Common Cause.

Ironically, this situation gets worse under certain conditions. When the manufacturing process is tightly controlled and consistent, the strength of identically manufactured products is similar. The probability distribution of strength has a small sigma (σ). When this happens, the likelihood of a common failure is increased (Figure 10-2). Taken to the extreme, if a redundant system could be built from identical units with identical strengths, all failures would be common cause! Of course, this is not realistic since strength varies due to many factors. If that situation were realistic, there would no benefit to redundancy.

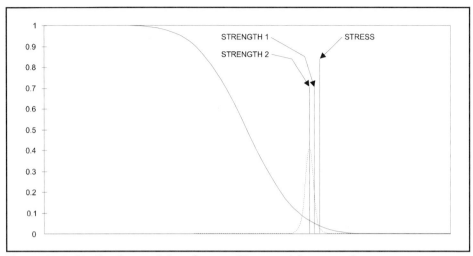

Figure 10-2. Similar Strength has Greater Chance of Common Cause.

Common-Cause Modeling

Several models are available to predict the effects of common-cause susceptibility. One of the models should be used in the safety and reliability analysis of redundant systems. Without proper consideration of common-cause effects, reliability and safety models can produce extremely optimistic results.

The Beta Model

One of the simplest models divides the failure rate of each component into common cause (two or more fail) and normal (one fails). A fractional multiplication factor known as the beta factor is used to subdivide the failure rates as shown in Figure 10-3.

The area within the rectangle of Figure 10-3 represents the total rate of stress events where stress is high enough to cause a failure. When only one unit is subjected to the stress, the stress rate equals the failure rate. Thus, the area within the rectangle represents the rate at which one or more units fail due to a stress event: the failure rate. Over a portion of the area, stress is high enough so that two or more units fail due to stress event. That portion is designated by β. The beta factor is used to divide the failure rate into the "common-cause" portion, λ^C, and the normal portion, λ^N. The following equations are used:

$$\lambda^C = \beta \times \lambda \qquad (10\text{-}1)$$

and

$$\lambda^N = (1 - \beta) \times \lambda \qquad (10\text{-}2)$$

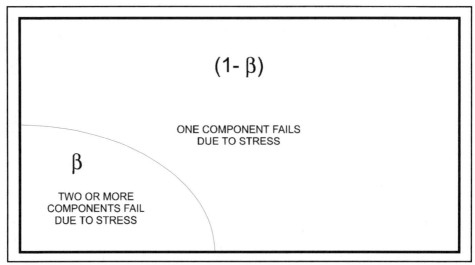

Figure 10-3. Venn Diagram of Failure Rate Showing Beta Factor.

Example 10-1

Problem: A system has two power supplies. The system is successful even if only one power supply operates. Assume that each power supply has a failure rate of 25,000 FITS (where one FIT equals one failure per billion hours), and that only one failure mode exists. A Markov reliability model for this system is shown in Figure 10-4.

State 0 represents the situation where both power supplies are successfully operating. In state 1, one power supply has failed but the system is successful. In state 2, both power supplies have failed and the system has failed. If common cause were not considered, $\lambda 1$ would be 50,000 FITS (25,000 FITS for each operating supply), $\lambda 2$ would be 25,000 FITS, and $\lambda 3$ would be zero. What would the failure rates be for $\lambda 1$, $\lambda 2$, and $\lambda 3$ if the beta factor is 0.1?

Solution: Using the beta model, the failure rates for each power supply are divided into normal and common cause. Using a beta factor of 0.1:

$$\lambda^C = \beta \times 25000$$
$$= 2500$$
$$\lambda^N = (1 - \beta) \times 25000$$
$$= 22500$$

Example 10-1 (continued)

For the beta factor of 0.1, the common-cause failure rate is 2,500 FITS. The normal mode failure rate is 22,500 FITS for each power supply. The failure rates in the Markov model are

$$\lambda 1 = 2 \times 22{,}500 = 45{,}000 \text{ FITS}$$

$$\lambda 2 = 25{,}000 \text{ FITS}$$

$$\lambda 3 = 2{,}500 \text{ FITS}$$

Note that no common-cause factor is applied to $\lambda 2$ since only one power supply is operating successfully in state 1. Common cause is a concept that applies only when multiple units are susceptible to failure. Note that the total failure rate from state 0 is 47,500 FITS and not 50,000 FITS. This seems counterintuitive since the "total" failure rate is "lowered" when common-cause modeling is applied. However, even though the overall failure rate from state 0 is lower, this does not improve system reliability. The reliability of a system with common-cause failure is actually lower because common-cause failure introduces additional pathways to the failure state.

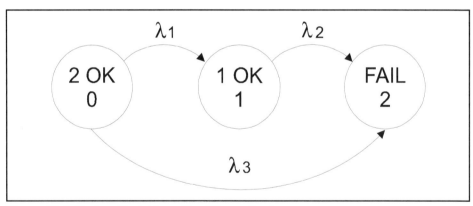

Figure 10-4. Markov Model for a Dual Power System.

Although the beta model is relatively understandable and easy to apply, it does not distinguish between two, three, or more failures due to a common stress. While this capability is not needed for most control architectures composed of dual systems or triple systems that tolerate only one failure, it is necessary to distinguish between two or three failures in a triple system that is successful when only one of three units is successful. For greater accuracy it is also necessary to separate common-cause failures of two, three, four, or more units in systems that use four or more units in a fault-tolerant scheme.

Multiple Error Shock Model (MESH)

The MESH model is more complicated than the beta model, but it does distinguish between failures of two, three, or more units. The MESH recognizes that failures occur when a system experiences a "shock." (A *shock* is defined as a stress that has exceeded an associated strength: a failure occurs.) Some shocks are of sufficient magnitude to cause multiple errors (more than one unit fails due to the shock). The model is described in detail in Ref. 1. The model presented here is modified in that it assumes that all failures are due to a shock (all failures occur when stress exceeds an associated strength: Chapter 3).

The objective of the MESH model is to calculate the failure rates of one, two, three, or more failures per shock. These failure rates are defined as:

$\lambda(1)$ = the failure rate where one unit fails per shock

$\lambda(2)$ = the failure rate where two units fail per shock

$\lambda(3)$ = the failure rate where three units fail per shock

$\lambda(n)$ = the failure rate where n units fail per shock

The calculation starts with an estimate of the probability that one, two, three, etc. units fail per shock. These probabilities are

$P1$ = the probability that one unit fails per shock

$P2$ = the probability that two units fail per shock

$P3$ = the probability that three units fail per shock

Pn = the probability that n units fail per shock

Note that the sum of all the probabilities must equal 1, since every shock will fail some quantity of units.

The *shock rate* is represented by the Greek letter nu, υ. If only one unit were exposed to the shocks, the shock rate would equal the failure rate, $\upsilon = \lambda$. If multiple units (n = the number of units) are exposed to the shock, then sometimes more than one unit will fail with each shock. Under such conditions the shock rate is less than n times the failure rate.

$$\upsilon < n \times \lambda \qquad (10\text{-}3)$$

The relationship between shock rate and individual failure rates is given by

$$\upsilon = \frac{n \times \lambda}{M} \qquad (10\text{-}4)$$

Common Cause

where M is the average number of units failed per shock. This can be calculated using the formula for the expected value of a discrete probability density function (Equation 2-8):

$$M = P1 + P2 \times 2 + P3 \times 3 + \ldots \qquad (10\text{-}5)$$

Once M is calculated, the shock rate can be calculated using Equation 10-4. The failure rates for one unit, two units, three units, and so forth are calculated using:

$$\lambda(1) = \upsilon \times P1$$

$$\lambda(2) = \upsilon \times P2$$

$$\lambda(3) = \upsilon \times P3$$

and so on until $\lambda(n) = \upsilon \times Pn$.

Example 10-2

Problem: Consider the dual power supply of Example 10-1. The system is successful if only one power supply operates. Each power supply has a failure rate of 25,000 FITS and only one failure mode. An expert estimates that $P1 = 0.95$ and $P2 = 0.05$. Calculate $\lambda(1)$ and $\lambda(2)$. Show all failure rates on a Markov model.

Solution: $\lambda = 25{,}000$ FITS and $n = 2$. First, calculate the average number of failures per shock:

$$M = P1 + 2 \times P2$$
$$= 1.05$$

Calculate the shock rate:

$$\upsilon = \frac{2 \times 25000 \text{ FITS}}{M}$$
$$= 47500 \times 10^{-9} \text{ shocks per hour}$$

$$\lambda(1) = P1 \times 47500 \times 10^{-9}$$
$$= 45000 \times 10^{-9} \text{ failures per hour}$$

$$\lambda(2) = P2 \times 47500 \times 10^{-9}$$
$$= 2500 \times 10^{-9} \text{ failures per hour}$$

These are shown in Figure 10-5. Note that the failure rates are the same as those calculated by the beta model (exact if $P1 = 0.947368421$). The MESH model will give the same results for $n = 2$ systems as the beta model if the parameters are estimated with the same considerations.

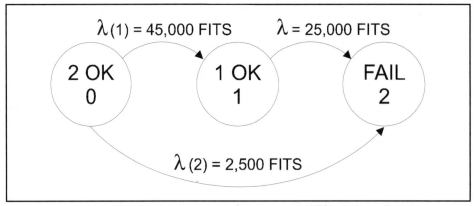

Figure 10-5. Markov Model of Dual Power System with MESH Failure Rates.

Example 10-3

Problem: Two sensors are used to measure an important process variable. The safety model should include common cause. Each sensor has a failure rate of 0.001 failures per hour. An expert estimates that $P1 = 0.8$ and $P2 = 0.2$. Calculate $\lambda(1)$ and $\lambda(2)$.

Solution: $\lambda = 0.001$ failures per hour and $n = 2$. First, calculate the average number of failures per shock:

$$M = P1 + 2 \times P2$$
$$= 1.2$$

Calculate the shock rate:

$$\upsilon = \frac{2 \times 0.001}{M}$$
$$= 0.001667 \text{ shocks per hour}$$

$$\lambda(1) = P1 \times \upsilon = 0.001334 \text{ failures per hour}$$

$$\lambda(2) = P2 \times \upsilon = 0.000334 \text{ failures per hour}$$

Example 10-4

Problem: What is the equivalent beta for Example 10-3?

Solution: Note that $\lambda(2)$ in the MESH model is equal to $\beta \times \lambda$ in the beta model (for $n = 2$ system). Therefore,

$$\beta = \frac{0.000334}{0.001} = 0.334$$

Common-Cause Avoidance

Control system designers at all levels must recognize that common-cause failures drastically reduce system safety and MTTF. Systems must be designed to achieve desired reliability goals even when common-cause failure rates are included in reliability models. Designers must recognize the failure sources that are responsible for common-cause failures. Specific solutions must be implemented to combat common-cause failures. The common-cause defense rules can be grouped into categories that result in three basic rules.

Rule 1: Physically Separate the Redundant Units

When redundant units are separated, there is less likelihood of a common stress. Most environmental stress factors vary nonlinearly as a function of physical distance. Both physical separation and electrical separation are important. Redundant units should not be physically mounted side by side, such that the physical and electrical common stress is nearly identical. In such situations, the beta factor will be higher.

Many abusive environmental failure sources could be segregated by putting redundant equipment in separate cabinets. Control systems that have redundant resources physically separated will be less susceptible to environmental common-cause failures simply because the common environment has been reduced.

In software, the chances of a common stress can be reduced by asynchronous operation (less chance of identical inputs with identical timing). Lock-step synchronization between processors should be avoided. The use of different operating modes between software in redundant processor units is another way to reduce common stressors for software. The data stored in each unit can be different.

Rule 2: Diversity

A second common-cause defense technique is "diversity." *Diversity* is a concept in which differently designed units are wired together in a redundant configuration. The motivation is simple: Different component designs will not suffer from the same common-cause faults. Components manufactured differently will not suffer from the same manufacturing faults.

The technique has been tested (Refs. 2 and 3) and has had some success in both hardware and software. However, serious trade-offs and cost considerations must be taken into account. The testing has shown that design diversity has not eliminated all common-cause design errors, including system-level design errors. In addition, many new problems having to do with synchronization, calibration, and data mismatch due to digital round-off have arisen.

In terms of environmental common cause, redundant components using different technologies may increase common-cause strength only if the designs respond differently to a common stress. For diversity to be effective, the redundant components must not respond to a common stress. The use of different manufacturers may eliminate some common manufacturing defects, but significant benefits are not realized if both units respond to the same stress. Using different manufacturers adds maintenance and operational complexity, which may override the benefits. Hardware diversity will work only if the redundant units respond differently to a common stress. A mechanical unit backing up an electrical unit would be a good use of hardware diversity.

Software diversity can be implemented by using different software written in different languages by different people. However, this can be very expensive and should not be done at the sacrifice of software strength (put the effort into software quality first).

At this point, there is a significant trade-off between the extra effort required to replicate a design more than once (extra design training, extra design documentation, extra maintenance training, extra spare parts, etc.) versus the time required to avoid faults during design. The extra complexity created when connecting diverse machines into a fault-tolerant system results in design faults. Given the new, inevitable problems caused by diversity, such systems should be approached with caution.

Rule 3: "Ruggedize" the Design for High Strength

Controllers that are designed to withstand tougher environments will work much better in fault-tolerant configurations, because the threshold of environmental stress required to fail multiple units is much higher. The system features that lower the component failure rate will also lower common-cause failure rate. Good heat sinking, coated circuit boards, rugged module covers, and secure mechanical connectors will lower the component failure rate because these features increase strength. If a module is less likely to fail due to a certain stress level, it is less likely to have a common-cause failure. All the factors that increase strength decrease common-cause susceptibility. The higher the design margin, the less likely is common-cause failure.

As complexity grows in control systems, common-cause design errors will increase in proportion. Of course, fault-avoidance design techniques should be used. A systematic software development process will reduce the chance of a software error and increase "software strength." This increases software common-cause strength. Third-party audit and inspection can help avoid faults. Such audits are typically included in the development process of safety-critical PLCs.

The operation and maintenance of a system can generate common-cause failures. Incorrect commands sent to synchronously operating controllers will cause both to fail. Complex operations should be automated whenever possible. Foolproofing techniques can be used for both operations and maintenance. Repairable assemblies should be keyed so that modules and connectors cannot be installed improperly. Preventive maintenance can also be helpful in maintaining high strength in some situations.

Estimating the Beta Factor

Common-cause strength as measured by the beta factor is highly dependent on physical and electrical implementation. The three rules for common-cause strength are the significant factors. Physically separated and electrically separated equipment is much less likely to be exposed to a common stress, which reduces the beta factor. Diverse design can reduce the beta factor, as can high strength. A combination of these parameters can reduce the beta factor significantly.

Published opinions of experts (Refs. 4-9) put the beta factor in the range of 0.005 to 0.11 for hardware failures. The beta factor has been estimated in the range of 0.05 to 0.6 for software failures. Reference 8, a study of redundant systems on the U.S. Space Shuttle, estimates the beta factor at 0.11 for hardware. This range of numbers in the models results in several orders of magnitude difference in reliability and safety metrics. Therefore, it is important to estimate the beta factor and include common-cause susceptibility in quantitative analysis.

A qualitative evaluation of the system implementation can be used to estimate the beta factor per Table 10-1. This table presents a simplified method of estimating that is consistent with the listed references. A more extensive method developed by R. A. Humphreys is published in Reference 9. Humphreys evaluates:

- *Separation*: physical separation (rule 1)
- *Similarity*: the level of diversity (rule 2)
- *Complexity*: a measure of "strength of design," assuming more complex designs are more likely to fail
- *Analysis*: a measure of the design testing
- *Operating procedures*: a measure of system strength against operational errors
- *Training*: a measure of strength against operational and maintenance errors
- *Environmental control*: a measure of access to the equipment. Strict access control would reduce the chance of an operational or maintenance error.
- *Environmental testing*: a measure of strength in the product (rule 3)

This method accounts for many legitimate human common-cause error sources that are not included in Table 10-1. The Humphreys method does a better job with "high-strength" systems. Table 10-1 does not give credit for a rugged design, which is probably less likely to experience a common-cause failure. The Humphreys method resulted in higher beta factors for several examples estimated by the author. It is recommended that Table 10-1 be used only for estimating. If rugged, high-strength components are being used, or if human factor failure rates are included in the models, then the Humphreys method is preferred.

	Evaluation Question	Add to the beta factor
1. Pick the level of physical separation:		
	Are redundant channels on the same PCB (printer circuit board)?	0.02
or	Are redundant channels in the same card rack or mounted side by side?	0.01
or	Are redundant channels in the same cabinet?	0.005
or	Redundant channels are in different cabinets.	0.002
2. Pick the level of electrical separation:		
	Are redundant channels on the same power common (nonisolated)?	0.01
or	Are redundant channels high-frequency isolated?	0.005
or	Are redundant channels galvanically isolated (DC and AC)?	0.002
3. Pick the level of diversity:		
	Are redundant channels identical technology, such as PE (programmable electronic)?	0.01
or	Are redundant channels similar technology such as PE and E (electronic)?	0.001
or	Are redundant channels diverse, such as PE/E and mechanical/hydraulic?	0.0001

Table 10-1. Beta Factor Evaluation Chart

Example 10-5

Problem: A redundant PLC is implemented with redundant controllers mounted side by side in one rack. They are electrically isolated via different power supplies and galvanically isolated communications circuits. They are identical in design. What is the estimated beta factor?

Solution: Start with $\beta = 0$. The redundant circuits are not on the same PCB, but they are in the same rack. Add 0.01 to the beta factor. The units are galvanically isolated, so add 0.002. Both modules are programmable electronic, so add another 0.01. The estimated beta factor is 0.022.

> **Example 10-6**
>
> **Problem:** A redundant PLC is implemented with redundant controllers mounted in different cabinets. They are electrically isolated via different power supplies and galvanically isolated communications circuits. They are identical in design. What is the estimated beta factor?
>
> **Solution:** Start with $\beta = 0$. The redundant circuits are physically separated in different cabinets, so add 0.002 to the beta factor. The units are galvanically isolated, so add 0.002. Both modules are programmable electronic, so add another 0.01. The estimated beta factor is 0.014.

> **Example 10-7**
>
> **Problem:** Redundant I/O circuits are provided within one module. They are not isolated. They are identical in design. What is the estimated beta factor?
>
> **Solution:** Start with $\beta = 0$. The redundant circuits are on the same PCB, so add 0.02 to the beta factor. The units are not isolated, so add 0.01. Both modules are programmable electronic, so add another 0.01. The estimated beta factor is 0.04.

Additional research is needed into the nature of common-cause failures. Methods such as common environment minimization and diversity can avoid some common-cause failures, but improvements are still required. System design techniques must be further refined. Methods of increasing component strength must be pursued. Architectures without common susceptibility must be refined. There is much room for advancement, but at least common cause is becoming recognized as a major consideration during the design of a system or in the evaluation of system reliability.

Including Common Cause in System Models

Common-cause effects can be modeled at the unit or system level using any of the techniques presented earlier in this book: reliability block diagrams, fault trees, and Markov models. Remember, common-cause failures occur when two or more devices fail at the same time. Additions can be made to each model type to show this event.

Common-Cause Modeling with Reliability Block Diagrams

A simple redundant system is shown in a reliability block diagram as two blocks in parallel. If either block operates, the system operates. If both units fail due to a common cause, the system fails. Thus, another block is added in series, as shown in Figure 10-6. The failure rates for non-common cause (normal) and common cause are used to calculate probabilities of successful operation. These probabilities are combined in the model.

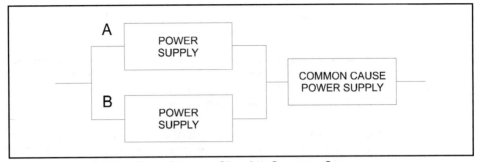

Figure 10-6. Reliability Block Diagram Showing Common Cause.

> **Example 10-8**
>
> **Problem:** A system has two power supplies. One power supply is required for the system to be successful. A power supply has a constant failure rate of 0.00005 failures per hour. We estimate a common-cause beta factor of 0.05. What is the system reliability for a time interval of 1,000 hr with and without consideration of the effects of common cause?
>
> **Solution:** Without common cause, the reliability of one power supply for 1,000 hr = exp − (0.0005)(1000) = 0.95123. The unreliability = 1 − 0.95123 = 0.04877. For a parallel system, the unreliability = 0.04877 × 0.04877 = 0.002379, and the reliability = 0.99762.

Including the effects of common cause, the block diagram of Figure 10-6 is used. The normal failure rate for the power supply is $(1 - \beta) \times \lambda$ = 0.95 × 0.00005 = 0.0000475 failures per hour. The common-cause failure rate = 0.05 × 0.0005 = 0.0000025.

$$\text{Reliability for block A} = \exp(-0.0000475 \times 1000)$$
$$= 0.95361$$

$$\text{Unreliability for block A} = 1 - 0.95361$$
$$= 0.04639$$

$$\text{Reliability for block B} = \exp(-0.0000475 \times 1000)$$
$$= 0.95361$$

$$\text{Unreliability for block A} = 1 - 0.95361$$
$$= 0.04639$$

$$\text{Reliability for the parallel combination of A and B}$$
$$= 1 - (0.046139^2) = 0.99785$$

Reliability for the common-cause block
$$= \text{EXP}\,(-\,0.0000025 \times 1000) = 0.99750$$

Reliability of power system including common cause
$$= 0.99785 \times 0.99750 = 0.99536$$

Common-Cause Modeling with Fault Tree Diagrams

The effects of common cause can be modeled using a fault tree in a manner similar to reliability block diagrams. A fault tree showing common cause is presented in Figure 10-7. Again, two power supplies are provided, either of which can operate the system. Without considering common cause, both must fail in order for the system to fail. When common cause is considered, this is the equivalent of another failure that will fail the system. This is shown as another input to an OR gate.

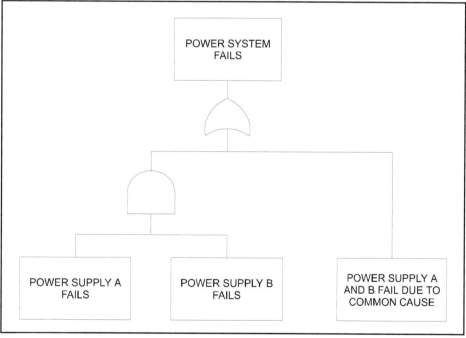

Figure 10-7. Fault Tree Diagram Showing Common Cause.

Example 10-9

Problem: A pressure sensor has a dangerous failure rate (λ^D) of 0.00001 failures per hour. Two sensors are used in an SIS such that if either sensor detects an overpressure, the SIS will trip and shut down the process. The sensor system will not fail dangerously unless both sensors fail dangerously. The common-cause beta factor is estimated to be 0.1. What is the PFD of the sensor system for a time interval of 8,760 hr with and without consideration of the effects of common cause? What are the risk reduction factors? Show the results on a fault tree.

Solution: Without common cause, the PFD of one sensor supply for a one-year time interval = $1 - \exp-(0.00001)(8760) = 1 - 0.916127 = 0.083873$. (Try the approximation, PFD = $\lambda^D \times t = 8760 \times 0.00001 = 0.0876$. Close, but obviously more pessimistic; use the more accurate value.) For an AND gate, the PFD = $0.083873 \times 0.083873 = 0.007035$. The risk reduction factor (RRF) = 142. This is shown in Figure 10-8.

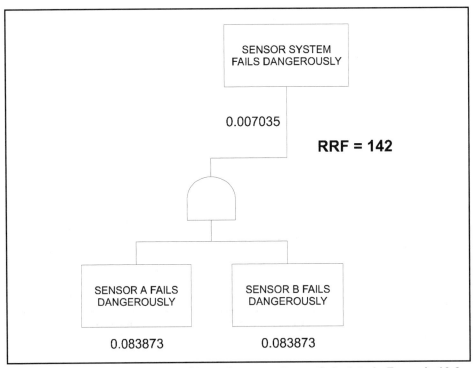

Figure 10-8. Fault Tree Diagram without Common Cause Calculated - Example 10-9.

Including the effects of common cause, the fault tree diagram in Figure 10-9 is used. The normal dangerous failure rate for a sensor is $(1 - \beta) \times \lambda^D = 0.9 \times 0.00001 = 0.000009$ failures per hour. The common-cause dangerous failure rate $= 0.1 \times 0.00001 = 0.000001$.

PFD for sensor A $= 1 - \exp(-0.000009 \times 8760) = 0.0758123$

PFD for sensor B $= 1 - \exp(-0.000009 \times 8760) = 0.0758123$

PFD for dangerous common-cause failure of A and B
$= 1 - \exp(-0.000001 \times 8760) = 0.008721743$

PFD for the AND gate $= 0.0758123 \times 0.0758123 = 0.005747492$

PFD for the sensor system with common cause
$= 0.005747492 + 0.008721743 - 0.005747492 \times 0.008721743$
$= 0.014419107$

$$\text{RRF} = \frac{1}{0.01441911} = 69$$

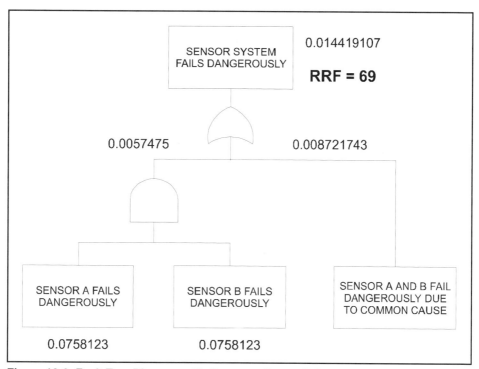

Figure 10-9. Fault Tree Diagram with Common Cause Calculated - Example 10-9.

Common-Cause Modeling with Markov Models

As demonstrated earlier, common-cause failures can be shown on a Markov model quite easily. Figure 10-10 shows a dual system with one failure mode. In state 0, both devices are operating and the system is successful. In state 1 the system has degraded; one device has failed but the system is still successful. State 2 represents the condition where both devices have failed and hence the system has failed. A common-cause failure takes the system directly from state 0 to state 2; this is shown by the directed arc labeled $\beta\lambda$. This technique can be extended to show various combinations of multiple failures as required.

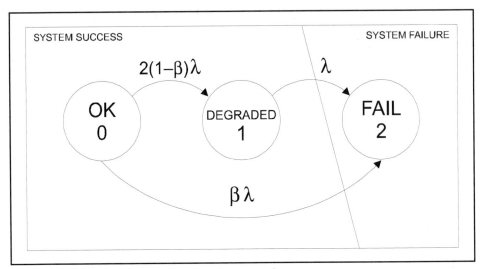

Figure 10-10. Markov Model Showing Common Cause.

The Markov model in Figure 10-11 shows a system consisting of three devices. The system operates successfully if only one of the three devices is operating. One failure mode is assumed. No common cause has been included. State 0 shows the condition where all three devices are successful. State 1 shows the condition where two of the devices are operating and one has failed. In state 2, two devices have failed and one is operating. In state 3, all three devices have failed and the system has failed.

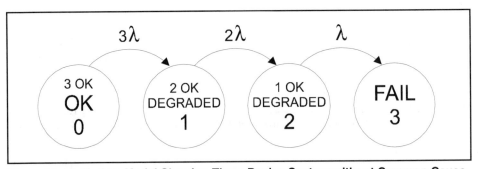

Figure 10-11. Markov Model Showing Three Device System without Common Cause.

Figure 10-12 shows the same system with possible common-cause failures included. From state 0, one, two, or three devices can fail per stress event. The arc marked λ3 (1) describes three devices exposed, one fails. The arc marked λ3 (2) means three devices are exposed, two fail. The arc marked λ3 (3) is the case where all three devices exposed fail due to a common stress. Other arcs are marked with the other possibilities for common-cause failure. The Markov modeling method is quite flexible when considering common cause.

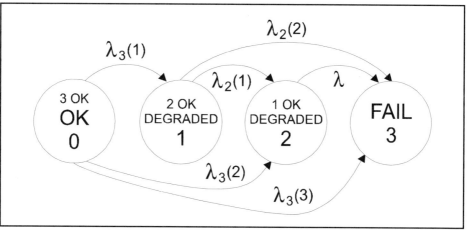

Figure 10-12. Markov Model Showing Three Device System with Common Cause.

EXERCISES

10.1 List sources of common-cause failures.

10.2 How can one software fault cause failure of two redundant controllers?

10.3 Describe the concept of diversity.

10.4 How would you achieve hardware and software diversity in a system design?

10.5 Redundant pressure sensors are used to measure pressure in a high-pressure protection system. They are fully isolated electrically. They are mounted several feet apart on the process vessel. They are identical in design. What is the estimated beta factor?

10.6 A sensor has a dangerous failure rate of 0.00005 failures per hour. Two of these sensors are used in a 1oo2 safety configuration. (The system will not fail dangerously unless both sensors fail.). The beta factor is estimated to be 0.025. What is the dangerous common-cause failure rate?

REFERENCES AND BIBLIOGRAPHY

1. Hokstad, P., and Bodsberg, L., "Reliability Model for Computerized Safety Systems," *1989 Proceedings of the Annual Reliability and Maintainability Symposium*, New York: IEEE, 1989.

2. Knight, J. C., and Leveson, N. G., "An Experimental Evaluation of the Assumption of Independence in Multi-Version Programming", *IEEE Transactions on Software Engineering*, Vol. SE-12, No. 1, 1986.

3. Brilliant, S. S., Knight, J. C., and Leveson, N. G., "The Consistent Comparison Problem in N-Version Software," *IEEE Transactions on Software Engineering*, Vol. 15, No. 11, 1989.

4. Gray, J., "A Census of Tandem Availability between 1985 and 1990," *IEEE Transactions on Reliability*, Vol. 39, No. 4, October 1990.

5. Siewiorek, D. P., "Fault Tolerance in Commercial Computers," *Computer*, July 1990.

6. Bukowski, J. V., and Goble, W. M., "Common Cause — Avoiding Control System Failures," *Proceedings of ICS94 Conference*, Research Triangle Park, NC: ISA—The Instrumentation, Systems, and Automation Society, 1994.

7. Beckman, L. V., "Match Redundant System Architectures with Safety Requirements," *Chemical Engineering Progress*, December 1995.

8. Rutledge, P. J., and Mosleh, A., "Dependent-Failures in Spacecraft: Root Causes, Coupling Factors, Defenses, and Design Implications," *1995 Proceedings of the Annual Reliability and Maintainability Symposium*, New York: IEEE, 1995.

9. Smith, D. J., *Reliability, Maintainability and Risk: Practical Methods for Engineers*, Oxford, U.K.: Butterworth-Heinemann, 1993.

10. Dhillon, B. S., and Rayapati, S. N., "Common-Cause Failures in Repairable Systems," *1988 Proceedings of the Annual Reliability and Maintainability Symposium*, New York: IEEE, 1988.

11. Page, L. B., and Perry, J. E., "A Model for System Reliability with Common-Cause Failures," *IEEE Transactions on Reliability*, Vol. 38, No. 4, October 1989.

12. Paula, H. M., Roberts, M. W., and Battle, R. E.,"Reliability Performance of Fault Tolerant Digital Control Systems," *Proceedings of the 24th Annual Loss Prevention Symposium*, New York: American Institute of Chemical Engineers, August 1990.

13. Common Cause Failure Models, Annex A, *Electrical (E)/Electronic (E)/Programmable Electronic Systems (PES) for Use in Safety Applications — Safety Integrity Evaluation Techniques, Draft 4*, Research Triangle Park, NC: ISA—The Instrumentation, Systems, and Automation Society, June 1995.

14. Bukowski, J. A., and Lele, A., "The Case for Architecture-Specific Common Cause Failure Rates and How They Affect System Performance," *1997 Proceedings of the Annual Reliability and Maintainability Symposium*, New York: IEEE, 1997.

11
Software Reliability

Software Failures

Some system failures are caused by software. These failures are different from most other failure types, in that all software failures are inadvertently "designed" into the system. Software does not wear out. Software is "manufactured" with virtually no undetectable duplication error. There are no latent manufacturing defects.

This situation leads some to believe that software failures cannot be modeled using statistical techniques. The argument states that a computer program always fails given a particular sequence of input data. Therefore, reliability equals zero for that input sequence. Reliability equals one for other input sequences. The system is completely deterministic, not probabilistic.

The argument is correct, but it does ignore a reality of computer operation. The number of input sequences is very large and statistical in nature. Probability is a concept that works perfectly on a specific input sequence for a specific system. The input sequence is the "stress event" that may or may not cause a particular software system to fail. When we consider a large number of systems with a large number of input sequences, we realize that some will fail. When we consider how often they fail, the concept of failure rate applies.

An engineer was adding new control logic to a PLC. An analog input module had been added to the system and the input to that module was a slide wire that normally generated a 1 to 4 v signal. The slide wire was measuring mechanical motion. Startup went well; the additions passed all testing. Several months later, a mechanical stop failed and the wiper of the slide wire traveled beyond its normal range to 0 v. The PLC stopped operation. The PLC software failed with a "divide by zero" failure message. For this failure:

Title	Software failure: divide by zero
Root cause	Wiper travel beyond normal range
Failure type	Functional
Primary stressor	Unexpected input to math software
Secondary stressors	Mechanical stop failure; not enough software stress testing

An industrial operator console has been operating in the plant for two years with no problems. A new operator joined the staff; during one of his first shifts, the console stopped updating the CRT screen and stopped responding to operator commands shortly after an alarm acknowledgment. The unit was powered down and restarted perfectly. There were no hardware failures. Since the manufacturer had more than 500 units in the field with 12 million operating hours it was hard to believe that a significant software fault existed in such a mature product.

The problem could not be duplicated after many hours of testing, so a test engineer visited the site and interviewed the operator. During the visit it was observed that "this guy is very fast on the keyboard." That input allowed the problem to be traced. It was discovered that if an alarm acknowledgment key is struck within 32 msec of the alarm silence key, a software routine will overwrite an area of memory and "crash" the computer. For this failure:

Title	Software failure: rapid input
Root cause	Ack key struck within 32 msec of Sil key
Failure type	Functional
Primary stressor	Unexpected input timing to software
Secondary stressors	Not enough software stress testing

A computer failed after an operator requested that a data file be displayed on the CRT. This particular data file had been successfully displayed on the system many times before. The problem was traced to a software module that did not always append a terminating "null zero" to the end of the file name character string depending on file name size. Although the software routine never appended the zero onto this file name before storing it in memory, most of the time the file name was stored in memory that had been cleared: zeros were written into all locations. Under those circumstances, the operation was always successful and the software fault was masked. Occasionally, the dynamic memory allocation algorithm picked memory that had not been cleared. The system failure occurred only when the software module did not append the zero in combination with a memory allocation in an uncleared area of memory. For this failure:

Title	Software failure: file name size failure
Root cause	Software did not append null terminator for certain input character strings
Failure type	Functional
Primary stressor	Character string size
Secondary stressors	Dynamic memory allocation in uncleared area of memory

A computer stopped operating after a message was received on its communication network. The message used the correct "frame" format, but was from an incompatible operating system with different data formats. The computer did not check for compatible data format. The data bits within the frame were incorrectly interpreted, eventually causing the computer to "crash." For this failure:

Title	Software failure: comm message
Root cause	Connected incompatible computers on same network
Failure type	Functional
Primary stressor	Incompatible data input
Secondary stressors	Not enough testing; Not enough checking of input data

Many factors in addition to inputs cause software to fail. Consider the above cases. Timing of input data was involved. Size of input data caused a failure. Changing operation (dynamic memory allocation) contributed. There are more. With such a wide variety of failure sources, each of which can be treated statistically, there is a solid basis for the use of statistical reliability and safety analysis of software.

Software Failure Rate Trends

The systems used to control and safeguard our chemical processes are growing more complex as we learn better ways to automate and optimize. Powerful new software tools have become the enabler for much of this advancement. However, this means that most of the complexity resides in software based systems. We depend on this software, and our dependency is growing. Software reliability — the ability of the software to perform the expected function when needed — is essential. Yet how many times do we experience trouble? "The company network is down."

"My computer hung again." "I forgot to save my file and just blew away four hours of work." Our experience is far from perfect. As our software dependency increases, our incentive for higher levels of software reliability is greater.

Unfortunately, software failure rates are increasing (Figure 11-1) (Refs. 1 and 2). Although software fault-avoidance techniques (known as software engineering) are improving, the amount of software is increasing at a faster pace. A top-of-the-line personal computer purchased by the author in 1980 provided a maximum of 8K (8,096 bytes) of RAM memory. Its replacement, purchased in 1986, had 640K (655,360 bytes) of memory. Another personal computer purchased in 1991 had 8M (8,388,608 bytes) of memory. The current machine has 64M of RAM memory. The machines need more memory because the software available for each subsequent machine required the available memory. The fundamental problem is complexity. It appears that software size has grown by more than three orders of magnitude. If complexity has grown only a portion of that amount, the ability to avoid software faults has not kept pace. The "strength" of the software decreases with complexity.

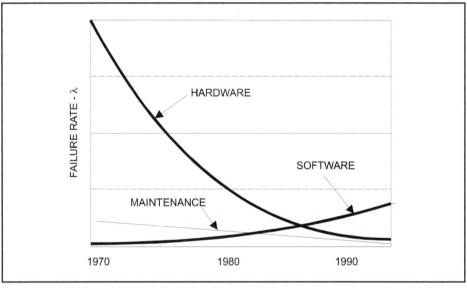

Figure 11-1. Software Failure Rate Trends.

Stress-Strength View of Software Failures

All failures occur when some "stress" is greater than the associated "strength." Like hardware, a software failure is caused when stress is greater than strength. This concept was presented in Chapter 3.

Intuitively, we guess that the software failure rate has some relationship to the number of faults (human design errors, or "bugs") in the software. Fault count is a "strength" factor. A program with few faults is stronger than one with many faults. Software strength is also affected by the amount of "stress rejection" designed into the software. Software that checks for valid inputs and rejects invalid inputs will fail much less frequently. Consider the communications example above. Although the software did check for correct data frame format, it did not check for correct data format. If it had, it is likely the failure would not have occurred.

We also guess that the software failure rate relates to the way in which the software is used. The usage "stress" to a software system is the combination of inputs, timing, and stored data seen by the CPU. Inputs and timing of inputs may be a function of other computer systems, operators, internal hardware, external hardware, or any combination of these factors.

Increasing Software Strength

Using the stress-strength concept as a guide, higher levels of software reliability are obtained by increasing software strength. Most of the effort in software reliability to date has focused on removing software faults (i.e., software process improvement). Software is created by human beings. The design process is not perfect; mistakes are made. Many companies expend considerable resources to establish and monitor the software development process (Refs. 3-8). This is intended to increase software strength by reducing the number of faults in the software.

These efforts vary significantly depending on implementation. Attempts are being made to audit software process effectiveness. The ISO 9000-3 standard establishes required practices. The Software Engineering Institute has a five-level software maturity model (Ref. 4), where level 5 is intended to represent the best process (Figure 11-2). Companies in regulated industries audit software vendors to internal standards. When the focus is on reducing the number of faults in software, these efforts do improve software reliability.

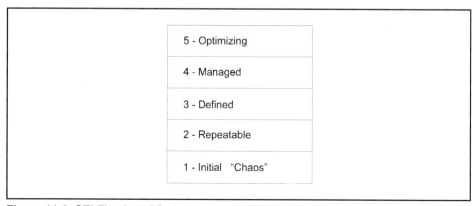

Figure 11-2. SEI Five Level Software Maturity Model.

The number of faults in a software system is also related to the testability of a system. Depending on the variability of execution, a software testing effort may or may not be effective. When software always executes in the computer in a consistent manner, tests can be done to verify correct execution. When software executes differently each time it is loaded, a test program cannot be complete. The number of test cases explodes to "virtual infinity." Execution variability is increased with dynamic memory allocation, number of CPU interrupts, number of tasks in a multitasking environment, and so on. After a software fault is discovered, it is easy to say "Not enough testing. What's wrong with the software testing department?" But under conditions of high execution variability, expensive software testing efforts are very ineffective. Software strength (and therefore reliability and safety) is improved when execution variability is reduced and testing can be more effective.

Several other important factors increase software strength. These are called "software diagnostics" and "stress rejection." Software diagnostics increase strength by doing on-line software verification. This not only helps prevent a software failure but also helps identify faults for subsequent repair.

Software diagnostics work much like hardware reference diagnostics. Certain conditions are expected, and the software checks to make sure these conditions are present. Some software diagnostic techniques, such as "program flow control," are specified in international standards for safety-critical software (Refs. 9 and 10). These diagnostics are required in safety-critical software approved by recognized third parties, such as FM (U.S.) and TUV (Germany).

Program flow control is a technique where each software component that must execute in sequence writes an indicator to memory. Each software component can check to verify that necessary predecessors have done their job. At the end of a sequence, checking can be done to ensure that all necessary software components have run. When time-critical operations are performed, the indicators can include time stamps. The time stamps can be checked to verify that maximum times have not been exceeded.

Many other software diagnostic techniques are in use, especially in safety-critical software. Testing can be done to check for certain patterns in memory. The software can measure the time required for execution of certain algorithms. The software may check to verify that memory has changed as expected.

Software *stress rejection* techniques also increase software strength. The concept is that potential stressors that might cause software failure (especially those that have not been tested during software quality assurance testing) are filtered out before they can do damage. One example is *plausibility assertions*, which are required in approved safety-critical software. The software must check inputs to software and stored data. Communication messages are checked for proper format and content before commands are executed. Data are checked to verify that they are within an expected range. Pointers to memory arrays are checked to verify they are within a valid range for the particular array. Many other techniques particular to special applications are used.

Software Complexity

Why are software failure rates going up? Why does the strength of software seem to be decreasing? Many think the answer is complexity — which is growing beyond the ability of the tools that help humans deal with complexity. To understand this growth in complexity, a view from the microprocessor unit (MPU) is needed. A microprocessor always starts the same way. It looks at a particular place in memory and expects to find a command. If that command is valid, the MPU begins execution of a long series of commands from that point, reading inputs and generating outputs. There are three ways to view this complexity: a series of digital "states," a sequence of digital states called a "path," or a series of inputs.

State Space

The first digital computers were easily recognized as state machines — digital circuits that moved from "state" to "state" depending on the input conditions and memory content. The "state" was represented by a number of binary bits stored on flip-flop (bistable latch) circuits. One group of flip-flop circuits was called the "program counter." Other groups of circuits were called "registers."

Outputs from the program counter selected a particular location in the memory. Logic circuits set or cleared bits in the flip-flops, depending on input switches and logic signals from the memory. The bit settings stored a calculation and determined the next state. The digital state machine moved from state to state, depending on contents of the memory and depending on input. Each state is defined as a permutation of bits from all the flip-flops in the machine.

The machine was successful if it moved from state to state as intended by the software engineer. However, if any bits were in error or if the input switches had some unanticipated input, the machine was not likely to follow the desired path or to perform its intended function. This constituted a failure.

The number of possible states in these early machines was extremely large. Consider a machine that has three 8-bit registers, one 16-bit register, and a 6-bit status register. A total of 46 bits are present. The total number of possible states (2^{46}) is over 70 trillion! In the many generations of machines developed since the first computers, complexity has certainly increased. Current-generation machines have many more states. The quantity has been called "virtual infinity"! Fortunately, if most bits are always set the same way, the number of states is quickly reduced. The collection of all possible states is called the *state space*.

Path Space

A sequence of states followed by a computer during execution of a program is called a path. The collection of all possible paths is called the *path space*. A particular path is determined by the contents of memory and the input received by the computer during the program execution. Simple microprocessors such as those installed in appliances execute only a few paths repeatedly.

Consider the simple program diagrammed in Figure 11-3. The program measures the time between two switch inputs and calculates the speed of a vehicle. How many paths can this simple program have? There are many possibilities:

1234567

121234567

12121234567

1212121234567

123454567

12345454567

1234545454567

121212345454567

... and many more.

Software Reliability   245

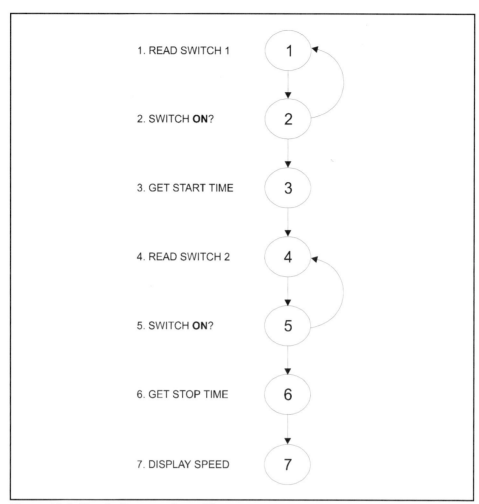

Figure 11-3. Simple Program State Diagram.

If we count only a single repetition through each loop, the program has three paths: 1234567, 121234567, and 123454567. A path that considers only a single loop iteration is called a "control flow" path. Control flow path structures associated with common program constructs are shown in Figure 11-4.

Paths are identified and counted in computer programs for many reasons, including the development of program test strategies. Theoretically, every path should be tested. If this is done, all software design faults should be discovered. "Test coverage" is a measure of the percentage of paths tested. In small programs and individual program modules, it is practical to test all paths. However, as the size of a program increases, the number of paths increases in a superlinear manner. In large programs the testing will take tens of years, even if only control flow paths are tested.

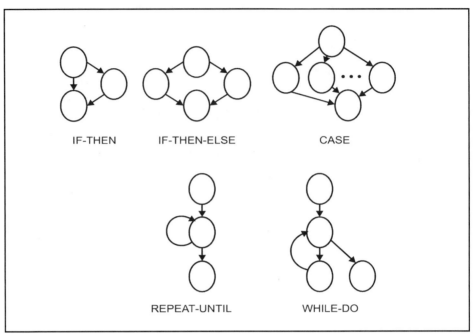

Figure 11-4. Control Flow Path Structures.

McCabe Complexity Metric

For any program that can be represented by a state diagram, one method of calculating the number of control flow paths is called the *McCabe complexity metric* (Ref. 6). Developed from graph theory, the number of paths is

$$NP = e - n + 2 \qquad (11\text{-}1)$$

The variable e represents the number of edges (which appear on the diagram as arrows). The variable n equals the number of nodes (represented on the diagram by circles). The program diagrammed in Figure 11-3 has a complexity measure of 3 ($8 - 7 + 2$). These three control flow paths are listed in the previous section.

EXAMPLE 11-1

Problem: How many control flow paths are present in the program diagrammed in Figure 11-5?

Solution: The number of nodes equals eight. The number of edges equals nine. Using Equation 11-1:

$$NP = 9 - 8 + 2$$
$$= 3$$

Software Reliability   247

Another simple program is diagrammed in Figure 11-5. Assume that this is the diagram of a program used in a toaster. The program scans the lever switch until a user pushes down the bread to be toasted. The heater is energized. The "heat time" is input from the "light-dark" switch on the side of the toaster. This time number should be an integer between 1 and 5. A value of 1 dictates a short heating time of 10 sec, resulting in light toast. Higher values dictate longer heating times. The program decrements the time number until the "remaining time" equals zero. The heater is then turned off and the toast is popped up. The number of program control flow paths equals three (Example 11-1).

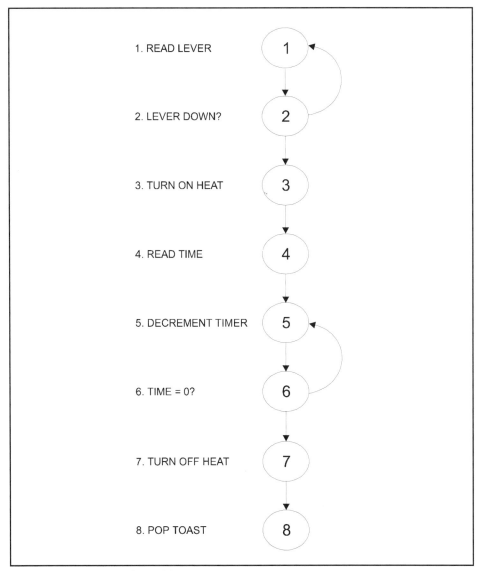

Figure 11-5. Program State Diagram.

These three paths are

> 12345678
>
> 1212345678
>
> 1234565678

All program control flow paths could be tested with three test runs.

Data-Driven Paths

Control flow path testing does not account for path variations due to input data. Whenever the number of loop iterations is controlled by input data, each loop iteration count should be considered a different path. The program from Figure 11-5 would have two paths for each possible data value of "time," an input obtained in step 4. For the input value of one, the paths are 12345678 and 1212345678, because the loop from step 6 to step 5 does not occur. Table 11-1 lists data-dependent paths for the valid data range of one through five.

Data	Path 1	Path 2
1	12345678	1212345678
2	1234565678	121234565678
3	123456565678	12123456565678
4	12345656565678	1212345656565678
5	1234565656565678	121234565656565678

Table 11-1. Data-Driven Program Paths

Even testing of these 10 paths may not find all software design faults. Certain "errors of omission" are not detected until the program is tested with unexpected inputs. What happens when an input of zero is received by our toaster program? Assume that the time value is stored in an 8-bit binary number. The program follows its path to step 5. The number is decremented. A zero is decremented to a binary "minus one," represented by eight binary ones (11111111). This is the same representation as the number 255. In step 6, the time number will not equal zero. The program will decrement 255 times! By the time the heater is turned off, the toast will be ashes and the kitchen will be full of smoke. Most consider this behavior to be product failure (a functional failure). In order to fully test this program, two paths must be tested for all possible inputs. With 256 input numbers, 512 paths need to be tested.

Many quality-conscious programmers will strengthen the design by adding a valid input data check, a "stress rejecter." After the time is input in step 5, it is checked against valid bounds. If the number is less than one or greater than five, the heat is turned off and the toast is popped up.

Invalid numbers are rejected in this manner. Figure 11-6 shows the modified state diagram. For each valid input, two paths exist as before. For all inputs above the limit, only two paths exist (total of 10). For all inputs below the limit, only two paths exist. The path space has been reduced. A total of only 14 paths need to be tested.

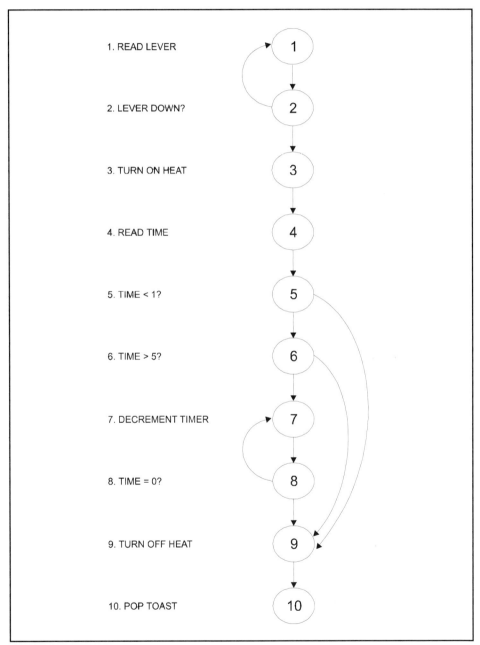

Figure 11-6. Modified Program State Diagram.

Asynchronous Functions

The path count goes up by orders of magnitude whenever the computer is designed to perform more than one function in an asynchronous manner. Most computers implement asynchronous functions with a feature known as an "interrupt." An electronic signal is wired to the computer. The computer is designed in such a way that whenever the signal activates, the computer stops following its path, saves enough information to return to the same spot in the old path, and then starts following a new path. In effect, every time an interrupt occurs, a new path is created.

Imagine that our new product development team has identified a need in the toaster market. We have estimated that thousands of additional units could be sold if a clock were added to the side of the toaster. In order to keep the cost down, we add only a digital display and a 1-sec timer. The timer will interrupt the microprocessor. The microprocessor will update the digital clock display. The state diagram for our enhanced toaster program is shown in Figure 11-7.

Since the interrupt that causes the computer to go to state B occurs at any time relative to the path execution, there are more paths. Consider the case when the time input equals one. We previously had two possible paths: 12123456789A and 123456789A. If state B can occur between any of the other states, we now have the following possibilities for path 12123456789A:

> 1B2123456789A, 12B123456789A, 121B23456789A,
>
> 1212B3456789A, 12123B456789A, 121234B56789A,
>
> 1212345B6789A, 12123456B789A, 121234567B89A,
>
> 1212345678B9A, 12123456789BA, and 12123456789AB

Similar possibilities exist for path 123456789A:

> 1B23456789A, 12B3456789A, 123B456789A, 1234B56789A,
>
> 12345B6789A, 123456B789A, 1234567B89A, 12345678B9A,
>
> 123456789BA, and 123456789AB

Two paths became 22! The path space increases by an order of magnitude.

When more than one asynchronous task is added to the computer, it is possible that the second task can interrupt the path of the first. The second task also can interrupt the main path. The number of paths goes up dramatically (we estimate one or more orders of magnitude for each asynchronous task, depending on program size). The possibility of testing all possible paths disappears.

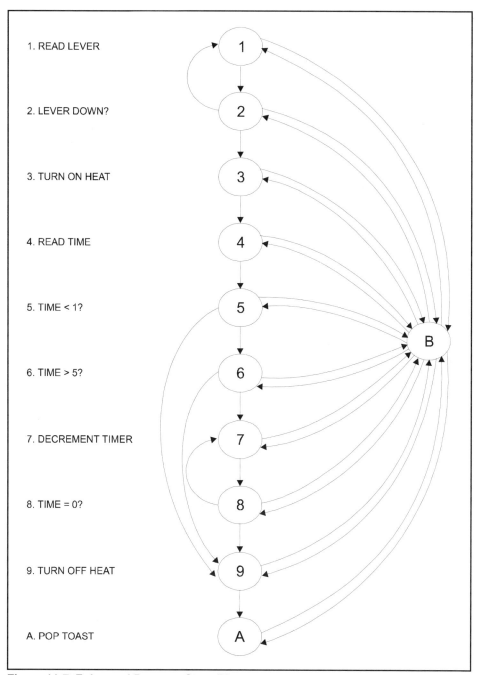

Figure 11-7. Enhanced Program State Diagram.

Not testing all paths is a problem only if one of those paths results in some unacceptable operation of the computer. This happens. Software designers often cannot foresee the interaction of these asynchronous tasks. The probability of interaction increases whenever common resources, such as memory, are shared by different tasks. Asynchronous tasks should not be used as a design solution unless necessary. When asynchronous tasks are necessary, shared resources such as memory or I/O devices should be eliminated.

We have decided to enhance our toaster with a computer communications port so that a voice-recognition system can automatically adjust the toast time according to each person's preferences. The communications port can interrupt program flow at any time. Once the program starts following the path required to receive an instruction from the other system, that path cannot be interrupted. To solve this problem, the computer logic gives the communication interrupt the highest priority. The communication protocol normally takes less than 1 sec to complete a message. Under error conditions, such as electrical noise, messages are repeated up to 10 times. Four messages are required to complete a transaction.

One morning Emily went into the kitchen and said, "Light toast," expecting a 10-sec heat time. She turned on the blender (generating lots of electrical noise) and pushed down the toast. Tyree walked in just then and said, "Medium toast next." The message from the voice-recognition system interrupted the toaster as its main program was decrementing the timer. Because of the blender noise, the messages took 40 sec. The toast, given 50 sec of heat instead of 10 sec, was burned. This functional failure occurred because the computer followed an unanticipated, untested path.

Input Space

So far, we have viewed computer operation in terms of state space and path space. The computer is successful if it stays in successful states. The computer is successful if it follows successful paths. Other states or other paths cause functional failure. In Ref. 11, Dr. J. V. Bukowski explains that there is a third way of looking at the problem: Consider the inputs. An input condition or sequence of input conditions will cause a particular path to be followed. Programs accept "input" from both the external world and its memory. The input space is the collection of all possible input sequences (Ref. 11). State space, path space, and input space are three different ways to view program operation.

Programs often perform many functions. A word processor does more than put text on a computer screen. Inputs received by a program vary depending on the function being performed. The input space view of program operation offers an advantage in that program usage can be estimated in terms of the functions being performed. This approach can be useful in explaining why some computer installations have software failure rates that are vastly different from others. The functions being performed are different. Usage is different. The paths utilized are different. A high software failure rate at one site does not indicate that every site will have a high failure rate. The input spaces at these sites are likely to be different. Such considerations can be very important when quantizing software failure rates.

Software Reliability Modeling

Many different methods of quantizing software reliability exist (Refs. 12-16). Early models measured "time between failures." Other models have counted failures during specific time intervals, and there are many variations. These models serve a number of useful purposes. For example, they are used to measure expected field failure rate. This information is needed for accurate system-level reliability evaluations. The models provide information to prospective buyers. They indicate the level of software quality. The models provide information useful in comparing new software development processes.

One of the primary concepts behind software reliability modeling is that the software design goes through a period of testing in which failures and failure times are recorded. The test period is typically during a formal test planned to uncover software design faults. Most models assume that design faults are repaired as soon as they are discovered. Thus, this test period represents a reliability growth process. The expected failure rate drops in some manner as reliability grows.

Basic Model

Versions of the basic model were developed independently by both Jelinski-Moranda and Shooman and were first published in 1972 (Ref. 12). The model depends on the measurement of time between failures during a software test. It assumes that there is some quantity of software design faults at the beginning of the test. These faults are assumed to be independent of each other, and to be equally likely to cause failure. The model assumes an ideal repair process in which a detected fault is repaired in negligible time and no new faults are introduced during the fault-removal process. Like many models, it assumes that the failure rate is proportional to the current number of faults in the program. A constant, k, relates failure rate to the number of faults that remain in the program. This is illustrated in Figure 11-8.

During the test period, the failure rate equals

$$\lambda(n_c) = k[N_0 - n_c(t)] \tag{11-2}$$

N_0 is the number of faults at the beginning of the test period. The cumulative number of faults that have been repaired is given by $n_c(t)$. Faults remaining in the software at any time during the test period are calculated by subtracting $n_c(t)$ from N_0. A proportionality constant, k, relates the remaining faults to the failure rate. After the test period, faults are not detected and repaired. The failure rate then remains constant.

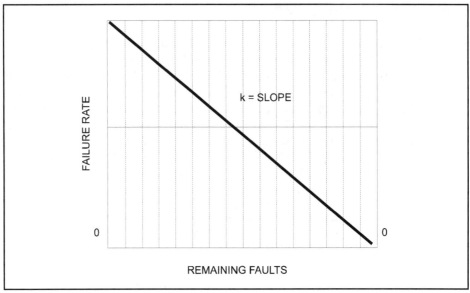

Figure 11-8. Failure Rate versus Faults.

EXAMPLE 11-2

Problem: A program has an estimated 100 software design faults. Previous programs have had a 0.01 ratio between remaining faults and field failure rate per hour. The following data were gathered during a formal software stress test:

First failure	5 hr
Second failure	12 hr
Third failure	21 hr
Fourth failure	38 hr
Fifth failure	55 hr

Plot the failure rate as a function of time during the test phase.

Solution: Using Equation 11-2, calculate the projected failure rate for each time interval between failure detection. Between failure discovery times, the failure rate is constant. After the first failure detection, the failure rate is

$$\lambda(1) = 0.01(100-1)$$
$$= 0.99$$

After the second failure detection, the failure rate is

$$\lambda(2) = 0.01(100-2)$$
$$= 0.98$$

EXAMPLE 11-2 (continued)

After the third failure detection, the failure rate is

$$\lambda(3) = 0.01(100 - 3)$$
$$= 0.97$$

After the fourth failure detection, the failure rate is

$$\lambda(4) = 0.01(100 - 4)$$
$$= 0.96$$

After the fifth failure detection, the failure rate is

$$\lambda(5) = 0.01(100 - 5)$$
$$= 0.95$$

These values are plotted in Figure 11-9.

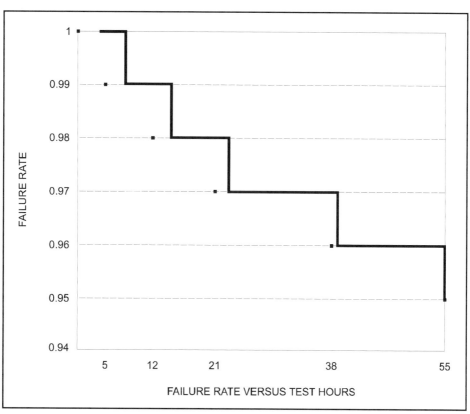

Figure 11-9. Failure Rate versus Test Hours.

Basic Execution Time Model

The basic execution time (BET) model is an enhanced version of the basic model. It was published in 1975 by J. D. Musa (Ref. 13). This model makes a careful distinction between computer execution time and calendar time. Although these times are equal when the computer spends all its time on one program, the times are not equal when a program is run on a time-shared system. The assumptions made by this model are similar to those of the basic model.

During the test phase of a project, the failure rate equals

$$\lambda(n_c) = k[N_0 - n_c(\tau)] \qquad (11\text{-}3)$$

where τ represents execution time. Normally, for control system computers, execution time equals calendar time, since the computers are dedicated to the control function.

During testing, the cumulative number of detected faults (failures) as a function of execution time is recorded. However, this information does not directly provide the values of N_0 and k that are needed in order to use Equation 11-3. To obtain the necessary parameters, an analytical formula for n_c as a function of execution time must be developed. Given such a formula, parameters are chosen that provide a best-fit curve for the data.

Proceeding with the model development, note that within both the basic model and the basic execution time model, it is assumed that the detection of any fault is equally likely. This means that in any given time period, a constant percentage of faults will be detected. To illustrate the concept, we estimate that a program has 1,000 faults (5,000 lines of source multiplied by 0.2 faults per thousand lines). Assume that each week we find a constant 25% of the remaining faults. Table 11-2 lists the results.

If we plot the cumulative number of faults found, we see a curve that appears to be an exponential rise. A plot of the faults remaining appears to be exponentially decreasing (Figure 11-10). Since the failure rate is proportional to the faults remaining, an exponentially decreasing failure rate would be expected.

Since the number of faults detected (and repaired) each period is a constant percentage of faults remaining, we can state:

$$N_{Detected} = kN_{Remaining} \qquad (11\text{-}4)$$

The number of faults remaining equals the total starting number of faults (N_0) minus the cumulative quantity of repaired faults. Therefore:

$$N_{Detected} = k[N_0 - n_c(\tau)] \qquad (11\text{-}5)$$

Software Reliability

Week	Remaining Faults	Faults Found	Cumulative Faults
0	1,000	0	0
1	750	250	250
2	563	187	437
3	423	140	577
4	318	105	682
5	239	79	761
6	180	59	820
7	135	45	865
8	102	33	898
9	77	25	923
10	58	19	942
11	44	14	956
12	33	11	967
13	25	8	975
14	19	6	981
15	15	4	985
16	12	3	988
17	9	3	991
18	7	2	993
19	6	1	994
20	5	1	995

Table 11-2. Program Formal Test Data

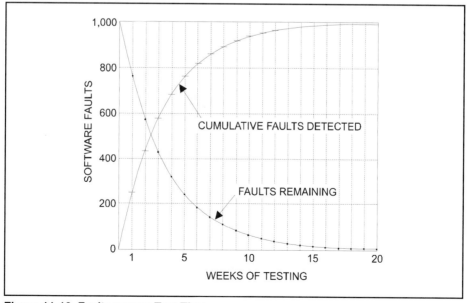

Figure 11-10. Faults versus Test Time.

The change in cumulative quantity of repaired faults in each time period equals the number detected during that time period. This change can be substituted into Equation 11-5, yielding:

$$\frac{dn_c}{d\tau} = k[N_0 - n_c(\tau)] \qquad (11\text{-}6)$$

This differential equation can be solved to obtain the cumulative quantity of repaired faults as a function of execution time:

$$n_c(\tau) = N_0[1 - e^{-k\tau}] \qquad (11\text{-}7)$$

Substituting Equation 11-7 into Equation 11-3 yields:

$$\lambda(\tau) = kN_0 e^{-k\tau} \qquad (11\text{-}8)$$

The failure rate is indeed an exponentially decreasing function of execution time.

The parameters of Equation 11-8 can be estimated by curve-fitting. The test data are plotted on a graph. A curve generated using Equation 11-8 is also plotted. The curve-fit can be done graphically. Several mathematical methods to fit data optimally also exist. Spreadsheet programs have such "data regression" methods available.

EXAMPLE 11-3

Problem: We are given test data obtained from a software test in which several testers exercised the software. Each day, the total number of test hours were recorded along with the total number of failures. Calculate the software failure rate using a BET model. Assume that test time equals execution time. The test data are given in Table 11-3.

Solution: The data are plotted in Figure 11-11. Using a simple spreadsheet, curves generated using Equation 11-8 are superimposed on the graph. The curve is varied until it seems to fit well.

The parameters used in the curve are $N_0 = 140$ and $k = 0.01$. Using these parameters along with our total execution time of 400 hours in Equation 11-8, a failure rate is obtained.

$$\lambda(400) = 1.4 e^{-0.01 \times 400}$$
$$= 0.026$$

Overall, the test data show that at the end of the 400-hr test period a field failure rate of 0.026 failures/hour is expected. It is estimated that 140 faults were originally in the program. Since we have discovered 121 faults, an estimated 19 faults remain in the program.

Software Reliability

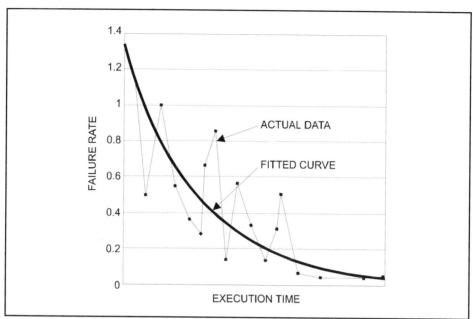

Figure 11-11. Failure Rate versus Test Time.

Day	Test Hours	Cumulative Test Hours	Failures	Failures/Hour
1	20	20	22	1.1
2	16	36	8	0.5
3	18	54	18	1
4	22	76	12	0.545455
5	25	101	9	0.36
6	14	115	4	0.285714
7	6	121	4	0.666667
8	14	135	12	0.857143
9	14	149	2	0.142857
10	16	165	9	0.5625
11	18	183	6	0.333333
12	22	205	3	0.136364
13	13	218	4	0.307692
14	4	222	2	0.5
15	30	252	2	0.066667
16	28	280	1	0.035714
17	60	340	2	0.033333
18	24	364	1	0.041667
19	22	386	0	. . .
20	14	400	0	. . .

Table 11-3. Data from Software Test

> **EXAMPLE 11-4**
>
> **Problem:** The boss will not accept our graphic curve-fitting technique. Redo Example 11-3 using numerical data regression.
>
> **Solution:** The spreadsheet we are using has a linear data regression capability. We can take the natural log of our data and convert it into a linear line. For example, if an exponential formula has the form:
>
> $$\lambda(\tau) = Ae^{B\tau} \qquad (11\text{-}9)$$
>
> then taking the log of both sides of the equation results in
>
> $$\ln \lambda(\tau) = \ln A + B\tau$$
>
> The slope of the line is calculated to be –0.0099 and the constant is 0.3299. Taking the exponent of 0.3299 results in an answer of 1.390834, which equals kN_0. The slope of the line is –0.0099, which is the value of k.

Logarithmic Poisson Model

One of the assumptions of the BET model is that all faults are equally likely to be detected and repaired. In many programs, especially those that are large in size with many asynchronous tasks, this assumption cannot be made. The author has many examples of software faults that have been present in programs for many years. Clearly, faults that are not found after hundreds of hours of stressful testing will be much harder to find than those quickly discovered.

The logarithmic Poisson (LP) model assumes that failure rate variation as a function of cumulative repaired faults is an exponential:

$$\lambda(n_c) = \lambda_0 e^{-\Theta n_c} \qquad (11\text{-}10)$$

where Θ is the failure intensity decay parameter, and λ_0 is the initial failure rate. Compare this equation with Equation 11-3. The difference between the LP model and the BET model is shown in Figure 11-12.

For the LP model, the graph indicates that some faults are more likely to cause failures. When they are removed, the failure rate drops quickly. Other faults are less likely to cause failure. As they are removed, the failure rate drops more slowly. We should also note that the rate does not drop significantly past a certain point in the fault-removal process. This is characteristic of a process in which some faults will be introduced when errors are repaired. It is realistic to expect some new faults to be introduced whenever changes are made to software.

Software Reliability

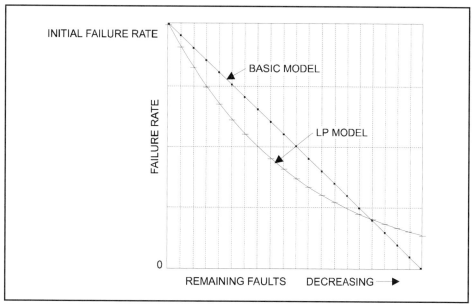

Figure 11-12. Failure Rate versus Faults.

The cumulative number of repaired faults for the LP model is given by:

$$n_c(\tau) = \frac{1}{\Theta}\ln(\lambda_0\Theta\tau + 1) \qquad (11\text{-}11)$$

Compare this with Equation 11-7 for the BET model. The cumulative number of repaired faults for both models is plotted in Figure 11-13.

The expected number of repaired faults tends toward N_0 for the BET model as execution time increases. For the LP model, however, the expected number of repaired faults tends toward infinity. This is characteristic of large, complex programs with a path space of "virtual infinity." If new faults are introduced when old faults are repaired, the cumulative number of repaired faults will tend toward infinity. It has been argued that programs used by many persons for a long period of time are more reliable. Programs that follow a LP reliability model will not support this argument. Given the large, complex path space, the faults will never be repaired beyond a certain point.

Failure rate, as a function of execution time, can be obtained for the LP model. That solution provides:

$$\lambda(\tau) = \frac{\lambda_0}{\lambda_0\Theta\tau + 1} \qquad (11\text{-}12)$$

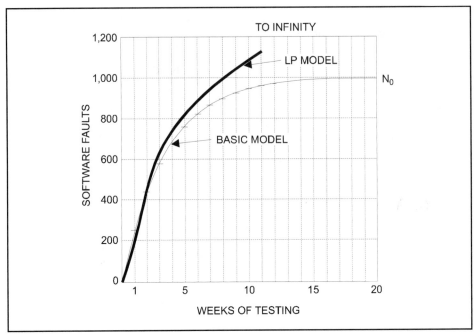

Figure 11-13. Faults versus Test Time.

The LP model is used in a manner similar to the BET model. Failure times are recorded during product tests. Parameters are estimated by best-case curve-fitting. Figure 11-14 shows the data from Table 11-2 fitted with both a BET curve and an LP curve.

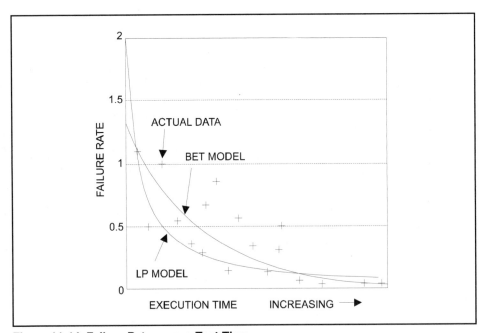

Figure 11-14. Failure Rate versus Test Time.

Operational Profile

Earlier in this chapter we learned the concept of an input space — the collection of all possible inputs. These inputs cause the computer to execute different paths through its program. Some failures occur only when certain inputs are received by the computer. Often, sets of inputs are grouped according to computer function. Commands that tell the computer to execute certain functions, followed by input data, represent a logical grouping within the input space.

Computer usage is different at different sites. This can be represented by a probability profile of input groupings, known as an *operational profile*. An example of an operational profile would be a scaled version of Figure 2-1, the computer usage histogram. Each computer function is characterized by a set of distinct inputs.

Estimates of the operational profile for a particular computer program can be used to direct testing efforts. More testing can be done on sections of the program that will get highest usage. Testing on each section of the program can be modeled separately. A failure rate can be calculated for each part. The overall failure rate could be calculated using a weighted average based on the operational profile. One strong advantage of this approach is that the overall failure rate could be recalculated for each customer with a distinctively different operational profile.

Software Reliability Model Assumptions

Many assumptions were mentioned when the concepts behind software reliability models were presented earlier in this chapter. Since many of these assumptions violate the common sense of professional programmers, the assumptions and their consequences need to be discussed in more detail. It will be seen that many assumptions are indeed violated during practical software testing. However, as A. L. Goel explains in Reference 14, none of the violations prevents us from using the models, especially if we understand the effects of assumption violation and account for them.

Faults Are Independent

Independence is generally assumed in software reliability models in order to simplify the model. Dependent probabilities are not required. A study of how faults are introduced reveals many reasons: misunderstood functionality, design error, coding error, etc. These reasons generally result in independent faults. Occasionally, dependent faults occur, but we have no strong reason to doubt this assumption.

Times between Failures Are Independent

This assumption would be valid if the input data set (or path) for each test were chosen randomly. Some testing is done randomly, but a majority of software testing is done by following carefully made plans. Whenever a failure occurs, testing in that area is often intensified.

We conclude, then, that this assumption is not valid for most testing processes. Two effects have been attributed to this assumption violation. First, the data are generally "noisy." The actual data in Figures 11-11 and 11-14, for example, bounce around quite a bit. This could result from nonrandom testing.

"Restarts" also result from nonrandom testing. As a new area of software is tested, the failure rate suddenly goes up (see Figure 11-15). A similar result has been observed when new test personnel are added to a project. These results do not prevent us from using the models; we must merely realize that our plots will be noisy.

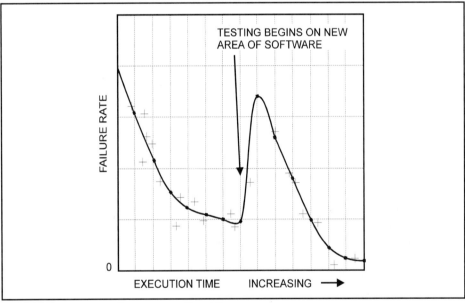

Figure 11-15. Failure Rate versus Test Time.

Detected Fault Removed in Negligible Time

It does require time to recode or redesign software to remove faults. This time can even be quite variable. The assumption of detected fault removal in negligible time is typically violated in real projects. This is not a serious problem, however. Testing can usually continue with little effect. In the worst case, duplicate faults will be reported. These can be deleted from the database.

No New Faults Are Introduced

The BET model assumes that new faults are not introduced when repairs are made to the software. Experience has shown that some new faults can be introduced when software repairs are made. This is more likely to be true in larger, more complicated programs and programs with poor documentation.

New fault introduction does not prevent us from using the model; it merely changes the parameters. If the new fault introduction rate is large, the failure rate will not decline. This is an indication that something more drastic needs to be done. Redesign or abandonment of the program is in order.

Faults Are Equal

The BET model also assumes that all faults are equally likely to cause failure. Experience has shown this not to be true. Some key faults can cause failure quickly. Others may remain hidden for years.

Although the assumption is not met in practice, again the effect is minimal. One such effect is that actual data do not correlate well with the best-fit curve. When this happens, alternatives exist. We can switch to the LP model or model each major element of the operational profile as a separate program.

Model Usage

Several studies have been done using these models, as well as a few others, on actual software test data. The results have been moderately successful. Usually, one model fits better than others for a particular piece of software. The characteristics of the software that allow one model or another to work best have not yet clearly been identified. Some patterns seem to be forming.

The LP model seems to fit better for larger, more complicated programs, especially when the operational profile is not uniform. The BET model seems to fit better when program size is changing. Research continues in this important area.

EXERCISES

11.1 List several system failures that are caused by software.

11.2 Describe how software failures can be modeled statistically.

11.3 What "strength" factors can be attributed to a software design?

11.4 What "stress" factors cause a program to fail?

11.5 Estimate how much software complexity has grown in the last decade.

11.6 Calculate the number of control flow paths in the program diagrammed in Figure 11-16.

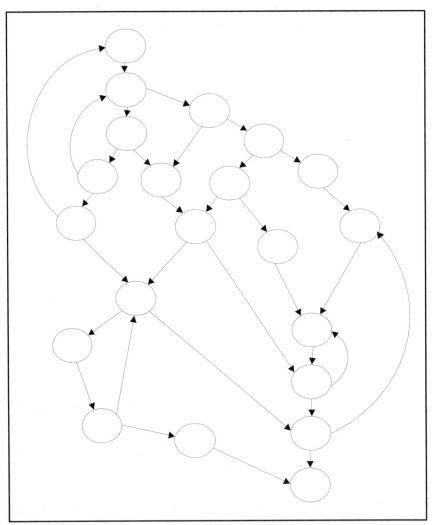

Figure 11-16. Program States.

11.7 Are all software design faults detected by executing all control paths?

11.8 How many control flow paths are present in the program diagrammed in Figure 11-7?

11.9 We have tested a program for 500 execution hours. A BET model shows a good correlation using parameters of $k = 0.01$ and $N_0 = 372$. If we continue testing and repairing problems for another 200 hr, what field failure rate would be expected?

11.10 The program in Exercise 11.9 has a customer-specified maximum software failure rate of 0.0001 failures per hour. How many hours of testing and repair will be required to meet the specification?

11.11 We have tested a program for 1,000 execution hours. An LP model shows a good correlation using parameters of $\lambda_0 = 2$ and $\Theta = 0.03$. What is the expected field failure rate?

11.12 We have a software reliability goal of 0.0001 failures per hour for the program in Exercise 11.11. How many test hours are required?

11.13 Is it reasonable to expect the program in Exercise 11.11 to ever achieve the reliability goal of 0.0001 failures per hour if new faults are added when old faults are repaired?

REFERENCES AND BIBLIOGRAPHY

1. Gray, J., "A Census of Tandem Availability between 1985 and 1990," *IEEE Transactions on Reliability*, Vol. 39, No. 4, October 1990.

2. Siewiorek, D. P., "Fault Tolerance in Commercial Computers," *Computer*, July 1990.

3. Crawford, S. G., and Fallah, M. H., "Software Development Process Audits — A General Procedure," *Proceedings of the 8th International Conference on Software Engineering*, New York: IEEE, 1985.

4. Humphrey, W. S., "Characterizing the Software Process: A Maturity Framework," *IEEE Software*, March 1988.

5. Johnson, D. A., Goble, W. M., and Stover, B. D., "Software Reliability in Control System Instrumentation," *Proceedings of the 14th Annual Advanced Industrial Control Conference*, West Lafayette, IN: Purdue University, 1988.

6. McCabe, T. J., "A Complexity Measure," *IEEE Transactions on Software Engineering*, Vol. SE-2, 1976.

7. Musa, J. D., and Everett, W. W., "Software-Reliability Engineering: Technology for the 1990's," *IEEE Software*, November 1990.

8. Nuefelder, A. M., *Ensuring Software Reliability*, New York: Marcel Dekker, 1993.

9. DIN V VDE 0801, *Principles for Computers in Safety-Related Systems*, October 1991.

10. IEC61508, *Functional Safety — Safety Related Systems*, 1997.

11. Bukowski, J. V., "Evaluating Software Test Results: A New Approach," *1987 Proceedings of the Annual Reliability and Maintainability Symposium*, New York: IEEE, 1987.

12. Shooman, M. L., "Software Reliability: A Historical Perspective," *IEEE Transactions of Reliability*, Vol. R-33, No. 1, 1984.

13. Musa, J. D., "A Theory of Software Reliability and Its Application," *IEEE Transactions on Software Engineering*, Vol. SE-1, No. 3, September 1971.

14. Goel, A. L., "Software Reliability Models: Assumptions, Limitations, and Applicability," *IEEE Transactions on Software Engineering*, Vol. SE-11, December 1985.

15. Musa, J. D., "Tools for Measuring Software Reliability," *IEEE Spectrum*, February 1989.

16. Musa, J. D., Iannino, A., and Okumoto, K., *Software Reliability: Measurement, Prediction, Application*, New York: McGraw-Hill, 1987.

17. Suydam, W. E., "Approaches to Software Testing Embroiled in Debate," *Computer Design*, November 15, 1986.

18. Ehrlich, W. K., Stampfel, J. P., and Wu, J. R., "Application of Software Reliability Modeling to Product Quality and Test Process," *Proceedings of the 12th International Conference on Software Engineering*, New York: IEEE, 1990.

19. Downs, T., and Garrone, P., "Some New Models of Software Testing with Performance Comparisons," *IEEE Transactions on Reliability*, Vol. 40, No. 3, 1991.

20. Gill, G. K., and Kemerer, C. F., "Cyclomatic Complexity Density and Software Maintenance Productivity," *IEEE Transactions on Software Engineering*, Vol. 17, No. 12, December 1991.

21. Lakshmanan, K. B., Jayaprakash, S., and Sinha, P. K., "Properties of Control-Flow Complexity Measures," *IEEE Transactions on Software Engineering*, Vol. 17, No. 12, December 1991.

22. Laprie, J. C., and Kanoun, K., "X-Ware Reliability and Availability Modeling," *IEEE Transactions on Software Engineering*, Vol. 18, No. 12, February 1992.

23. Elbert, M., Howe, R. C., and Weyant, T., "Software Reliability Measurement," *Proceedings of the 29th Annual Spring Reliability Seminar*, Boston Chapter of the Reliability Society, IEEE, 1991.

24. Avizienis, A., "Systematic Design of Fault Tolerant Computers," *SAFECOMP '96, Proceedings of the 15th International Conference on Computer Safety, Reliability and Security*, New York: Springer-Verlag, October 1996.

12 Modeling Detail

Key Issues

The amount of detail included in a safety and reliability model depends on the objectives of the modeling. The amount of effort (cost!) is also affected by the level of detail in the model, but it should be noted that costs depend much more on available computer tools. Key issues to be considered include:

1. Number of relevant failure modes
2. Availability of on-line diagnostics in the equipment being modeled
3. Level of detail for system components
4. Amount of redundancy
5. Common cause
6. Imperfect inspection and repair

The objective of the modeling is to account for the important factors and to eliminate the rest in order to keep the model to a manageable size. This may sound easy, but it is not. Sometimes the important factors are not obvious.

The objective is to simplify and approximate the appropriate amount of detail. Simplifications and approximations are especially useful when they reduce complexity and allow a model to become understandable. Simplifications and approximations are reasonable when they produce accurate or even pessimistic results — higher failure probabilities. If the pessimistic numbers meet the goal, the work is finished. One must be especially careful with simplifications and approximations, however, because they can produce optimistic and possibly misleading results.

In this chapter, a simple system consisting of two switches wired in series (1oo2, normally energized) will be analyzed for the probability of failing dangerously (both switches must fail dangerously, i.e., short circuit). In safety instrumented systems (SIS), this measure is called probability of failure on demand (PFD). Another way to express this measure is called the risk reduction factor (RRF). The RRF equals one divided by the PFD. Each switch has two failure modes: open circuit (safe, de-energized) and short circuit (dangerous, energized). Various levels of detail will be added using different modeling techniques and the answers can be compared.

Probability Approximations

When the failure rate of a component is known, the probability of failure for a given time interval is approximated by multiplying the failure rate times the time interval (see the section "A Useful Approximation" in Chapter 4). In SISs it is good practice to inspect the system for failures periodically. In this situation, the time interval used in the calculation is the inspection period. While this method is an approximation, errors tend to be in the pessimistic direction; therefore, the method is conservative. The failure probabilities for system components can be used in reliability block diagrams or fault tree diagrams to calculate system failure probabilities.

Figure 12-1 shows a simple fault tree for the 1oo2 series wired switch system. The probability of short-circuit (dangerous) failure for one switch is approximated by multiplying the short-circuit failure rate, λ^D, times the periodic inspection interval, TI (the mission time). The system fails short circuit only if both switches A and B fail short circuit. Therefore, the probability of the system failing short circuit (dangerous) is given by

$$\text{PFD} = (\lambda^D \times TI)^2 \qquad (12\text{-}1)$$

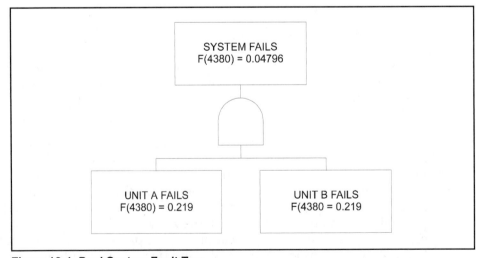

Figure 12-1. Dual System Fault Tree.

Modeling Detail

This simple fault tree assumes that there is no common cause. It assumes that perfect repairs are made at each inspection period. It does not account for more rapid repairs made if diagnostics detect the failure. The model assumes constant failure rates.

Example 12-1

Problem: A system consists of two switches wired in series (1oo2). Both must fail short circuit (dangerous) for the system to fail short circuit (dangerous). The constant failure rate for short-circuit failures is 0.00005 failures per hour. The system operates for time intervals (*TI*) of six months and is then inspected and restored to full operation if any failures are found. Using a probability approximation technique, what is the probability that the system will fail dangerously during the six-month operating time interval? What is the RRF?

Solution: The approximate probability of dangerous failure for one switch is found by using Equation 4-20:

$$PFD_{one}(4380) = \lambda^D \times TI$$
$$= 0.00005 \times 4380$$
$$= 0.219$$

Since both units must fail for the system to fail, system probability of failure using Equation 12-1 is

$$PFD(4380) = (\lambda^D \times TI)^2$$
$$= 0.219 \times 0.219$$
$$= 0.048$$

$$RRF = \frac{1}{PFD}$$
$$= 20.8$$

Example 12-2

Problem: The 1oo2 system of Example 12-1 is built using switches of much higher strength (lower failure rate). The failure rate for short-circuit (dangerous) failures of this higher-strength switch is 0.000005 failures per hour ($MTTF^D$ = 22.8 years). Using a probability approximation technique, what is the probability that this system will fail short circuit (dangerous) during the six-month time interval?

Solution: Using Equation 12-1,

$$PFD(4380) = (\lambda^D \times TI)^2$$
$$= 0.0219 \times 0.0219$$
$$= 0.00048$$

$$RRF = \frac{1}{PFD}$$
$$= 2085$$

The average PFD over the time interval may be the required calculation. The average approximation is given by:

$$PFD_{avg}(t) = \frac{1}{t}\int_0^t (\lambda^D t')^2 dt'$$

substituting $t = TI$

$$PFD_{avg}(TI) = \frac{1}{TI}\int_0^{TI} (\lambda^D t')^2 dt$$

and integrating

$$PFD_{avg}(TI) = \frac{1}{TI}\left[(\lambda^D)^2 \frac{t^3}{3}\right]_0^{TI}$$

which evaluates to

$$PFD_{avg} = \frac{(\lambda^D)^2 \times TI^2}{3} \tag{12-2}$$

Example 12-3

Problem: Using a probability approximation technique, what is the PFD$_{avg}$ and the RRF using the data from Example 12-1?

Solution: Using Equation 12-2,

$$PFD_{avg} = \frac{(0.00005 \times 4380)^2}{3}$$

$$= 0.01599$$

$$RRF = 62$$

Example 12-4

Problem: Using a probability approximation technique, what is the PFD$_{avg}$ and the RRF using the data from Example 12-2?

Solution: Using Equation 12-2,

$$PFD_{avg} = \frac{(0.000005 \times 4380)^2}{3}$$

$$= 0.0001599$$

$$RRF = 6255$$

Probability Approximation with Common Cause

The two switches are identical technology and are mounted side by side. In such circumstances, common-cause failures must be considered (Chapter 10). The dangerous failure rate is divided into normal (independent) failures and common-cause failures. The simplest model to use for this is the beta model. The failure rate is multiplied times β to get the common-cause failure rate. The failure rate is multiplied times $1 - \beta$ to get the normal (independent) failure rate.

$$\lambda^{DN} = (1 - \beta)\lambda^D \tag{12-3}$$

and

$$\lambda^{DC} = \beta\lambda^D \tag{12-4}$$

A new fault tree is shown in Figure 12-2. The equation for PFD using this fault tree is

$$\text{PFD} = (\lambda^{DN})^2 \times TI^2 + \lambda^{DC} \times TI \tag{12-5}$$

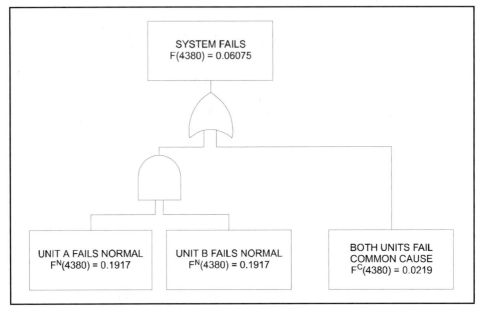

Figure 12-2. Dual System Fault Tree with Common Cause.

Example 12-5

Problem: Using a probability approximation technique, recalculate the PFD and the RRF with data from Example 12-1 and accounting for common cause. Assume a beta factor of 0.1.

Solution: The failure rate of 0.00005 failures per hour must be divided into normal-mode failures and common-cause failures. Using Equations 12-3 and 12-4,

$$\lambda^{DN} = (1 - \beta)\lambda^D$$
$$= 0.000045$$

$$\lambda^{DC} = \beta\lambda^D$$
$$= 0.000005$$

Modeling Detail

Example 12-5 (continued)

The probability of normal-mode switch failure equals $\lambda^{DN} \times t = 0.000045 \times 4380 = 0.1971$, and the probability of a common]-cause failure equals $\lambda^{DC} \times t = 0.000005 \times 4380 = 0.0219$. Using Equation 12-5,

$$PFD = 0.1971 \times 0.1971 + 0.0219$$
$$= 0.06075$$

$$RRF = \frac{1}{PFD}$$
$$= 16.4$$

Example 12-6

Problem: Using a probability approximation technique, recalculate the PFD and the RRF with data from Example 12-2 and accounting for common cause using a beta factor of 0.1.

Solution: The failure rate of 0.000005 failures per hour must be divided into normal-mode failures and common-cause failures. Using Equations 12-3 and 12-4,

$$\lambda^{DN} = (1-\beta)\lambda^{D}$$
$$= 0.0000045$$

$$\lambda^{DC} = \beta\lambda^{D}$$
$$= 0.0000005$$

The probability of normal-mode switch failure equals $\lambda^{DN} \times t = 0.0000045 \times 4380 = 0.01971$, and the probability of a common-cause failure equals $\lambda^{DC} \times t = 0.0000005 \times 4380 = 0.00219$. Using Equation 12-5,

$$PFD = 0.01971 \times 0.01971 + 0.00219$$
$$= 0.00258$$

$$RRF = \frac{1}{PFD}$$
$$= 387$$

This result should be compared with Example 12-2, where RRF was calculated without common cause for the same failure-rate data. That result was 2,085. This shows the importance of including common cause in the RRF evaluation, especially in higher safety integrity levels.

Modeling Detail

The equation for PFD$_{avg}$ can be derived by integrating the PFD equation and dividing by the time interval. The average approximation with common cause is given by:

$$\text{PFD}_{avg}(TI) = \frac{1}{TI} \int_0^{TI} [(\lambda^{DN} t')^2 + \lambda^{DC} t'] dt'$$

and integrating

$$\text{PFD}_{avg}(TI) = \frac{1}{TI} \left[(\lambda^{DN})^2 \frac{t^3}{3} + \lambda^{DC} \frac{t^2}{2} \right]_0^{TI}$$

which evaluates to

$$\left[(\lambda^{DN})^2 \frac{t^2}{3} + \lambda^{DC} \frac{t}{2} \right]_0^{TI}$$

Substituting the integration limits:

$$\text{PFD}_{avg} = \frac{(\lambda^{DN})^2 \times TI^2}{3} + \frac{\lambda^{DC} \times TI}{2} \tag{12-6}$$

Example 12-7

Problem: Using a probability approximation technique, calculate the PFD$_{avg}$ and the RRF using data from Example 12-5 and accounting for common cause with a beta factor of 0.1.

Solution: The short-circuit (dangerous) failure rate of the switch is 0.00005 failures per hour. The dangerous normal failure rate of the switch is 0.000045 failures per hour, and the common-cause failure rate is 0.000005 failures per hour. Using Equation 12-6,

$$\text{PFD}_{avg} = \frac{(0.1971 \times 0.1971)}{3} + \frac{0.0219}{2}$$

$$= 0.0239$$

$$\text{RRF} = \frac{1}{\text{PFD}_{avg}}$$

$$= 41.8$$

> **Example 12-8**
>
> **Problem:** Using a probability approximation technique, calculate the PFD_{avg} and the RRF using data from Example 12-6 and accounting for common cause using a beta factor of 0.1.
>
> **Solution:** The failure rates are $\lambda^{DN} = (1 - \beta)\lambda^D = 0.0000045$ and $\lambda^{DC} = \beta\lambda^D = 0.0000005$. Using Equation 12-6,
>
> $$PFD_{avg} = \frac{(0.01971 \times 0.01971)}{3} + \frac{0.00219}{2}$$
>
> $$= 0.001225$$
>
> $$RRF = \frac{1}{PFD}$$
>
> $$= 816$$

Simple Markov Model

The dual-switch system can also be modeled using a simple Markov model, as shown in Figure 12-3. Both switches are working in state 0. One switch has failed in state 1. Both switches have failed in state 2 and the system has failed dangerously. A "single repairman" is assumed. The **P** matrix for this model is

$$\mathbf{P} = \begin{bmatrix} 1 - 2\lambda^D & 2\lambda^D & 0 \\ \mu & 1 - (\lambda^D + \mu) & \lambda^D \\ 0 & \mu & 1 - \mu \end{bmatrix}$$

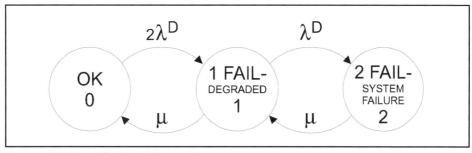

Figure 12-3. Dual System Markov Model.

Markov models can be solved for time-dependent probabilities or steady-state probabilities. To solve for steady state probabilities for this Markov model (a fully repairable system), a row matrix, $\mathbf{S}^L$, equals $\mathbf{S}^L \times \mathbf{P}$ (see Chapter 8, starting with Equation 8-5). In this case:

$$\begin{bmatrix} S_0^L & S_1^L & S_2^L \end{bmatrix} \begin{bmatrix} 1-2\lambda^D & 2\lambda^D & 0 \\ \mu & 1-(\lambda^D+\mu) & \lambda^D \\ 0 & \mu & 1-\mu \end{bmatrix} = \begin{bmatrix} S_0^L & S_1^L & S_2^L \end{bmatrix}$$

Multiplying (see Appendix B for matrix math) yields the following equations:

$$S_0^L \times (1-2\lambda^D) + S_1^L \times \mu = S_0^L$$

$$S_0^L \times (2\lambda^D) + S_1^L \times [1-(\lambda^D+\mu)] + S_2^L \times \mu = S_1^L$$

$$S_1^L \times \lambda^D + S_2^L \times (1-\mu) = S_2^L$$

and since state probabilities must always add to one:

$$S_0^L + S_1^L + S_2^L = 1$$

These can be simplified to:

$$-2\lambda^D S_0^L + \mu S_1^L = 0$$

$$\lambda^D S_1^L - \mu S_2^L = 0$$

and

$$S_0^L + S_1^L + S_2^L = 1$$

Note that one of the equations from the matrix multiplication was discarded. The equations resulting from a matrix multiplication are not all independent. Solving these equations for steady-state probabilities, the steady-state probability of being in state 2 (PFD$_{ss}$) at time equals infinity is

$$S_2^L = \text{PFD}_{SS} \tag{12-7}$$

$$= \frac{2(\lambda^D)^2}{2(\lambda^D)^2 + 2\lambda^D \mu + \mu^2}$$

Modeling Detail

The simple Markov model technique assumes constant failure rates, constant repair rates, no diagnostics, no common cause, a single failure mode, and perfect repair. In addition, it should be remembered that this equation presents an average probability at steady state, possibly a long time period.

The repair rate (μ) is the inverse of MTTR. In this case, the average time to repair is the sum of detection time and actual repair time. A failure could occur right before an inspection, right after an inspection, or any time in between. If this is equally likely at any moment of time, the average detection time is the inspection interval divided by two. The MTTR would then be

$$\text{MTTR} = \frac{TI}{2} + RT \qquad (12\text{-}8)$$

where TI equals the time interval for inspections and RT equals the actual time to repair the device.

Example 12-9

Problem: A normally energized dual-switch system wired in series (1oo2) uses two switches, each of which have a short-circuit (dangerous) failure rate of 0.00005 failures per hour. It is operated for six months, then inspected and repaired if problems are found. Actual repair time averages 8 hr. Using the steady-state solution for a simple Markov model, calculate the PFD$_{SS}$ and the RRF.

Solution: Using Equation 12-8, the repair rate is

$$\mu = \frac{1}{(4380/2) + 8}$$
$$= 0.000455$$

Substituting the values into Equation 12-7,

$$\text{PFD}_{SS} = \frac{2(0.00005)^2}{2 \times (0.00005)^2 + 2 \times 0.00005 \times 0.000455 + (0.000455)^2}$$
$$= 0.0194$$

$$\text{RRF} = \frac{1}{\text{PFD}}$$
$$= 51.5$$

> **Example 12-10**
>
> **Problem:** A normally energized dual-high strength switch system wired in series (1oo2) uses two switches, each of which have a dangerous failure rate of 0.000005 failures per hour. It is operated for six months, then inspected and repaired if problems are found. Actual repair time averages 8 hr. Using a steady-state solution from a simple Markov model, calculate the PFD_{SS} and the RRF.
>
> **Solution:** Using Equation 12-8, the repair rate is 0.000455. Substituting the values into Equation 12-7,
>
> $$PFD_{SS} = \frac{2(0.000005)^2}{2 \times (0.000005)^2 + 2 \times 0.000005 \times 0.0000455 + (0.0000455)^2}$$
>
> $$= 0.000236$$
>
> $$RRF = \frac{1}{PFD}$$
>
> $$= 4231$$

Time-dependent probabilities can also be calculated from a Markov model. However, in systems where repair is made after a periodic inspection, the modeling of repair must be done differently. In the time period during inspection, there is no repair. At the time of inspection, repair is made. If it is assumed that the repair is perfect, all components are operating and the system returns to state 0. The PFD is then zero. Imperfect repair can also be assumed, and the PFD can be lowered to some value greater than zero.

Time-dependent probabilities are calculated using the techniques in Chapter 8, where a starting-state probability row matrix is repeatedly multiplied by the **P** matrix. Since no repair occurs between inspections, the repair arc is removed from the model. The resulting **P** matrix is

$$\mathbf{P} = \begin{bmatrix} 1 - 2\lambda^D & 2\lambda^D & 0 \\ 0 & 1 - \lambda^D & \lambda^D \\ 0 & 0 & 1 \end{bmatrix}$$

When this matrix is multiplied by a starting row matrix of [1,0,0] repeatedly, a probability of being in state 2 (fail short circuit, dangerous) can be calculated as a function of time. The average PFD can also be numerically calculated. Figure 12-4 shows a plot of the PFD and the

PFD$_{avg}$ for several years using the data from Example 12-1. The PFD$_{avg}$ numeric calculation equals 0.0136.

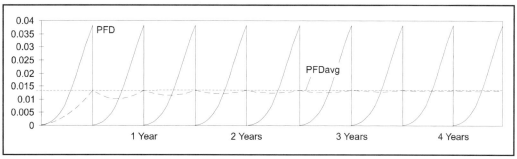

Figure 12-4. Time Dependent Probabilities.

Simple Markov Model with Common Cause

Common cause can be included in a simple Markov model. This is shown in Figure 12-5. The arc between state 0 and state 1 has two normal (independent) failures. The arc between state 0 and state 2 has the common-cause failure. The **P** matrix for this model is

$$\mathbf{P} = \begin{bmatrix} 1-(2\lambda^{DN}+\lambda^{DC}) & 2\lambda^{DN} & \lambda^{DC} \\ \mu & 1-(\lambda^{D}+\mu) & \lambda^{D} \\ 0 & \mu & 1-\mu \end{bmatrix}$$

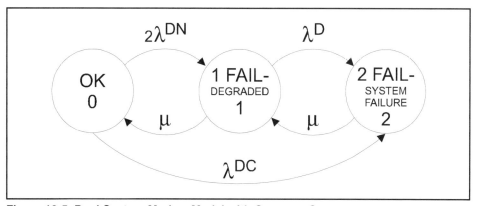

Figure 12-5. Dual System Markov Model with Common Cause.

This model can be solved for steady-state probabilities using the methods described in Chapter 8. The solution for steady-state probability of being in state 2 (fail-danger) is given by

$$\text{PFD}_{SS} = \frac{2\lambda^D \lambda^{DN} + \lambda^D \lambda^{DC} + \mu \lambda^{DC}}{\mu^2 + 2\mu\lambda^{DN} + 2\mu\lambda^{DC} + 2\lambda^D \lambda^{DN} + \lambda^D \lambda^{DC}} \qquad (12\text{-}9)$$

Note that if β goes to zero, $\lambda^{DC} = 0$ and $\lambda^{DN} = \lambda^D$. When these values are substituted into Equation 12-9, it reduces to Equation 12-7.

Example 12-11

Problem: Recalculate the PFD$_{SS}$ and RRF of Example 12-9 accounting for common cause with a beta factor of 0.1 using a steady-state solution to a simple Markov model with common cause.

Solution: Using Equation 12-9 and substituting the values $\lambda^D = 0.00005$, $\lambda^{DN} = 0.000045$, $\lambda^{DC} = 0.000005$, and $\mu_P = 0.000455$,

$$\text{PFD}_{SS} = 0.02731$$

$$\text{RRF} = 36.6$$

Example 12-12

Problem: Recalculate the PFD$_{SS}$ and RRF of Example 12-10 (high-strength switch) including the effects of common cause using a beta factor of 0.1.

Solution: Using Equation 12-9 and substituting the values, $\lambda^D = 0.000005$, $\lambda^{DN} = 0.0000045$, $\lambda^{DC} = 0.0000005$, and $\mu_P = 0.000455$,

$$\text{PFD}_{SS} = 0.0012996$$

$$\text{RRF} = 769$$

Multiple Failure Modes

The Markov model of Figure 12-3 assumed only one failure mode. The switch can fail open circuit (safe) or short circuit (dangerous). A more detailed Markov model that includes the two failure modes and no common cause is shown in Figure 12-6. This model accounts for both failure modes of the system. A Markov model can show multiple failure modes in one drawing. The **P** matrix for Figure 12-6 is

$$\mathbf{P} = \begin{bmatrix} 1 - (2\lambda^S + 2\lambda^D) & 2\lambda^D & 2\lambda^S & 0 \\ \mu_P & 1 - (\lambda^S + \lambda^D + \mu_P) & \lambda^S & \lambda^D \\ \mu_{SD} & 0 & 1 - \mu_{SD} & 0 \\ \mu_P & 0 & 0 & 1 - \mu_P \end{bmatrix}$$

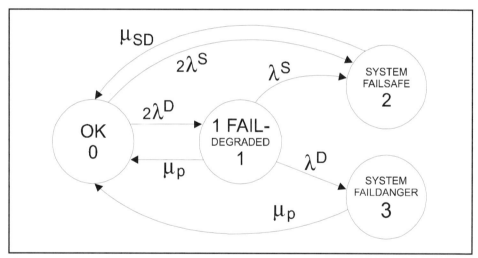

Figure 12-6. Dual System Markov Model with Multiple Failure Modes.

The system is operating correctly in state 0. If one switch fails dangerously, the system degrades into state 1, where the system is still successful. State 2 is a condition where the two-switch system has failed open circuit (safe) and shuts down the process. The system fails dangerously in state 3. The model shows two repair rates. After a shutdown (transition to state 2), we assume that the system is repaired and restarted immediately without waiting for a periodic inspection.

The model can be solved for steady-state probabilities. Solving for probability of being in state 3:

$$\text{PFD}_{SS} = \frac{2\mu_{SD}(\lambda^D)^2}{\mu_{SD}\mu_P\lambda^S + 3\mu_{SD}\mu_P\lambda^D + \mu_{SD}\mu_P^2 + 2(\lambda^S)^2\mu_P + 2\lambda^S\mu_P^2 + 4\lambda^S\lambda^D\mu_P + 2(\lambda^D)^2\mu_{SD}} \quad (12\text{-}10)$$

While this analytical equation is complex and thus does not provide much insight, note that the PFD_{SS} depends on a number of variables, including both repair rates, the safe failure rate, and the dangerous failure rate.

Example 12-13

Problem: A dual-switch system wired in series (1oo2) uses two switches, each of which has a dangerous failure rate of 0.00005 failures per hour and a safe failure rate of 0.00005 failures per hour. It is operated for six months, then inspected and repaired if problems are found. Actual repair time averages 8 hr. After a shutdown, the system is repaired and restarted in an average time period of 24 hr. What is the PFD_{SS}? What is the RRF?

Solution: Using Equation 12-8, the repair rate is calculated to be 0.000455. Substituting the values, $\lambda^D = 0.00005$, $\lambda^S = 0.00005$, $\mu_P = 0.000455$, and $\mu_{SD} = 1/24 = 0.041667$ into Equation 12-10,

$$\text{PFD}_{SS} = 0.016467$$

$$\text{RRF} = \frac{1}{\text{PFD}} = 60.7$$

Example 12-14

Problem: Repeat Example 12-13 using two high-strength switches, each of which has a short-circuit (dangerous) failure rate of 0.000005 failures per hour and an open-circuit (safe) failure rate of 0.000005 failures per hour. What is the PFD_{SS}? What is the RRF?

Solution: Substituting the values $\lambda^D = 0.000005$, $\lambda^S = 0.000005$, $\mu_P = 0.000455$, and $\mu_{SD} = 1/24 = 0.041667$ into Equation 12-10,

$$\text{PFD}_{SS} = 0.0002313$$

$$\text{RRF} = \frac{1}{\text{PFD}} = 4323$$

Multiple Failure Modes with Time-Dependent PFD

The Markov model in Figure 12-6 can be solved for time-dependent state probabilities using matrix multiplication techniques as described in Chapter 8. Again, since no periodic repair is done between inspections, the periodic repair rate is removed from the model. The modified Markov model is shown in Figure 12-7. The **P** matrix is

$$\mathbf{P} = \begin{bmatrix} 1-(2\lambda^S + 2\lambda^D) & 2\lambda^D & 2\lambda^S & 0 \\ 0 & 1-(\lambda^S + \lambda^D) & \lambda^S & \lambda^D \\ \mu_{SD} & 0 & 1-\mu_{SD} & 0 \\ 0 & 0 & 0 & 1 \end{bmatrix}$$

If it assumed that the model starts in state 0 (everything is working when we start the system), the initial starting-state matrix is

$$\mathbf{S}_0 = [\,1, 0, 0, 0\,]$$

The starting-state matrix is multiplied by the **P** matrix to obtain the state probabilities at time increment one. This is done repeatedly, obtaining state probabilities as a function of time.

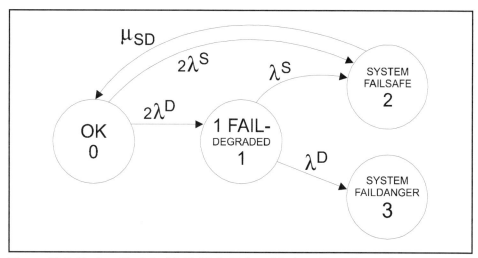

Figure 12-7. Markov Model with Multiple Failure Modes - Time Dependent.

Example 12-15

Problem: The dual-switch system of Example 12-13 is operated for six months, then inspected and repaired if problems are found. What is the time-dependent PFD? What is the RRF at 4,380 hr? What is the PFD_{avg} at 4,380 hr?

Solution: The problem can be quickly solved using a spreadsheet. Substituting the failure rates and repair rates into the **P** matrix gives:

$$\mathbf{P} = \begin{array}{c|cccc} & 0 & 1 & 2 & 3 \\ \hline 0 & 0.9998 & 0.0001 & 0.0001 & 0 \\ 1 & 0 & 0.9999 & 0.00005 & 0.00005 \\ 2 & 0.0416667 & 0 & 0.958333 & 0 \\ 3 & 0 & 0 & 0 & 1 \end{array}$$

Assuming that the system is operating correctly at startup (it starts in state 0), the **P** matrix is multiplied repeatedly by the **S** matrix, creating a table of values for PFD as a function of time. Figure 12-8 shows a table of data from a spreadsheet for the first few time increments. Spreadsheet size can be reduced by using larger time increments. The **P** matrix can be multiplied by itself several times to get a matrix that represents transition probabilities for larger time increments. For this example, the 16-hr increment is represented by $\mathbf{P}^{16}$:

$$\mathbf{P}^{16} = \begin{bmatrix} 0.997219 & 0.001597 & 0.001184 & 5.99\text{E-}07 \\ 0.000207 & 0.998401 & 0.000592 & 0.000799 \\ 0.493112 & 0.000414 & 0.506474 & 1.02\text{E-}07 \\ 0 & 0 & 0 & 1 \end{bmatrix}$$

Time	S0 Prob.	S1 Prob.	S2 Prob.	S3 Prob.	
0	1	0	0	0	Starting row vector - SRV
1	0.9998	0.0001	0.0001	0	S1={mmult(SRV, P)}
2	0.999604	0.0002	0.000196	5E-09	S2={mmult(S1, P)}
3	0.999412	0.0003	0.000288	1.5E-08	S3={mmult(S2, P)}
4	0.999225	0.0004	0.000376	3E-08	S4={mmult(S3, P)}
5	0.99904	0.0005	0.00046	5E-08	S5={mmult(S4, P)}
6	0.99886	0.0006	0.000541	7.5E-08	S6={mmult(S5, P)}

Figure 12-8. Spreadsheet Results for Time Step Probabilities.

Modeling Detail

The PFD$_{avg}$ can be calculated as a straight average of PFD values. In the spreadsheet, all PFD values are added and divided by the total number of values. This can be done as function of time on the spreadsheet with two additional columns. The PFD at 4,384 hr equals 0.036256. The RRF at 4,384 hr equals 27.6. The PFD$_{avg}$ at 4,384 hr equals 0.0129. The RRF using PFD$_{avg}$ equals 77. A plot of PFD and PFD$_{avg}$ as a function of time with no periodic inspection for a two-year time interval is shown in Figure 12-9.

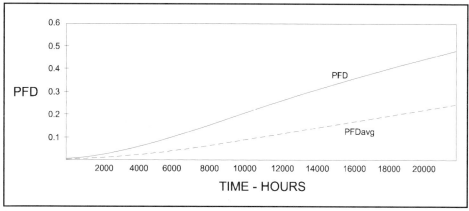

Figure 12-9. Time Dependent PFD.

Example 12-16

Problem: Repeat the dual-switch system of Example 12-15 using the high-strength switches of Example 12-14. What is the time-dependent PFD? What is the RRF at 4,380 hr? What is the PFD$_{avg}$ at 4,380 hr?

Solution: The problem can be quickly solved using a spreadsheet. Substituting the failure rates and repair rates into the **P** matrix gives:

		0	1	2	3
	0	0.99998	0.00001	0.00001	0
P =	1	0	0.99999	0.000005	0.000005
	2	0.0416667	0	0.958333	0
	3	0	0	0	1

The **P** matrix is multiplied repeatedly by the **S** matrix, creating a table of values for PFD as a function of time. The PFD at 4,384 hr equals 0.000467. The RRF at 4,380 hr equals 2,143. The PFD$_{avg}$ at 4,384 hr equals 0.0001568. The RRF using PFD$_{avg}$ equals 6,379.

Including Common Cause

In past examples, it has been shown that more conservative results are obtained when the potential effect of common cause is included in the reliability and safety model. These effects should be included in the multiple failure mode model. Since our model uses both safe and dangerous failure rates, both should be partitioned into normal-mode and common-cause failures. Using the beta model:

$$\lambda^{SN} = (1-\beta)\lambda^{S} \tag{12-11}$$

$$\lambda^{SC} = \beta\lambda^{S} \tag{12-12}$$

$$\lambda^{DN} = (1-\beta)\lambda^{D} \tag{12-13}$$

and

$$\lambda^{DC} = \beta\lambda^{D} \tag{12-14}$$

The revised Markov model is shown in Figure 12-10. The state conditions are the same as in Figure 12-7, but failure rates are changed. The failure rate from state 0 to state 2 now accounts for common cause. The most significant change in the model is the added failure rate from state 0 to state 3. A dangerous common-cause failure fails the entire system.

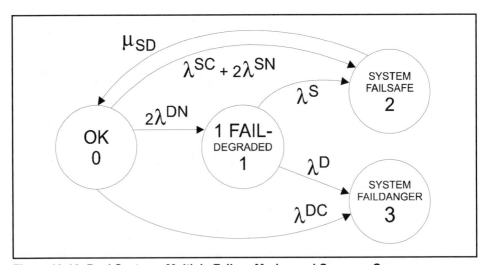

Figure 12-10. Dual System - Multiple Failure Modes and Common Cause.

Modeling Detail

The **P** matrix for this Markov model is

$$\mathbf{P} = \begin{bmatrix} 1-(\lambda^{SC}+2\lambda^{SN}+\lambda^{DC}+2\lambda^{DN}) & 2\lambda^{DN} & \lambda^{SC}+2\lambda^{SN} & \lambda^{DC} \\ 0 & 1-(\lambda^{S}+\lambda^{D}) & \lambda^{S} & \lambda^{D} \\ \mu_{SD} & 0 & 1-\mu_{SD} & 0 \\ 0 & 0 & 0 & 1 \end{bmatrix}$$

A time-dependent numerical solution is the fastest and most accurate way to solve the model if a spreadsheet program on a personal computer is available. Failure rates and repair rates can be calculated and entered into the **P** matrix. A starting matrix is entered indicating that the system starts in state 0. The **S** matrix can be repeatedly multiplied by the **P** matrix, creating a time-dependent column of probabilities (see Chapter 8).

Example 12-17

Problem: Recalculate the time-dependent PFD and time-dependent PFD_{avg} of Example 12-15, accounting for the effects of common cause using a beta factor of 0.1.

Solution: Substituting the failure rates and repair rates into a spreadsheet provides the numeric values of the **P** matrix:

		0	1	2	3
P =	0	0.99981	0.00009	0.000095	0.000005
	1	0	0.9999	0.00005	0.00005
	2	0.041666667	0	0.958333	0
	3	0	0	0	1

Starting in state 0 and multiplying the **P** matrix by the **S** matrix repeatedly gives the PFD as a function of time. The PFD at 4,384 hr is 0.051. The RRF is 19.6. The PFD_{avg} at 4,384 hr is 0.0214 and the RRF using PFD_{avg} is 46.8.

> **Example 12-18**
>
> **Problem:** Recalculate the time-dependent PFD and time-dependent PFD$_{avg}$ of Example 12-16 accounting for the effects of common cause using a beta factor of 0.1.
>
> **Solution:** The high-strength switch has a dangerous failure rate of 0.000005 failures per hour and a safe failure rate of 0.000005 failures per hour. Substituting these failure rates and the repair rates into a spreadsheet provides the numeric values of the **P** matrix:
>
> $$\mathbf{P} = \begin{array}{c|cccc} & 0 & 1 & 2 & 3 \\ \hline 0 & 0.999981 & 0.000009 & 9.5\text{E-}06 & 5\text{E-}07 \\ 1 & 0 & 0.99999 & 0.000005 & 0.000005 \\ 2 & 0.041666667 & 0 & 0.958333 & 0 \\ 3 & 0 & 0 & 0 & 1 \end{array}$$
>
> Starting in state 0 and multiplying the **P** matrix by the **S** matrix repeatedly gives the PFD as a function of time. The PFD at 4,380 hr is 0.002567. The RRF is 389. The PFD$_{avg}$ at 4,384 hr is 0.001222, and the RRF using PFD$_{avg}$ is 818.

On-line Diagnostics

System reliability and safety can be substantially improved when automatic diagnostics are programmed into a system to detect component, module, or unit failures. This can benefit the system by reducing repair time. Diagnostics can identify and annunciate failures. Repairs can be made quickly without waiting until failures are discovered during periodic inspections.

Imagine that the two switches used in the 1oo2 example had small microcomputers that open the switch for a few microseconds and check to verify that the current flowing through the switch begins to drop. With such a diagnostic it is possible to detect many of the short-circuit failure modes in the switch. If diagnostic capability exists in a system, the model should account for that capability. To do that, the failure rates are divided into those detected by the on-line diagnostics and those not detected by the on-line diagnostics. If the failure rates are obtained from a FMEDA (Chapter 9), the failure rates are already separated into these categories. Alternatively, a "diagnostic coverage factor" may be given. This can also be used to divide the failure rates.

For the two failure modes, safe and dangerous, four failure rates result:

SU: Safe, undetected

SD: Safe, detected

DU: Dangerous, undetected

DD: Dangerous, detected

The appropriate diagnostic coverage factor is used as follows:

$$\lambda^{SU} = (1 - C^S)\lambda^S \tag{12-15}$$

$$\lambda^{SD} = C^S \lambda^S \tag{12-16}$$

$$\lambda^{DU} = (1 - C^D)\lambda^D \tag{12-17}$$

and

$$\lambda^{DD} = C^D \lambda^D \tag{12-18}$$

where C^S is the safe coverage factor and C^D is the dangerous coverage factor.

Example 12-19

Problem: A switch with electronic diagnostic circuitry is used in a dual-series wired system (1oo2). The switch can detect most short-circuit (dangerous) failures and has a short-circuit mode (dangerous) coverage factor of 0.95. The short-circuit failure rate is 0.00005. What is the dangerous detected failure rate? What is the dangerous undetected failure rate?

Solution: Using Equation 12-17, the dangerous undetected failure rate equals 0.00005 × (1 − 0.95) = 0.0000025. Using Equation 12-18, the dangerous detected failure rate equals 0.00005 × 0.95 = 0.0000475.

Fault Tree Analysis with Diagnostics

Figure 12-11 shows a fault tree diagram for the 1oo2 dual-switch system that accounts for the two different diagnostic failure modes. If each switch has two possible dangerous failure modes, the system has four combinations of dangerous failures. These are shown as the four AND gates. The system fails when one combination of the two switches fail. The OR gate accounts for this.

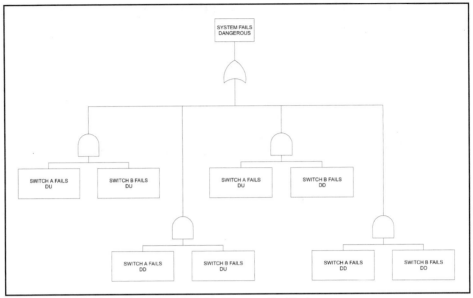

Figure 12-11. Fault Tree Accounting for Diagnostic Modes.

An equation can be developed from the fault tree. The probability that a switch will fail dangerously if the failure is undetected is approximated by multiplying the dangerous undetected failure rate times the inspection interval ($\lambda^{DU} \times TI$). The probability that a switch will fail dangerously when the failure is detected actually depends on maintenance policy. The switch will fail dangerously only until it is repaired. The faster the repair time, the lower the probability of dangerous failure. This is approximated by multiplying the dangerous detected failure rate times the actual average repair time ($\lambda^{DD} \times RT$). The full equation for system dangerous failure is given by:

$$\text{PFD} = (\lambda^{DU} \times TI)^2 + 2(\lambda^{DD} \times RT \times \lambda^{DU} \times TI) + (\lambda^{DD} \times RT)^2 \qquad (12\text{-}19)$$

The equation for PFD_{avg} can be derived by integrating the PFD equation and dividing by the time interval. This average approximation is given by:

$$\text{PFD}_{avg} = \frac{1}{TI} \int_0^{TI} [(\lambda^{DU} t)^2 + 2\lambda^{DD}\lambda^{DU} RT t + (\lambda^{DD} RT)^2] dt$$

which evaluates to

$$\text{PFD}_{avg} = \frac{1}{TI} \left[(\lambda^{DU})^2 \frac{t^3}{3} + 2\lambda^{DD}\lambda^{DU} RT \frac{t^2}{2} + (\lambda^{DD} RT)^2 t \right]_0^{TI}$$

Substituting the integration limits:

$$\text{PFD}_{avg} = \frac{(\lambda^{DU})^2 \times TI^2}{3} + \lambda^{DD} \times \lambda^{DU} \times RT \times TI + (\lambda^{DD} \times RT)^2 \qquad (12\text{-}20)$$

Example 12-20

Problem: A 1oo2 dual-switch system operates for six months between periodic inspections. During this period, when failures are detected and annunciated, repairs are made within 8 hr. Calculate the PFD, RRF, PFD_{avg}, and RRF using PFD_{avg} for the dual-switch system using a dangerous failure rate of 0.00005 and a dangerous coverage factor of 0.95.

Solution: The dangerous failure rate of the switch is divided into dangerous detected and dangerous undetected:

$$\lambda^{DD} = 0.0000475 \text{ failures per hour}$$

$$\lambda^{DU} = 0.0000025 \text{ failures per hour}$$

Substitute failure rate and repair rate data into Equation 12-19 to obtain PFD and into Equation 12-20 to obtain PFD_{avg}:

$$\text{PFD} = 0.0001284, \text{ RRF} = 7{,}790$$

$$\text{PFD}_{avg} = 0.0000443, \text{ RRF} = 22{,}587$$

Example 12-21

Problem: The first term of Equation 12-20 appears to dominate the result. Recalculate the PFD for Example 12-20 accounting only for the dangerous undetected failures and check the accuracy level.

Solution: The first term of Equation 12-20 is

$$\text{PFD}_{avg} = \frac{(\lambda^{DU} \times TI)^2}{3}$$

Substituting our values, $\text{PFD}_{avg} = 0.00003997$. This represents a little over 90% of the previous value (0.0000443). This result depends on the coverage factor. For low coverage factors (less than 50%), this simplification will reach 99% accuracy. For systems with a higher coverage factor, this simplification is not recommended.

Markov Model Analysis with Diagnostics

Figure 12-12 presents a Markov model for the dual-series wired switch that shows the effect of on-line diagnostics. Note that the Markov model also accounts for multiple failure modes, but does not account for common cause. The system is successful in states 0, 1, and 2. The system fails open circuit (safe) in state 3. The system has failed short circuit (dangerous) in states 4 and 5. The model has two degraded states, state 1 and state 2. In state 1, one switch has failed dangerously, but this failure is detected by the diagnostics. An on-line repair rate from state 1 to state 0 assumes that the system can be repaired without a shutdown. In state 2, one switch has failed dangerously and the failure is not detected by any on-line diagnostics; therefore, no repair rate from this state is shown.

In state 4 the system has failed dangerously, but the failure in one or both of the units has been detected by diagnostics. The model assumes that the repair technician will inspect and repair both units if a service visit is made. The model also assumes that maintenance policy allows on-line repair of the system without shutting down the process. State 5 represents the condition where the system has failed dangerously and the failures are not detected by diagnostics.

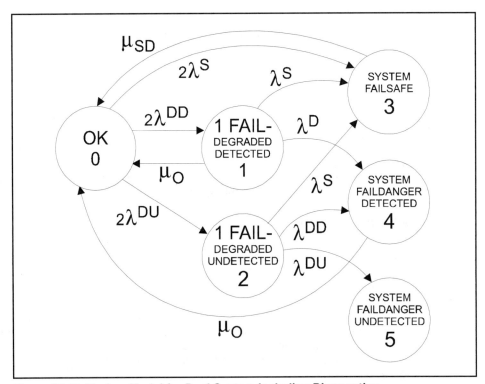

Figure 12-12. Markov Model for Dual System Including Diagnostics.

Two different repair rates are shown. State 1 has a repair rate labeled μ_O, on-line repair. The repair time has the potential to be much lower from state 1 if detected failures are repaired quickly. State 3 has a repair rate labeled μ_{SD}. It is reasonable to assume that it would take a different amount of time to restart a process than to repair a system component, so the model allows for a different average "repair/restart" time. Repair from states with undetected failures depends on a periodic inspection, and since the model will calculate time-dependent probabilities, no repair rate is shown from those states. Note that perfect inspection and repair is assumed, as repair rates lead back to state 0 — where everything is working correctly.

The **P** matrix for this Markov model is

$$\mathbf{P} = \begin{bmatrix} 1-(2\lambda^D+2\lambda^S) & 2\lambda^{DD} & 2\lambda^{DU} & 2\lambda^S & 0 & 0 \\ \mu_O & 1-(\lambda^D+\lambda^S+\mu_O) & 0 & \lambda^S & \lambda^D & 0 \\ 0 & 0 & 1-(\lambda^D+\lambda^S) & \lambda^S & \lambda^{DD} & \lambda^{DU} \\ \mu_{SD} & 0 & 0 & 1-\mu_{SD} & 0 & 0 \\ \mu_O & 0 & 0 & 0 & 1-\mu_O & 0 \\ 0 & 0 & 0 & 0 & 0 & 1 \end{bmatrix}$$

Example 12-22

Problem: A dual-series wired switch system (1oo2) uses smart switches with built-in diagnostics. The switch has two failure modes: open circuit (safe) and short circuit (dangerous). The safe failure rate is 0.00005 failures per hour. The dangerous failure rate is 0.00005 failures per hour. Diagnostics detect 95% of the dangerous failures. The switches are also wired to a digital fieldbus used to annunciate the failures. When failures are detected, the average system repair time is 8 hr. The system is operated for six-month time intervals, after which an inspection is made and any undetected failures are repaired. If the system fails safely, it is restarted in an average time period of 24 hr. What is the system PFD, RRF, PFD$_{avg}$, and RRF using PFD$_{avg}$ for a time interval of 4,380 hr?

> **Example 12-22 (continued)**
>
> **Solution:** The Markov model of Figure 12-12 is appropriate. Substituting the failure and repair rates into a spreadsheet provides the numeric values of the **P** matrix:
>
> | | 0 | 1 | 2 | 3 | 4 | 5 |
> |---|---|---|---|---|---|---|
> | 0 | 0.999800 | 0.000095 | 0.000005 | 0.000100 | 0.000000 | 0.000000 |
> | 1 | 0.125000 | 0.874900 | 0.000000 | 0.000050 | 0.000050 | 0.000000 |
> | 2 | 0.000000 | 0.000000 | 0.999900 | 0.000050 | 0.000048 | 0.000003 |
> | 3 | 0.041667 | 0.000000 | 0.000000 | 0.958333 | 0.000000 | 0.000000 |
> | 4 | 0.125000 | 0.000000 | 0.000000 | 0.000000 | 0.875000 | 0.000000 |
> | 5 | 0.000000 | 0.000000 | 0.000000 | 0.000000 | 0.000000 | 1.000000 |
>
> **P =**
>
> Starting in state 0 and multiplying the **P** matrix by the **S** matrix repeatedly gives the PFD as a function of time. The PFD at 4,392 hr is 0.000111. The RRF is 9,042. The PFD_{avg} at 4,392 hr is 0.00004, and the RRF using PFD_{avg} is 24,997.

Diagnostics and Common Cause

In all previous models, common cause was a significant factor in the results of our modeling. Because of its importance, another model must be created that shows the combination of on-line diagnostics and common cause.

The failure rates must be partitioned into eight categories:

> SDN: Safe, detected, normal
>
> SDC: Safe, detected, common cause
>
> SUN: Safe, undetected, normal
>
> SUC: Safe, undetected, common cause
>
> DDN: Dangerous, detected, normal
>
> DDC: Dangerous, detected, common cause
>
> DUN: Dangerous, undetected, normal
>
> DUC: Dangerous, undetected, common cause

Modeling Detail

The four failure rates SU, SD, DU, and DD are divided using the beta model:

$$\lambda^{SDN} = (1 - \beta) \lambda^{SD} \tag{12-21}$$

$$\lambda^{SDC} = \beta \lambda^{SD} \tag{12-22}$$

$$\lambda^{SUN} = (1 - \beta) \lambda^{SU} \tag{12-23}$$

$$\lambda^{SUC} = \beta \lambda^{SU} \tag{12-24}$$

$$\lambda^{DDN} = (1 - \beta) \lambda^{DD} \tag{12-25}$$

$$\lambda^{DDC} = \beta \lambda^{DD} \tag{12-26}$$

$$\lambda^{DUN} = (1 - \beta) \lambda^{DU} \tag{12-27}$$

$$\lambda^{DUC} = \beta \lambda^{DU} \tag{12-28}$$

Example 12-23

Problem: A switch has a safe failure rate of 0.00005 failures per hour and a dangerous failure rate of 0.00005 failures per hour. On-line diagnostics detect 95% of the dangerous failures and 90% of the safe failures. Two identical switches are closely coupled in a system. The beta factor is estimated to be 0.1. What are the eight failure rates required to model multiple failure modes, diagnostics, and common cause?

Solution: First, the failure rates are divided by diagnostic capability using Equations 12-15 to 12-18:

$$\lambda^{SD} = 0.00005 \times 0.9 = 0.000045$$
$$\lambda^{SU} = 0.00005 \times (1 - 0.9) = 0.000005$$
$$\lambda^{DD} = 0.00005 \times 0.95 = 0.0000475$$
$$\lambda^{DU} = 0.00005 \times (1 - 0.95) = 0.0000025$$

These failure rates are multiplied by β using Equations 12-21 to 12-28. The following failure rates result:

$$\lambda^{SDN} = 0.0000405$$
$$\lambda^{SDC} = 0.0000045$$
$$\lambda^{SUN} = 0.0000045$$
$$\lambda^{SUC} = 0.0000005$$
$$\lambda^{DDN} = 0.00004275$$
$$\lambda^{DDC} = 0.00000475$$
$$\lambda^{DUN} = 0.00000225$$
$$\lambda^{DUC} = 0.00000025$$

Fault Tree Analysis with Diagnostics and Common Cause

Common cause can be incorporated into a fault tree. Figure 12-13 shows a fault tree for a 1oo2 dual-switch system; compare it with Figure 12-11. A dangerous undetected common-cause failure has been added to the drawing along with a dangerous detected common-cause failure.

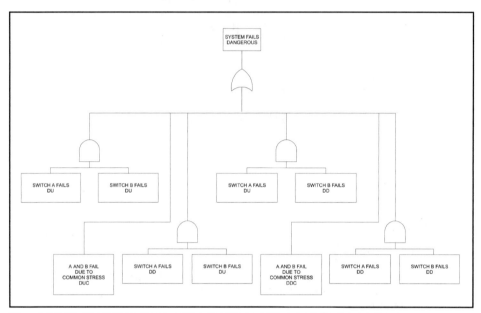

Figure 12-13. Fault Tree Accounting for Diagnostics and Common Cause.

This fault tree can be analytically evaluated. The probability of a dangerous detected common-cause failure is approximated by multiplying λ^{DDC} times RT. The probability of a dangerous undetected common-cause failure is approximated by multiplying λ^{DUC} times TI. The complete equation is

$$\text{PFD} = (\lambda^{DUN} \times TI)^2 + 2(\lambda^{DDN} \times RT \times \lambda^{DUN} \times TI) + (\lambda^{DDN} \times RT)^2 + \lambda^{DDC} \times RT + \lambda^{DUC} \times TI \quad (12\text{-}29)$$

The equation for PFD_{avg} can be derived by integrating the PFD equation and dividing by the time interval. This average approximation is given by:

$$\text{PFD}_{avg} = \frac{1}{TI}\int_0^{TI}\left[(\lambda^{DUN}t)^2 + 2\lambda^{DDN}\lambda^{DUN}RTt + (\lambda^{DDN}RT)^2 + \lambda^{DDC}RT + \lambda^{DUC}t\right]dt$$

which evaluates to

$$\text{PFD}_{avg} = \frac{1}{TI}\left[(\lambda^{DUN})^2\frac{t^3}{3} + 2\lambda^{DDN}\lambda^{DUN}RT\frac{t^2}{2} + (\lambda^{DDN}RT)^2 t + \lambda^{DDC}RTt + \lambda^{DUC}\frac{t^2}{2}\right]_0^{TI}$$

Substituting the integration limits:

$$PFD_{avg} = \frac{(\lambda^{DUN})^2 \times TI^2}{3} + \lambda^{DDN} \times \lambda^{DUN} \times RT \times TI + (\lambda^{DDN} \times RT)^2 + \lambda^{DDC} \times RT + \lambda^{DUC} \times \frac{TI}{2} \quad (12\text{-}30)$$

Example 12-24

Problem: Repeat the fault tree of Example 12-20, accounting for common cause using a beta factor of 0.1. Calculate the PFD, RRF, PFD_{avg}, and RRF using PFD_{avg}.

Solution: The failure rates are calculated in Example 12-23. Using an approximation technique, substitute failure rate and repair rate data into Equation 12-29 to obtain PFD and into Equation 12-30 to obtain PFD_{avg}.

$$PFD = 0.00124, RRF = 808$$

$$PFD_{avg} = 0.00062, RRF = 1{,}609$$

Example 12-25

Problem: Repeat Example 12-24 using a high-strength switch with a safe failure rate of 0.000005 failures per hour and a dangerous failure rate of 0.000005 failures per hour. Account for common cause using a beta factor of 0.1. Calculate the PFD, RRF, PFD_{avg}, and RRF using PFD_{avg}.

Solution: Using an approximation technique, substitute failure rate and repair rate data into Equation 12-29 to obtain PFD and into Equation 12-30 to obtain PFD_{avg}.

$$PFD = 0.0001143, RRF = 8{,}746$$

$$PFD_{avg} = 0.0000589, RRF = 16{,}975$$

Markov Model Analysis with Diagnostics and Common Cause

The Markov model that shows a combination of multiple failure modes, diagnostics, and common cause is presented in Figure 12-14. It has the same state combinations as Figure 12-12, but additional failure rates are added. Two of the most significant are the dangerous detected common-cause failure rate from state 0 to state 4 and the dangerous undetected common-cause failure rate from state 0 directly to state 5. The **P** matrix for this Markov model is

$$\mathbf{P} = \begin{bmatrix} 1-(\lambda^{DC}+2\lambda^{DN}+\lambda^{SC}+2\lambda^{SN}) & 2\lambda^{DDN} & 2\lambda^{DUN} & \lambda^{SC}+2\lambda^{SN} & \lambda^{DDC} & \lambda^{DUC} \\ \mu_O & 1-(\lambda^S+\lambda^D+\mu_O) & 0 & \lambda^S & \lambda^D & 0 \\ 0 & 0 & 1-(\lambda^S+\lambda^D) & \lambda^S & \lambda^{DD} & \lambda^{DU} \\ \mu_{SD} & 0 & 0 & 1-\mu_{SD} & 0 & 0 \\ \mu_O & 0 & 0 & 0 & 1-\mu_O & 0 \\ 0 & 0 & 0 & 0 & 0 & 1 \end{bmatrix}$$

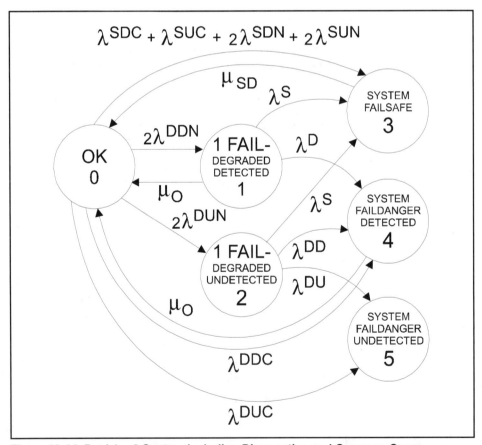

Figure 12-14. Dual 1oo2 System Including Diagnostics and Common Cause.

Modeling Detail

> **Example 12-26**
>
> **Problem:** Using the Markov model of Figure 12-14 and the failure rates of Example 12-22, calculate the PFD, RRF, PFD_{avg}, and RRF using PFD_{avg} of the dual-switch system for a time interval of six months.
>
> **Solution:** When the failure rates and repair rates are substituted into the **P** matrix, the following numeric values result:
>
> $$\mathbf{P} = \begin{array}{c|cccccc} & 0 & 1 & 2 & 3 & 4 & 5 \\ \hline 0 & 0.999810 & 0.000086 & 0.000005 & 0.000095 & 0.00000475 & 0.00000025 \\ 1 & 0.125000 & 0.874900 & 0.000000 & 0.000050 & 0.00005000 & 0.00000000 \\ 2 & 0.000000 & 0.000000 & 0.999900 & 0.000050 & 0.00004750 & 0.00000250 \\ 3 & 0.041667 & 0.000000 & 0.000000 & 0.958333 & 0.00000000 & 0.00000000 \\ 4 & 0.125000 & 0.000000 & 0.000000 & 0.000000 & 0.87500000 & 0.00000000 \\ 5 & 0.000000 & 0.000000 & 0.000000 & 0.000000 & 0.00000000 & 1.00000000 \end{array}$$
>
> When this matrix is repeatedly multiplied times an **S** matrix starting on state 0, the PFD at 4,392 hr is 0.00122. The RRF is 818. The PFD_{avg} at 4,392 hr is 0.000621, and the RRF is 1,811.

> **Example 12-27**
>
> **Problem:** Using the Markov model of Figure 12-14 and the high-strength failure rates of Example 12-25, calculate the PFD and RRF of the dual-switch system for a time interval of six months.
>
> **Solution:** When the failure rates and repair rates are substituted into the **P** matrix and repeatedly multiplied times an **S** matrix starting on state 0, the PFD at 4,392 hr is 0.0001146. The RRF is 8,726. The PFD_{avg} at 4,392 hr is 0.0000593, and the RRF is 16,852.

Comparing the Techniques

A range of answers have been obtained for the same fundamental problem. This result alone shows the importance of modeling the proper level of detail. A comparison will give some insight to the value of model detail.

Common Cause

Table 12-1 compares a number of different levels of detail with and without the effects of common cause. Common cause has less of an effect on the results of the approximate models with high failure rates. The results become much more significant when failure rates are lower and the models are more detailed. In effect, common cause must be considered in the higher safety integrity levels. If the beta factor is relatively high (as is the case in our examples), RRF results can be an order of magnitude in error if common cause is ignored.

Without Common Cause			With Common Cause					
Example	PFD/PFD$_{avg}$	RRF	Example	PFD/PFDavg	RRF	Method*	λ^D	Diff. %
12-1	0.04769	21.0	12-5	0.06075	16.5	Fault Tree Approx.	0.00005	-21
12-3	0.01579	63.3	12-7	0.0239	41.8	Fault Tree Approx.	0.00005	-34
12-2	0.00048	2083.3	12-6	0.00258	387.6	Fault Tree Approx.	0.000005	-81
12-4	0.0001599	6255.0	12-8	0.001225	816.3	Fault Tree Approx.	0.000005	-87
12-9	0.0213	46.9	12-11	0.02731	36.6	Simple Markov	0.00005	-22
12-10	0.0002388	4187.6	12-12	0.001299	769.8	Simple Markov	0.000005	-82
12-15	0.01296	77.2	12-17	0.021347	46.8	Markov, MFM	0.00005	-39
12-16	0.0001568	6379.3	12-18	0.001222	818.3	Markov, MFM	0.000005	-87
12-20	0.0000443	22588.7	12-24	0.0006214	1609.3	Fault Tree, Diagnostics	0.00005	-93
12-22	0.00004	24996.9	12-26	0.0006205	1611.7	Markov, Diagnostics	0.00005	-94
* MFM, multiple failure modes								

Table 12-1. Common Cause

Method Type Comparison

The two methods used — fault trees with probability approximation and Markov models — can be compared. Example 12-1, a simple fault tree approximation, gave the result PFD = 0.04769. This can be compared with the Markov results from Example 12-15, which gave a result of PFD = 0.03625. Example 12-20 (fault tree) and Example 12-22 (Markov) use the same level of detail and give similar results. A similar comparison is valid for Example 12-24 (fault tree) and Example 12-26 (Markov).

Markov Model Comparison

The differences between the various Markov models show the effect of including detail in the models. Markov modeling provides a unique way to account for time dependency and the interaction of multiple failure modes. The advantage of this is shown in Table 12-2, as the results are improved when these details are included in the model.

Example	PFD$_{avg}$	RRF	Method*
12-10	0.000236	4,231	Simple Markov, SS, λ^D = 0.000005
12-14	0.000231	4,323	MFM, SS
12-16	0.000157	6,379	TD, MFM
* MFM, multiple failure modes; SS, steady state; TD, time dependent			

Table 12-2. Markov Model Comparison

Modeling Detail

The comparison in Table 12-3 of steady-state Markov solutions versus time-dependent solutions shows that the steady-state solutions provide an "average" result. Those solutions are closer to PFD_{avg} than PFD. The results are not exact, however, with the time-dependent solutions typically providing better results.

Example	PFD_{avg}	RRF	Method
12-13	0.016467	60.7	Markov, MFM, SS, $\lambda^D = 0.00005$
12-15	0.01296	77.2	Markov, MFM, TD, $\lambda^D = 0.00005$

Table 12-3. Markov Model Time Comparison

Approximations

Example 12-21 shows a PFD_{avg} approximation of Example 12-20. It produced optimistic results. The results are especially optimistic when compared with Example 12-24, where common cause was taken into account. This is shown in Table 12-4. While simplifications do save on engineering time, they must be treated with great caution.

Example	PFD_{avg}	RRF	Method
12-21	0.0000399	25,020	Fault Tree Approx., Diagnostics, $\lambda^D = 0.00005$
12-20	0.000044	22,587	Fault Tree, Diagnostics, $\lambda^D = 0.00005$
12-24	0.0001512	6,612	Fault Tree, Diagnostics, Common Cause, $\lambda^D = 0.00005$

Table 12-4. Approximation Comparison of PFD_{avg}

EXERCISES

12.1   Example 12-24 uses a fault tree that accounts for both common cause and on-line diagnostic capability. Repeat that example using a smaller beta factor of 0.02.

12.2   Compare the result of Exercise 12.1 with the result of Example 12-20. What is the percentage difference?

12.3   Repeat Example 12-20 for a periodic inspection interval of one year (8,760 hr) and a diagnostic coverage of 50%.

12.4   Using the conditions of Exercise 12.3, approximate the answer using the technique in Example 12-21. Compare these results with those of Exercise 12.3.

BIBLIOGRAPHY

1. Brombacher A.C., Spiker R.Th.E., "Evaluating Safety Aspects of Safeguarding Systems," *Annual Conference of the American Society of Chemical Engineers*, 1991.

2. Stavrianidis, P., "Inspection, Testing and Repair/Replace (ITR) Policies as Part of Reliability Certification for Programmable Electronic Systems," *Proceedings of the Probabilistic Safety Assessment and Management Conference*, June 1996.

3. Rouvroye, J. L., Goble, W. M., Brombacher, A. C., and Spiker, R. Th. E., "A Comparison Study of Qualitative and Quantitative Analysis Techniques for the Assessment of Safety in Industry," *Proceedings of the Probabilistic Safety Assessment and Management Conference*, June 1996.

4. Gruhn, P., "Safety System Risk Analysis Needs to Be More Quantitative," *Control*, August 1991.

13
Reliability and Safety Model Construction

System Model Development

A successful system of reliability and safety evaluation depends on the process used to define the model for the system. Knowledge of proper system operation is essential. Perhaps more important is an understanding of system operation under failure conditions. What happens when the process sensor fails? What happens if the impulse line gets plugged? What happens if a controller fails with its output energized? What happens if the valve jams open? One of the best tools to systematically answer these questions is the failure modes and effects analysis (FMEA).

A series of steps, including an FMEA, can be used to help ensure the construction of an accurate reliability and safety model. The following steps are recommended:

1. Define a failure.
2. Complete a system-level FMEA.
   a. Identify and list all system components.
   b. For each system component, identify all failure modes and system effects.
3. Classify failures according to effect.
4. Determine the level of model detail.
5. Develop the model.
   a. List all failure rates.
   b. Build a model that accounts for all failure rates.
6. Solve for the needed reliability and safety measurements.

Step 1: What Is a Failure?

The lack of a clear definition of "system failure" has confused many reliability and safety analysis efforts. This may be difficult, especially in situations where a *system* runs in a degraded mode when a *component* fails. Is this degraded mode a system failure? This depends on how the system is being used.

Fortunately, for industrial control systems, the answer is usually clear. *Needed functions must be accomplished in a needed time period.* If a degraded mode does this, that mode is not a system failure. This simple definition, "Needed functions must be accomplished in a needed time period," is also comprehensive; it includes hardware failures, software failures, and human failures.

Step 2: System Level FMEA

An FMEA is done on all components within the system (Chapter 5). The scope of the system must be identified. For control and safety systems, this scope usually includes process connections, sensors, controllers, actuators, and valves. The system-level FMEA is a critical qualitative procedure. The accuracy of the information obtained in this step influences the model that is formed and the reliability and safety measures that are obtained.

Consider a safety system that consists of a pressure switch, two single-board controllers, and a valve (Figure 13-1). When the process is operating normally, pressure is low and the pressure switch is closed (energized). The controller keeps its outputs energized, and the valve is closed. Whenever pressure exceeds a preset limit, the switch opens (de-energizes).

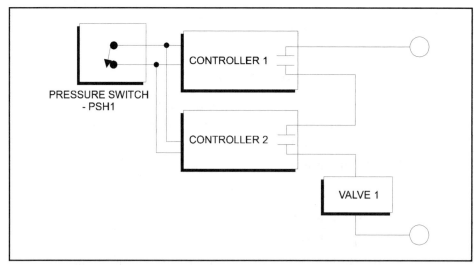

Figure 13-1. Safety System.

The controller reads the pressure switch status and executes a timing and filtering function. When the pressure switch is open for a sufficient period of time and the process is active (not in startup or inactive), the controller will de-energize its output. When the controller de-energizes its output, the valve opens and relieves pressure.

If the system fails such that the controller output is always energized or the valve is jammed closed, that is called a "dangerous" failure because the safety system cannot relieve the pressure. If the system fails such that the controller output is de-energized or the valve fails open, that is called a "safe" failure because pressure is inadvertently relieved.

The FMEA chart is presented in Table 13-1. Each system-level component is listed along with its failure modes. The system effect for each failure is listed and categorized. The failure rates are typically obtained from the component manufacturers. These are given in units of FITS (failures per 10^{-9} hr).

1	2	3	4	5	6	7	8	9
Name	Code	Function	Mode	Cause	Effect	Criticality	λ	Remarks
Valve	VALVE1	Opens to relieve pressure	Fail open	Power, corrosion	False trip	Safe	1000	Valve opens for safety mitigation
			Jam closed	Corrosion, dirt	Cannot trip	Dangerous	100	
			Coil open	Electrical surge	False trip	Safe	50	If coil opens, valve opens
			Coil short	Corrosion, wire	False trip	Safe	50	If coil shorts, valve opens
Pressure switch	PSH1	Sense overpressure	Short	Power surge	System output energized, cannot trip	Dangerous	100	
			Open	Many	False trip	Safe	400	
			Ground fault	Corrosion	Cannot trip	Dangerous	100	Assume grounding of positive side
Controller 1	C1	Logic solver	No comlink	Many	No effect	. . .	145	No effect on safety function
			Output energize	Surge, heat	Cannot trip	Dangerous	230	
			Output open	Many	False trip	Safe	950	
Controller 2	C2	Logic solver	No comlink	Many	No effect	. . .	145	No effect on safety function
			Output energize	Surge, heat	Cannot trip	Dangerous	230	
			Output open	Many	False trip	Safe	950	

Table 13-1. FMEA Chart

Step 3: Categorize Failures

The failures are categorized as directed by the FMEA. The FMEA chart (Table 13-1) shows two significant categories of system-level failures: safe (valve open) and dangerous (valve closed).

Starting with the two primary failure categories, all failure rates can be placed into one of the two categories using the FMEA chart (see Figure 13-2). For example, a short-circuit failure of a pressure switch belongs in the dangerous (valve closed) category. The total failure rate is divided:

$$\lambda_{TOT} = \lambda^S + \lambda^D \qquad (13\text{-}1)$$

where the superscript S represents a "safe" failure and the superscript D represents a "dangerous" failure.

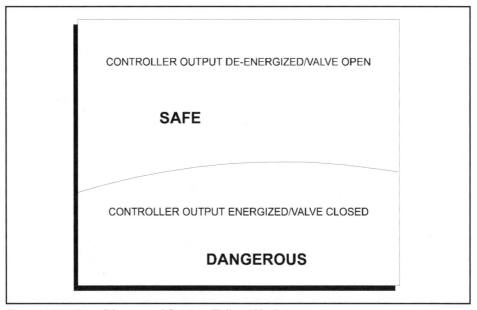

Figure 13-2. Venn Diagram of System Failure Modes.

Example 13-1

Problem: Divide the failure rates of Table 13-1 into the two categories — safe and dangerous — based on the FMEA table.

Solution: For each component, add the failure rates for each mode:

Valve	Safe	Dangerous
	1000	100
	50	
	50	

$\lambda^S_{VALVE1} = 1100 \times 10^{-9}$ failures per hour

$\lambda^D_{VALVE1} = 100 \times 10^{-9}$ failures per hour

Pressure Switch	Safe	Dangerous
	400	100
		100

$\lambda^S_{PSH1} = 400 \times 10^{-9}$ failures per hour

$\lambda^D_{PSH1} = 200 \times 10^{-9}$ failures per hour

Controller	Safe	Dangerous
	950	230

$\lambda^S_C = 950 \times 10^{-9}$ failures per hour

$\lambda^D_C = 230 \times 10^{-9}$ failures per hour

Note that the communications failure mode has no effect on safety, so that failure rate has been discarded. This would not be valid if the communications were required as part of the system functionality.

Step 4: Determine the Level of Model Detail

The level of detail chosen for a model will have an effect on the result. Generally, more accurate models will have more detail. For this example, multiple failure modes, common cause, and on-line diagnostics will be included. The model will be solved using time-dependent Markov solution techniques.

Common cause is a factor when redundant devices exist in a system. This example has redundant controllers. Common-cause failures can result in safe or dangerous system failures, so the controller failure rates are divided into two mutually exclusive categories — safe common-cause failures and dangerous common-cause failures — using a simple beta model (Chapter 10). The superscript SC is used to designate safe common-cause failures, and the superscript SN is used to designate safe normal-stress failures. The dangerous failures must also be divided into two groups: dangerous common-cause failures (DC) and dangerous normal-stress failures (DN). The failure rates are divided into two mutually exclusive groups where:

$$\lambda^S = \lambda^{SC} + \lambda^{SN} \qquad (13\text{-}2)$$

and

$$\lambda^D = \lambda^{DC} + \lambda^{DN} \qquad (13\text{-}3)$$

The beta factor is based on the chances of multiple redundant units failing due to a common stress. The factors to be considered are physical location, electrical separation, inherent strength of the components versus the environment, and any diversity in the redundant components. Though there may be different beta factors for safe failures and dangerous failures, the considerations are the same; therefore, the same beta is typically used.

The failure rates must be further classified into those that are detected by on-line diagnostics ("detected") and those that are not ("undetected"). Both safe and dangerous normal failures are classified, as are safe and dangerous common-cause failures.

EXAMPLE 13-2

Problem: The system of Figure 13-1 has redundant controller components; therefore, common cause must be modeled. Divide the controller failure rates into normal and common-cause categories.

Solution: This example has two identical controllers mounted in one cabinet with no electrical connection between units except I/O wiring. Past records indicate that high-stress events occur approximately 10% of the time. Therefore, a beta of 0.1 is chosen. The controller failure rates given in the FMEA chart are divided into four categories:

$$\lambda_C^{SC} = 0.1(950)$$
$$= 95$$

$$\lambda_C^{SN} = 0.9(950)$$
$$= 855$$

$$\lambda_C^{DC} = 0.1(230)$$
$$= 23$$

$$\lambda_C^{DN} = 0.9(230)$$
$$= 207$$

"Coverage" is the measure of the built-in test capability of a system. It is represented by a number from 0 to 1 and denoted by C. If the failure rates are not already provided by an FMEDA, a coverage factor must be obtained for each component in the system in order to separate the detected failures from the undetected failures. The eight failure rate categories are calculated as follows:

$$\lambda^{SDN} = C^S \lambda^{SN} \tag{13-4}$$

$$\lambda^{SUN} = (1 - C^S) \lambda^{SN} \tag{13-5}$$

$$\lambda^{DDN} = C^D \lambda^{DN} \tag{13-6}$$

$$\lambda^{DUN} = (1 - C^D) \lambda^{DN} \tag{13-7}$$

$$\lambda^{SDC} = C^S \lambda^{SC} \tag{13-8}$$

$$\lambda^{SUC} = (1 - C^S)\lambda^{SC} \tag{13-9}$$

$$\lambda^{DDC} = C^D\lambda^{DC} \tag{13-10}$$

$$\lambda^{DUC} = (1 - C^D)\lambda^{DC} \tag{13-11}$$

> **EXAMPLE 13-3**
>
> **Problem:** Calculate the failure rates for the system components of Figure 13-1.
>
> **Solution:** The controller manufacturer provides a safe coverage factor of 0.97 and a dangerous coverage factor of 0.99. Our own testing has shown that one failure was undetected from a series of 100 failure simulations. This is consistent with the data provided, so the manufacturer's numbers for the controller are used. Using Equations 13-4 to 13-11 for the controller,
>
> $\lambda^{SDN} = 0.97 \times 855 = 829$
>
> $\lambda^{SUN} = (1 - 0.97) \times 855 = 26$
>
> $\lambda^{DDN} = 0.99 \times 207 = 205$
>
> $\lambda^{DUN} = (1 - 0.99) \times 207 = 2$
>
> $\lambda^{SDC} = 0.97 \times 95 = 92$
>
> $\lambda^{SUC} = (1 - 0.97) \times 95 = 3$
>
> $\lambda^{DDC} = 0.99 \times 23 = 22.7$, round to 22.7 as a worst case
>
> $\lambda^{DUC} = (1-0.99) \times 23 = 0.23$, round up to 0.3 as a worst case

The pressure switch has no diagnostics, but the controller is capable of detecting ground fault failures in the switch, so those failures are categorized as dangerous detected. Other pressure switch failures are categorized as undetected. The valve has no diagnostics, but the controller can detect shorted or open coils. The valve failure rates are divided accordingly. The results are listed in Table 13-2; as a check, add up the failure rate categories for each component. The total should equal the starting number. For the controller, the total equals 1,180. All totals match.

Component	Coverage	SDC	SUC	DDC	DUC	SDN	SUN	DDN	DUN
Switch	...	...	...	...	...	0	400	100	100
Controller safe	0.97	92	3	...	...	829	26	...	...
Controller Dangerous	0.99	...	...	22.7	0.3	...	...	205	2
Valve	...	...	...	...	...	100	1000	...	100
Failure rates are in FITS.									

Table 13-2. Calculation Results

Step 5: Develop the Model

Model construction should begin by creating a checklist of all failure rates that must be included in the model (see Table 13-3). This will help verify that the model is complete. Regardless of the modeling technique, the checklist should be used. Once a checklist is complete, the construction process depends on the type of model (reliability block diagram, fault tree, Markov model).

Nonredundant	SD	SU	DD	DU
Switch	0	400	100	100
Valve	100	1000	. . .	100
Common-cause controller	SDC	SUC	DDC	DUC
	92	3	22.7	0.3
Normal controller	SDN	SUN	DDN	DUN
Controller 1	829	26	205	2
Controller 2	829	26	205	2
Failure rates are in FITS.				

Table 13-3. Failure Rate Checklist

Fault Tree Model Construction

Each fault tree shows a failure mode. In order to solve this model for both failure modes, two fault trees must be created. The process is the same:

1. Draw top-level event showing system failure mode.
2. Draw all lower events that will trigger the top event.
3. Continue the process with all failure rates.

The first step in fault tree construction for the dangerous failure mode is shown in Figure 13-3. Three major events will fail the system dangerously: a dangerous switch failure, a dangerous failure in both controllers, and a dangerous valve failure.

The process continues using the failure rate checklist. A switch can fail dangerously if it fails dangerous detected or if it fails dangerous undetected. An OR gate is added to the fault tree showing these events. The only dangerous valve failure is dangerous undetected. A basic fault showing this condition is added to the fault tree.

Both controllers can fail dangerously in a number of ways. These include a detected common-cause failure or an undetected common-cause failure. Both controllers can fail dangerously if any of four combinations of dangerous detected and dangerous undetected occur. These are also added to the fault tree. The complete model is shown in Figure 13-4.

316   Reliability and Safety Model Construction

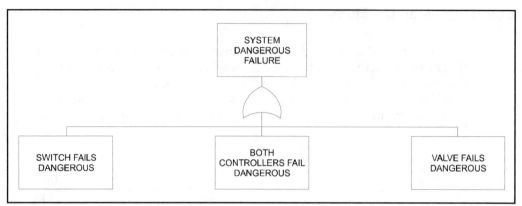

Figure 13-3. First Step, PFD Fault Tree.

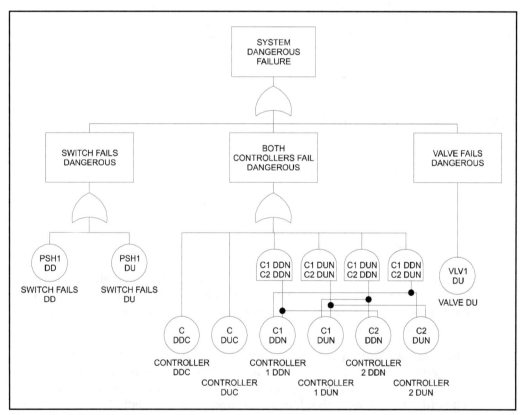

Figure 13-4. PFD Fault Tree.

Step 6: Solve the Fault Tree Model

The fault tree of Figure 13-4 can be solved for the probability that the system will fail dangerously. This term is called *probability of failure on demand* (PFD). Probability approximations can be used to determine basic fault probabilities. These are combined using the methods of Chapter 6 to solve for PFD.

Markov Model Construction

If the Markov modeling technique is chosen, different steps are used to construct the model. Markov model construction begins from a state in which all components are successful. This state is typically numbered 0. When building the Markov model, follow this rule: "For any successful state, list all failure rate categories for all successful components." In this case, all components are successful. Therefore, the failure rate categories from the pressure switch, the two controllers, and the valve should exit from state 0. The same checklist (Table 13-3) is redrawn in Figure 13-5.

PRESSURE SWITCH - PSH1						SU	DD	DU
VALVE - VALVE1					SD	SU		DU
CONTROLLER - C1	SDC	SUC	DDC	DUC	SDN	SUN	DDN	DUN
CONTROLLER - C2					SDN	SUN	DDN	DUN

Figure 13-5. Failure Rate Category Checklist.

Failures from a number of these failure rate categories will result in an open valve (fail safe). These are circled in Figure 13-6. The Markov model will have a failure state in which the valve has failed open. The initial Markov model resulting from our first step is shown in Figure 13-7.

The arc drawn from state 0 to the new failure state is labeled with the symbol l_1. This failure rate includes all those circled in Figure 13-6:

$$\lambda_1 = \lambda_C^{SDC} + \lambda_C^{SUC} + \lambda_{PSH1}^{SU} + \lambda_{C1}^{SDN} + \lambda_{C1}^{SUN} + \lambda_{C2}^{SDN} + \lambda_{C2}^{SUN} + \lambda_{VALVE1}^{SD} + \lambda_{VALVE1}^{SU} \quad (13\text{-}12)$$

Next, system behavior during a dangerous controller failure is examined. One controller can fail with its output energized without causing a system failure. This happens because the controller outputs are wired in series. Although one controller has failed with output energized, the other controller can still de-energize the valve if needed. This situation requires that more states be added to the diagram.

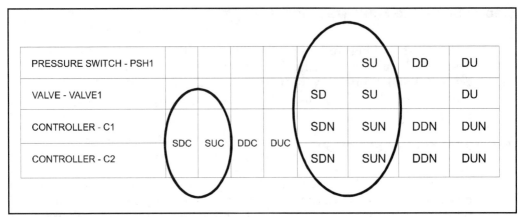

Figure 13-6. System Fail-Safe Failure Rate Categories.

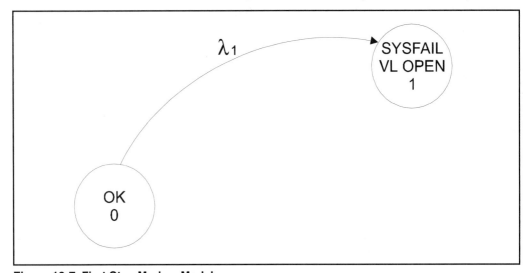

Figure 13-7. First Step Markov Model.

Four failure rates cause new states. These are circled in Figure 13-8. When the additional system success states are added to the Markov model, the interim diagram looks like Figure 13-9. Since states 2 and 3 represent failures detected by on-line diagnostics, an on-line repair rate goes from these states to state 0. The failure rates from Figure 13-9 are

$$\lambda_2 = \lambda_{C1}^{DDN} \tag{13-13}$$

$$\lambda_3 = \lambda_{C2}^{DDN} \tag{13-14}$$

$$\lambda_4 = \lambda_{C1}^{DUN} \tag{13-15}$$

$$\lambda_5 = \lambda_{C2}^{DUN} \tag{13-16}$$

Reliability and Safety Model Construction

PRESSURE SWITCH - PSH1						SU	DD	DU
VALVE - VALVE1					SD	SU		DU
CONTROLLER - C1	SDC	SUC	DDC	DUC	SDN	SUN	DDN	DUN
CONTROLLER - C2					SDN	SUN	DDN	DUN

Figure 13-8. Controller Fail-Danger Failure Rate Categories.

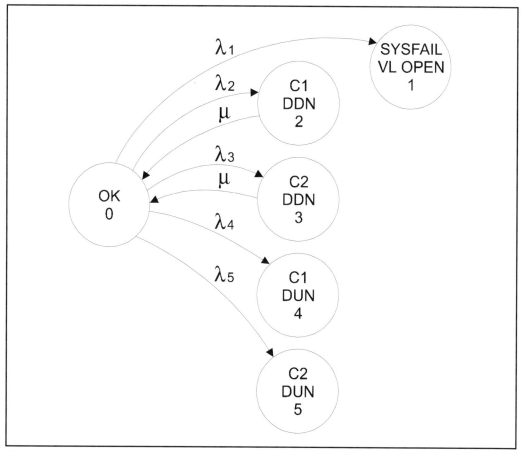

Figure 13-9. Second Step Markov Model.

The remaining failure rates from Figure 13-5 cause the system to fail with the valve closed. These are circled in Figure 13-10. To show this system behavior, two more failure states are added to the model. State 6 represents a condition where the system has failed with the valve closed (dangerous), but the failure is detected. In state 7, the system has failed dangerously and the failure is not detected. Assuming maintenance policy allows, an on-line repair could be made from state 6. This is indicated with a repair arc from state 6 to state 0. The new states are shown in Figure 13-11, where:

$$\lambda_6 = \lambda_C^{DDC} + \lambda_{PSH1}^{DD} \qquad (13\text{-}17)$$

$$\lambda_7 = \lambda_C^{DUC} + \lambda_{PSH1}^{DU} + \lambda_{VALVE1}^{DU} \qquad (13\text{-}18)$$

A check of the failure rates originally listed in Figure 13-5 will show that all failure rate categories have been included as transition rates from state 0 in the Markov model.

The Markov model continues by remembering the rule: "For any successful system state, list all failure rate categories for all successful components." The interim model now has four successful system states that must be considered. Construction continues from state 2. In this state, the system has one controller, the pressure switch, and the valve working successfully. A list of failure rates is shown in Figure 13-12.

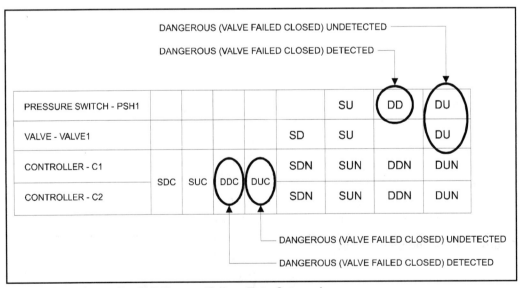

Figure 13-10. Controller Fail-Danger Failure Rate Categories.

Reliability and Safety Model Construction 321

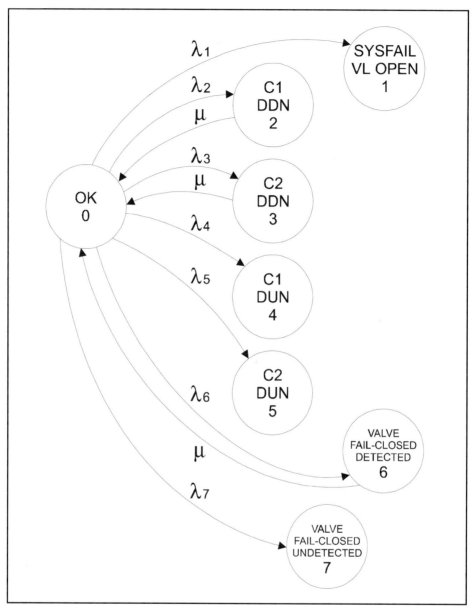

Figure 13-11. Step Three Markov Model.

322   Reliability and Safety Model Construction

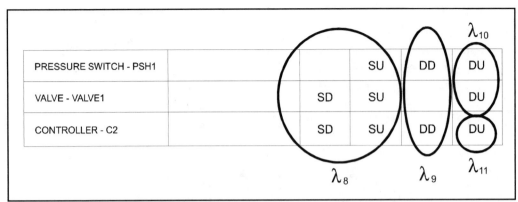

Figure 13-12. State 2 Failure Rates.

Examination shows that these failures will send the system to either state 1 (the valve open failure state), state 6 (the valve closed with all failures detected state), or new system failure states where some failures are detected and some are not. These will be modeled with two new states in order to show the effects of on-line repair. Four arcs are added as exit rates from state 2 as shown in Figure 13-13, where:

$$\lambda_8 = \lambda_{PSH1}^{SU} + \lambda_C^{SD} + \lambda_C^{SU} + \lambda_{VALVE1}^{SD} + \lambda_{VALVE1}^{SU} \qquad (13\text{-}19)$$

$$\lambda_9 = \lambda_{PSH1}^{DD} + \lambda_C^{DD} \qquad (13\text{-}20)$$

$$\lambda_{10} = \lambda_{PSH1}^{DU} + \lambda_{VALVE1}^{DU} \qquad (13\text{-}21)$$

$$\lambda_{11} = \lambda_C^{DU} \qquad (13\text{-}22)$$

Reliability and Safety Model Construction

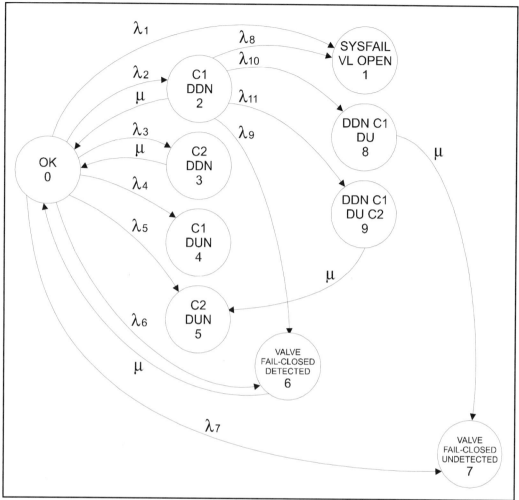

Figure 13-13. Step Four Markov Model.

State 3 is also a system success state. An examination of failure rates from state 3 shows that they are the same as from state 2. Therefore, the same arcs are added to state 3, creating two additional new states — as shown in Figure 13-14 along with associated repair arcs. The new states are added to the Markov model in Figure 13-15.

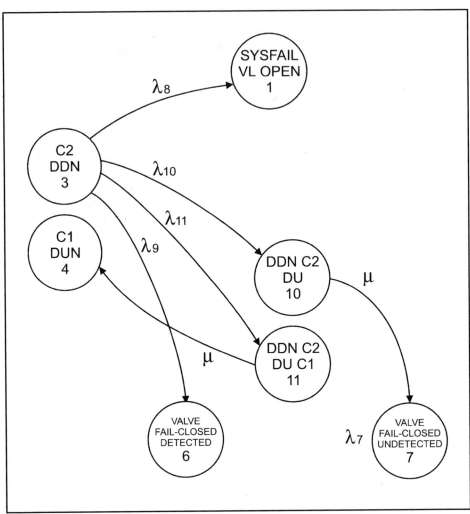

Figure 13-14. Exit Rates from State 3.

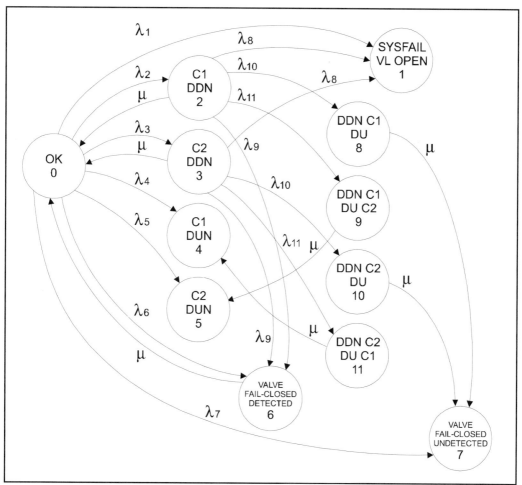

Figure 13-15. Step Five Markov Model.

The model is still not complete. States 4 and 5 are system success states where additional failures will fail the system. Figure 13-16 shows the list of failure rates in state 4. Any safe failure rate will take the Markov model to state 1. That group of failure rates is the same group as in states 2 and 3. Any dangerous detected failure rate will take the system to a new failure state where there is a combination of detected and undetected components. A repair from that new state will return to state 4. Any dangerous undetected failure will take the system to state 7 where all dangerous component failures are undetected. These are labeled:

$$\lambda_{12} = \lambda_{PSH1}^{DU} + \lambda_{VALVE1}^{DU} + \lambda_C^{DU} \qquad (13\text{-}23)$$

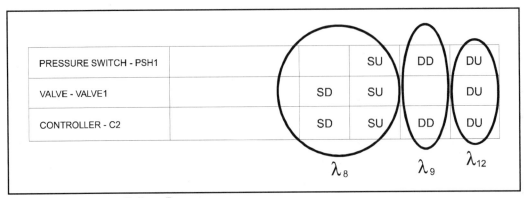

Figure 13-16. State 4 Failure Rates.

State 5 has the same failure and repair rates. Figure 13-17 shows the arcs from states 4 and 5. Figure 13-18 shows these additions to the completed Markov model. The system is successful in states 0, 2, 3, 4, and 5. The system has failed safely in state 1. The system has failed dangerously in states 6, 7, 8, 9, 10, 11, 12, and 13.

Note that the repair arcs from dangerous system failure states where component failures are detected (states 6, 8, 9, 10, 11, 12, and 13) are valid only if repairs are made to the system without shutting it down. In some companies, operators are instructed to shut down the process if a dangerous detected failure occurs. The model could be changed to show this by replacing the current repair arc with an arc from those states to state 1 with the average shutdown rate (1/average shutdown time).

The Markov model could also change depending on maintenance policy. If it is policy that whenever a service call is made all system components are inspected and repaired, then repair arcs from states 6, 8, 9, 10, 11, 12, and 13 will all go to state 0.

Reliability and Safety Model Construction

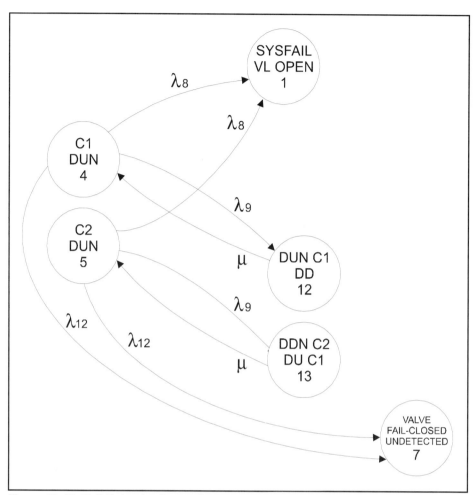

Figure 13-17. Exit Rates for States 4 and 5.

328 Reliability and Safety Model Construction

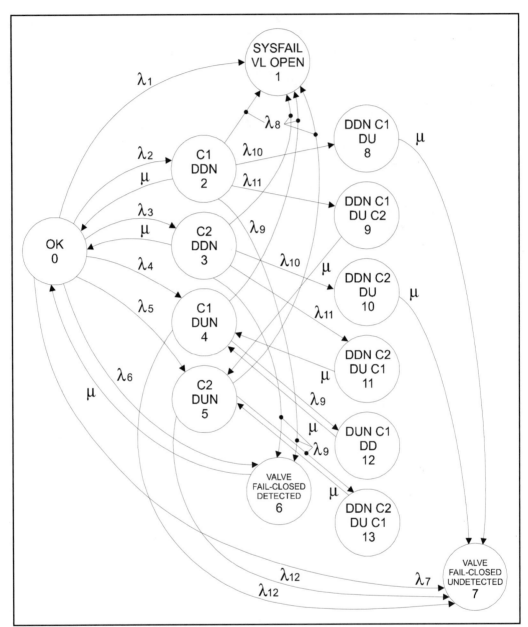

Figure 13-18. Completed Markov Model.

Simplify the Markov Model

In some Markov models, states can be merged (Ref. 1). The simple rule is: "When two states have transition rates that are identical to common states, those two states can be merged into one. Entry rates are added. Exit rates remain the same." Within the model, one obvious simplification can be made when it is assumed that all components are inspected and repaired when a service call is made. States 6, 8, 9, 10, 11, 12, and 13 can all be merged into state 6. This is shown in Figure 13-19.

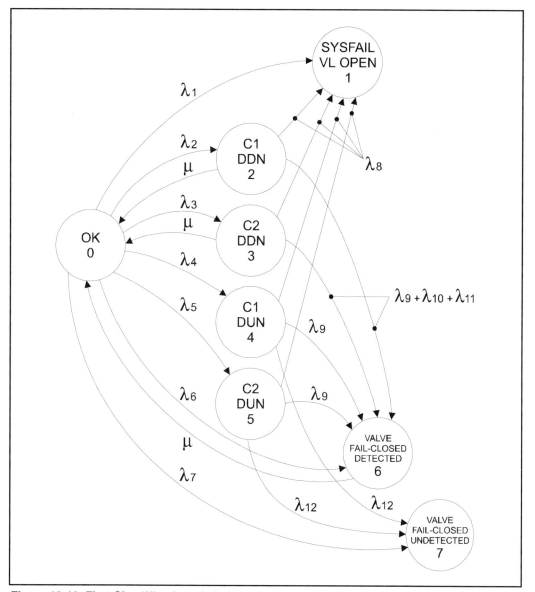

Figure 13-19. First Simplification of Markov Model.

Further state merging is possible. Both state 2 and state 3 have identical transition rates to state 0. These two states also have the same exit rate to states 1, 6, and 7. There are no more exit rates to check; these two states can be merged. A similar situation exists for states 3 and 4, which also can be merged. A repair rate may be added from state 1 to state 0 if a process is restarted after a shutdown. The simplified Markov model is redrawn in Figure 13-20.

Other simplification techniques involve decomposition of the system into stages and model truncation (Ref. 2). These techniques can be used by the experienced engineer, especially when only upper and lower bounds on appropriate reliability indices are necessary.

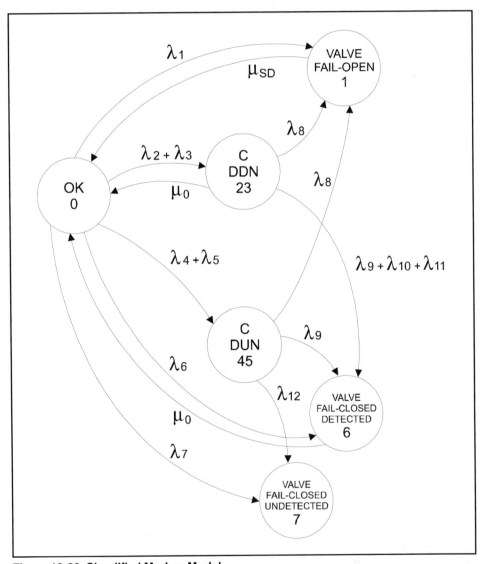

Figure 13-20. Simplified Markov Model.

Step 7: Solve for Reliability Indices

Using the techniques developed in Chapter 8, we can solve for the needed measures of system reliability. All calculations start with the **P** matrix. For Figure 13-20, the transition matrix, **P**, is

$$\mathbf{P} = \begin{bmatrix} 1-\left(\sum_{i=1}^{7}\lambda_i\right) & \lambda_1 & \lambda_2+\lambda_3 & \lambda_4+\lambda_5 & \lambda_6 & \lambda_7 \\ \mu_{SD} & 1-(\mu_{SD}) & 0 & 0 & 0 & 0 \\ \mu_O & \lambda_8 & 1-(\mu_O+\lambda_8+\lambda_9+\lambda_{10}+\lambda_{11}) & 0 & \lambda_9+\lambda_{10}+\lambda_{11} & 0 \\ 0 & \lambda_8 & 0 & 1-(\lambda_8+\lambda_9+\lambda_{12}) & \lambda_9 & \lambda_{12} \\ \mu_O & 0 & 0 & 0 & 1-\mu_O & 0 \\ 0 & 0 & 0 & 0 & 0 & 1 \end{bmatrix}$$

From this matrix, many reliability and safety measures can be calculated, including MTTF, MTTFS, MTTFD, time-dependent PFD, time-dependent PFD$_{avg}$, availability, safety availability, PFS, and others. All calculations can be done on a personal computer spreadsheet.

Sensors and Final Elements

In many systems the sensors, final elements, and even power supplies can be modeled as subsystems. Care must be taken, however. If the components are wired such that the failure of one device will affect another, the failures are not independent. Consider the system of Figure 13-21. Another sensor is added. Each sensor is wired directly into the corresponding controller. If pressure switch 1 fails dangerously, controller 1 cannot perform any safety function — although it can still fail with its outputs de-energized and cause a false trip.

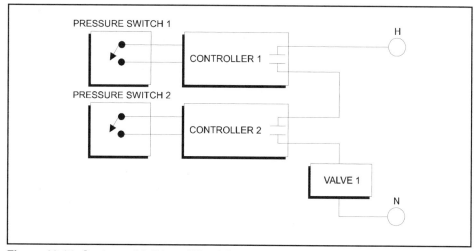

Figure 13-21. System with Direct Wired Sensors.

The dual-sensor system of Figure 13-22 operates very differently. In this configuration, sensors are cross-wired. The failure of one pressure switch will not affect operation of one controller. In this type of configuration, the sensor system can be modeled as an independent subsystem. In general, independence of subsystems depends on how the components are wired.

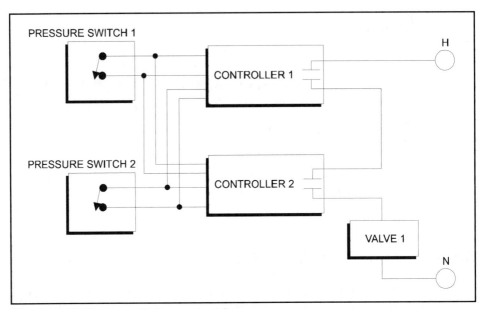

Figure 13-22. System with Crosswired Sensors.

EXERCISES

13.1 Why is a system-level FMEA necessary to model control system reliability accurately?

13.2 List degraded modes of operation that are valid for a control system. Are these modes of operation considered failures?

13.3 A control system has no on-line diagnostics and no parallel (redundant) components. The control system can fail only in a de-energized condition. How many failure rate categories exist?

13.4 Under what circumstances are common-cause failures distinguished from normal-mode failures?

13.5 When must "detected" versus "undetected" failures be modeled differently?

13.6 Solve the fault tree model of Figure 13-4 using data from Table 13-3 for PFD. Assume an MTTR for detected failures of 8 hr. Assume a periodic inspection period of one year (8,760 hr). Use approximation techniques to simplify the calculation.

13.7 Solve the Markov model of Figure 13-20 using data from Table 13-3 for PFD at one year. Assume an MTTR for detected failures of 8 hr. Assume an MTTR for shutdowns of 24 hr. Assume a periodic inspection interval of one year.

REFERENCES

1. Shooman, M. L., and Laemmel, A. E., "Simplification of Markov Models by State Merging," *1987 Proceedings of the Annual Reliability and Maintainability Symposium*, New York: IEEE, 1987.

2. Laemmel, A. E., and Shooman, M. L., "Bounding and Approximating Markov Models," *1990 Proceedings of the Annual Reliability and Maintainability Symposium*, New York: IEEE, 1990.

3. Houtermans, M. J. M., and Brombacher, A. C., "Automatic Synthesis of Markov Models: Creating Markov Models for Quantitative Safety Assessment in Process Industry," *Proceedings of the Probabilistic Safety Assessment and Management Conference*, June 1996.

4. Brombacher, A. C., "New Trends in Reliability Analysis," *ISCAS 1994*, London, June 1994.

5. Babcock, P. S. IV, Rosch, G., and Zinchuk, J. J., "An Automated Environment for Optimizing Fault-Tolerant Systems Designs," *1991 Proceedings of the Annual Reliability and Maintainability Symposium*, New York: IEEE, 1991.

14
System Architectures

Introduction

There are many ways in which to arrange control-system components when building a system. Some arrangements have been designed to maximize the probability of successful operation (reliability or availability). Some arrangements have been designed to minimize the probability of failure with outputs energized. Some arrangements have been designed to minimize the probability of failure with outputs de-energized. Other arrangements exist to protect against other specific failure modes.

These various arrangements of control-system components are referred to as *system architectures*. This chapter presents a few of the common system architectures for programmable electronic systems (PES). They are representative of the many architectures that exist in practice. Actual systems may even be combinations of the common arrangements presented in this chapter. It should be remembered that field devices should also be added to a complete system analysis.

Controller Construction

The various architectures and their characteristics can be explained using a simple model for a programmable electronic controller. Figure 14-1 shows a simple model of a "single-board controller" (SBC) with input circuits, output circuits, and common circuitry. This level of detail is sufficient for architecture comparison. The single-board architecture is characteristic of many smaller programmable logic controllers (PLC) and single-loop controllers (SLC). These units typically have a power supply, a microprocessor, and some quantity of industrial process I/O points on a single printed circuit board (PCB) assembly. This board is a single repairable unit; therefore, we can assume that any component failure will cause the controller to fail. The failure rate for the controller is the sum of the failure rates for each circuit block.

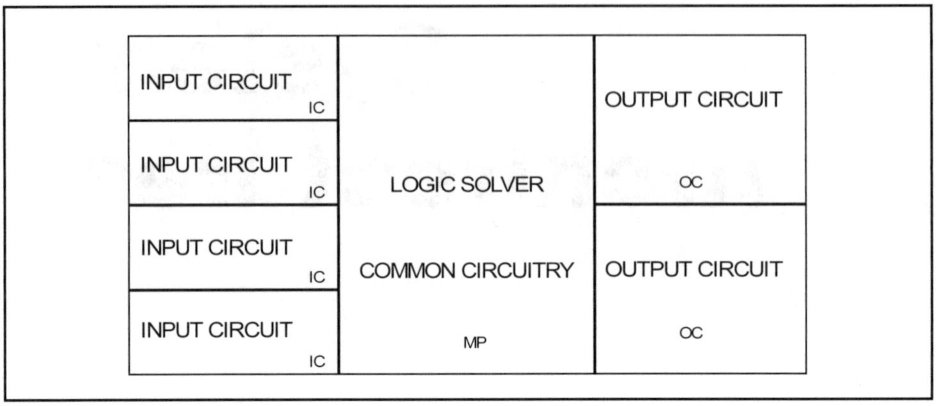

Figure 14-1. Single Board Controller Model.

EXAMPLE 14-1

Problem: An FMEDA analysis is done on a micro PLC with eight inputs and four outputs. The results of the input circuit analysis indicate the following failure rate categories (refer to Table 5-3 for FMEDA):

$$\lambda_{IC}^{SD} = 0$$

$$\lambda_{IC}^{SU} = 70$$

$$\lambda_{IC}^{DD} = 0$$

$$\lambda_{IC}^{DU} = 24$$

All failure rates are in units of FITS (failures per billion hours). The results of the output circuit FMEDA are

$$\lambda_{OC}^{SD} = 26$$

$$\lambda_{OC}^{SU} = 27$$

$$\lambda_{OC}^{DD} = 4$$

$$\lambda_{OC}^{DU} = 34$$

EXAMPLE 14-1 (continued)

The results of the common-circuitry FMEDA are

$$\lambda_{MP}^{SD} = 3790$$

$$\lambda_{MP}^{SU} = 2210$$

$$\lambda_{MP}^{DD} = 2890$$

$$\lambda_{MP}^{DU} = 2720$$

What are the failure rate categories for the single-board controller?

Solution: The unit has eight input circuits, four output circuits, and one common set of circuitry. It is assumed that this is a series system (where the failure of any circuit will cause failure of the entire unit), so the failure rates may be added per Equation 7-6. For the SBC:

$$\lambda_{SBC}^{SD} = 8 \times 0 + 4 \times 26 + 3790$$
$$= 3894 \text{ FITS}$$

$$\lambda_{SBC}^{SU} = 8 \times 70 + 4 \times 27 + 2210$$
$$= 2878 \text{ FITS}$$

$$\lambda_{SBC}^{DD} = 8 \times 0 + 4 \times 4 + 2890$$
$$= 2906 \text{ FITS}$$

$$\lambda_{SBC}^{DU} = 8 \times 24 + 4 \times 34 + 2720$$
$$= 3048 \text{ FITS}$$

Conventional	SD	SU	DD	DU	
IC	0	70	0	24	
MP	3790	2210	2890	2720	
OC	26	27	4	34	Total
Total	3894	2878	2906	3048	12,726

EXAMPLE 14-2

Problem: An FMEDA analysis is done on a small safety PLC with eight inputs and four outputs. This unit was designed specifically for critical applications. The results of the critical input circuit analysis indicate the following failure rate categories (refer to Table 5-4 for FMEDA):

$$\lambda_{IC}^{SD} = 87$$

$$\lambda_{IC}^{SU} = 1$$

$$\lambda_{IC}^{DD} = 22$$

$$\lambda_{IC}^{DU} = 1$$

The results of the critical output circuit FMEDA are

$$\lambda_{OC}^{SD} = 85$$

$$\lambda_{OC}^{SU} = 10$$

$$\lambda_{OC}^{DD} = 50$$

$$\lambda_{OC}^{DU} = 1$$

The results of the critical common-circuitry FMEDA are

$$\lambda_{MP}^{SD} = 8960$$

$$\lambda_{MP}^{SU} = 50$$

$$\lambda_{MP}^{DD} = 5580$$

$$\lambda_{MP}^{DU} = 100$$

What are the failure rate categories for the small safety PLC?

System Architectures

EXAMPLE 14-2 (continued)

Solution: The unit has eight input circuits, four output circuits, and one common set of circuitry. It is assumed that this is a series system, so the failure rates may be added per Equation 7-6. For the single safety controller (SSC):

$$\lambda_{SSC}^{SD} = 8 \times 87 + 4 \times 85 + 8960$$
$$= 9996 \text{ FITS}$$

$$\lambda_{SSC}^{SU} = 8 \times 1 + 4 \times 10 + 50$$
$$= 98 \text{ FITS}$$

$$\lambda_{SSC}^{DD} = 8 \times 22 + 4 \times 50 + 5580$$
$$= 5956 \text{ FITS}$$

$$\lambda_{SSC}^{DU} = 8 \times 1 + 4 \times 1 + 100$$
$$= 112 \text{ FITS}$$

Critical	SD	SU	DD	DU	
IC	87	1	22	1	
MP	8960	50	5580	100	
OC	85	10	50	1	Total
Total	9996	98	5956	112	16,162

EXAMPLE 14-3

Problem: Two units of SBC electronics (data from Example 14-1) will be wired together to form a two-channel system. Common cause should be modeled. Therefore, the failure rates must be split into common cause versus normal. The beta factor is estimated at 0.03. What are the common-cause and normal failure rates for the conventional micro PLC?

Solution: The four failure rate categories are divided using Equations 12-21 to 12-28:

$$\lambda_{SBC}^{SDC} = 3894 \times 0.03$$
$$= 116.82 \text{ FITS}$$

$$\lambda_{SBC}^{SUC} = 2878 \times 0.03$$
$$= 86.34 \text{ FITS}$$

$$\lambda_{SBC}^{DDC} = 2906 \times 0.03$$
$$= 87.18 \text{ FITS}$$

$$\lambda_{SBC}^{DUC} = 3048 \times 0.03$$
$$= 91.44 \text{ FITS}$$

$$\lambda_{SBC}^{SDN} = 3894 \times (1 - 0.03)$$
$$= 3777.18 \text{ FITS}$$

$$\lambda_{SBC}^{SUN} = 2878 \times (1 - 0.03)$$
$$= 2791.66 \text{ FITS}$$

$$\lambda_{SBC}^{DDN} = 2906 \times (1 - 0.03)$$
$$= 2818.82 \text{ FITS}$$

$$\lambda_{SBC}^{DUN} = 3048 \times (1 - 0.03)$$
$$= 2956.56 \text{ FITS}$$

Beta	0.03								
Conventional	SDC	SUC	DDC	DUC	SDN	SUN	DDN	DUN	Total
	116.82	86.34	87.18	91.44	3777.18	2791.66	2818.82	2956.56	12,726

To check for errors, the total is calculated. It should match the previous total; it does.

> **EXAMPLE 14-4**
>
> **Problem:** Two units of small safety controller electronics (data from Example 14-2) will be wired together to form a dual-channel system. Common cause should be modeled. Therefore, the failure rates must be split into common cause versus normal. The beta factor is estimated at 0.03. What are the common-cause and normal failure rates for the safety PLC?
>
> **Solution:** The four failure rate categories are divided using Equations 12-21 to 12-28.
>
> | Beta | 0.03 | | | | | | | | |
> |---|---|---|---|---|---|---|---|---|---|
> | Critical | | SDC | SUC | DDC | DUC | SDN | SUN | DDN | DUN | Total |
> | | | 299.88 | 2.94 | 178.68 | 3.36 | 9696.12 | 95.06 | 5777.32 | 108.64 | 16,162 |

System Configurations

A number of controller configurations exist in real implementations of control systems. Some simplified configurations are representative. For these simplified configurations, fault trees and Markov models are developed. A number of assumptions have been made in these models. For example, it is assumed that an SBC assembly is used, and that any component failure on the SBC will cause the entire controller to fail. Constant failure rates and repair rates are assumed, as are two failure modes labeled "safe" and "dangerous" in accordance with a de-energize-to-trip system. The models will account for on-line diagnostic capability and common cause. The Markov models will be solved using time-dependent solutions, and no periodic repair rates will be used so that the model will show state probabilities without inspection.

It is assumed that maintenance policies allow the quick repair of detected dangerous system failures without shutting down the process. Perfect inspection and repair is assumed. The models also assume that during a repair call, all pieces of equipment are inspected and repaired if failures exist. This may or may not be realistic, depending on training and maintenance policy. In some architectures (1oo1D, 2oo2D, and 1oo2D), the self-diagnostic circuitry can directly control an output switch. The Markov models for these systems assume that the self-diagnostic detection time is much less than system repair time. This assumption is very realistic since most self-diagnostic detection times are in the range of 1 to 5 sec.

1oo1: Single-Channel System

The single controller with single microprocessing unit (MPU) and single I/O (Figure 14-2) represents a minimum system. No fault tolerance is provided by this system, nor is failure mode protection. The electronic circuits can fail safely (outputs de-energized, open circuit) or dangerously (outputs frozen or energized, short circuit). Since the effects of on-line diagnostics should be modeled, four failure categories are included: DD, dangerous detected; DU, dangerous undetected; SD, safe detected; and SU, safe undetected.

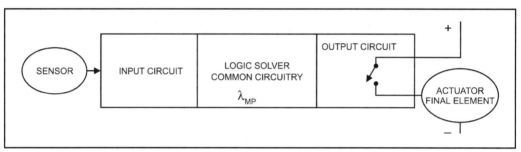

Figure 14-2. 1oo1 Architecture.

PFD Fault Tree for 1oo1

Figure 14-3 shows the fault tree for dangerous failures. The system can fail dangerously if the unit fails DD or DU. Using rough, first-order approximation techniques, a simple formula can be generated from the fault tree for the probability of dangerous failure, or probability of failure on demand (PFD):

$$\text{PFD}_{1oo1} = \lambda^{DD} \times RT + \lambda^{DU} \times TI \qquad (14\text{-}1)$$

where RT is the actual repair time and TI is the test interval for a periodic inspection. It should be pointed out that the approximation techniques are valid only for very small system failure rates.

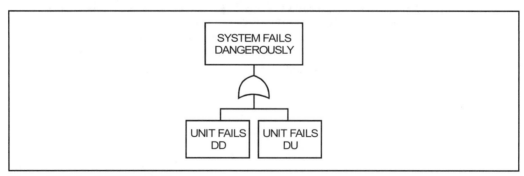

Figure 14-3. 1oo1 PFD Fault Tree.

System Architectures

EXAMPLE 14-5

Problem: An SBC is used in a 1oo1 architecture system. The failure rates are obtained from an FMEDA (Example 14-1). The average repair time is 8 hr. A periodic inspection is performed every year (8,760 hr). The average time to restart the process after a shutdown is 24 hr. What is the approximate PFD?

Solution: Using Equation 14-1,

$$\text{PFD}_{1oo1} = 0.291 \times 10^{-5} \times 8 + 0.305 \times 10^{-5} \times 8760$$

$$= 0.0267$$

EXAMPLE 14-6

Problem: A small safety controller is used in a 1oo1 architecture system. The failure rates are obtained from an FMEDA (Example 14-2). The average repair time is 8 hr. A periodic inspection is performed every year (8,760 hr). The average time to restart the process after a shutdown is 24 hr. What is the approximate PFD?

Solution: Using Equation 14-1,

$$\text{PFD}_{1oo1} = 0.596 \times 10^{-5} \times 8 + 0.11 \times 10^{-6} \times 8760$$

$$= 0.001$$

PFS Fault Tree for 1oo1

Figure 14-4 shows that the system will fail safely if the unit fails with SD or SU failures. The same rough approximation techniques can be used to generate a formula for the probability of failing safely (PFS):

$$\text{PFS}_{1oo1} = \lambda^{SD} \times SD + \lambda^{SU} \times SD \tag{14-2}$$

where SD is the time required to restart the process after a shutdown.

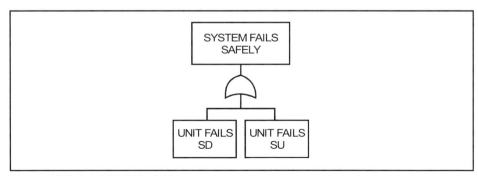

Figure 14-4. 1oo1 PFS Fault Tree.

> **EXAMPLE 14-7**
>
> **Problem:** A system consists of a 1oo1 SBC with failure rates from Example 14-1. The average repair time is 8 hr. A periodic inspection is performed once a year (8,760 hr). The average time to restart the process after a shutdown is 24 hr. What is the approximate PFS?
>
> **Solution:** Using Equation 14-2,
>
> $$PFD_{1oo1} = 0.389 \times 10^{-5} \times 24 + 0.288 \times 10^{-5} \times 24$$
> $$= 0.00016$$

> **EXAMPLE 14-8**
>
> **Problem:** A small safety controller is used in a 1oo1 architecture system. The failure rates are obtained from an FMEDA (Example 14-2). The average repair time is 8 hr. A periodic inspection is performed every year (8,760 hr). The average time to restart the process after a shutdown is 24 hr. What is the approximate PFS?
>
> **Solution:** Using Equation 14-2,
>
> $$PFD_{1oo1} = 0.1 \times 10^{-4} \times 24 + 1 \times 10^{-7} \times 24$$
> $$= 0.00024$$

Markov Model for 1oo1

The 1oo1 architecture can also be modeled using a Markov model (Figure 14-5). In the Markov model for this configuration, state 0 represents the condition where there are no failures. From this state, the controller can reach three other states. State 1 represents the fail-safe condition. In this state, the controller has failed with its outputs de-energized. State 2 represents the fail-danger condition with a detected failure. In this state, the controller has failed with its outputs energized, but the failure is detected by diagnostics and can be repaired. The system has also failed dangerously in state 3, but the failure is not detected by on-line diagnostics.

System Architectures

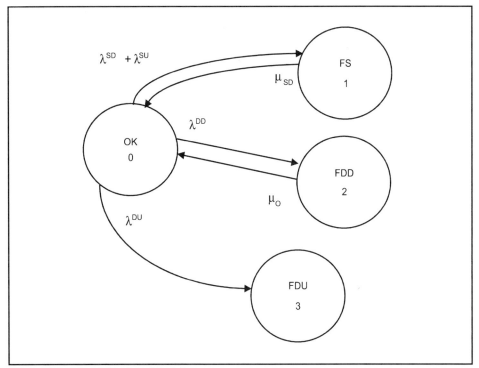

Figure 14-5. 1oo1 Markov Model.

The transition matrix, **P**, for the 1oo1 system is

$$\mathbf{P} = \begin{bmatrix} 1-(\lambda^S+\lambda^D) & \lambda^{SD}+\lambda^{SU} & \lambda^{DD} & \lambda^{DU} \\ \mu_{SD} & 1-\mu_{SD} & 0 & 0 \\ \mu_O & 0 & 1-\mu_O & 0 \\ 0 & 0 & 0 & 1 \end{bmatrix}$$

To derive the MTTF formula, use the techniques described in Chapter 8. As a first step, the transition matrix failure state rows and columns are truncated. This operation yields the **Q** matrix:

$$\mathbf{Q} = [1-(\lambda^S+\lambda^D)]$$

This is subtracted from the identity matrix:

$$\begin{aligned}\mathbf{I}-\mathbf{Q} &= 1-[1-(\lambda^S+\lambda^D)] \\ &= \lambda^S+\lambda^D\end{aligned}$$

The **N** matrix is obtained from $[\mathbf{I} - \mathbf{Q}]^{-1}$. In this case:

$$\mathbf{N} = \frac{1}{\lambda^S + \lambda^D}$$

Since MTTF is a sum of the row elements of the **N** matrix for a given starting state, it is simply:

$$\text{MTTF} = \frac{1}{\lambda^S + \lambda^D} \tag{14-3}$$

EXAMPLE 14-9

Problem: Using failure rates for an SBC as given in Example 14-1, calculate MTTF (any failure, safe or dangerous) for a 1oo1 configuration built from that SBC.

Solution: The total safe failure rate is 6,772 FITS. The total dangerous failure rate is 5,954 FITS. Using Equation 14-3,

$$\text{MTTF} = \frac{1}{(12726 \times 10^{-9})}$$
$$= 78{,}579 \text{ hr}$$

EXAMPLE 14-10

Problem: Using failure rates for a small safety controller as given in Example 14-2, calculate MTTF for a 1oo1 configuration built from that small safety PLC.

Solution: The total safe failure rate is 10,094 FITS. The total dangerous failure rate is 6,068 FITS. Using Equation 14-3,

$$\text{MTTF} = \frac{1}{16162 \times 10^{-9}}$$
$$= 61{,}873 \text{ hr}$$

The time-dependent probabilities can be calculated by multiplying the **P** matrix times a row matrix, **S**, per the methods described in Chapter 8. Assuming that the unit is working properly when started, the system starts in state 0 and the **S** matrix is

$$\mathbf{S} = [1\ 0\ 0\ 0]$$

System Architectures

EXAMPLE 14-11

Problem: Calculate the PFS and PFD from the Markov model of a 1oo1 small safety controller using failure rates from Example 14-2. The average repair time is 8 hr. A periodic inspection is performed every year (8,760 hr). The average startup time after a shutdown is 24 hr.

Solution: The failure rate and repair rate numbers are substituted into a **P** matrix. The numeric matrix is

	0	1	2	3
0	0.9999838	0.0000101	0.0000060	0.0000001
1	0.0416667	0.9583333	0.0000000	0.0000000
2	0.1250000	0.0000000	0.8750000	0.0000000
3	0.0000000	0.0000000	0.0000000	1.0000000

The PFS at 8,760 hr is 0.0002419. The PFD is the sum of state 2 and state 3 probabilities. At 8,760 hr the PFD is 0.001.

The PFD for the single-board controller could be improved by the addition of an external "watchdog timer." As discussed in Chapter 9, many failures are detected by timing methods. Figure 14-6 shows a 1oo1 architecture with an external retriggerable timer added to enhance diagnostics. The controller must be programmed to generate a periodic pulse via one of its outputs.

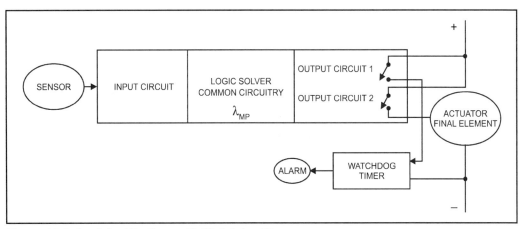

Figure 14-6. 1oo1 Architecture with Watchdog Timer.

1oo2: Dual-Channel System

Two controllers can be wired to minimize the effect of dangerous failures. For de-energize-to-trip systems, a series connection of two output circuits requires that both controllers fail in a dangerous manner for the system to fail dangerously. The 1oo2 configuration typically utilizes two independent main processors with their own independent I/O (see Figure 14-7). The system offers low probability of failure on demand, but it increases the probability of a fail-safe failure. The "false trip" rate is increased in order to improve the ability of the system to shut down the process.

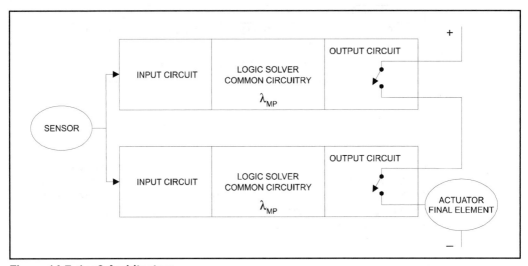

Figure 14-7. 1oo2 Architecture.

PFD Fault Tree for 1oo2

Figure 14-8 shows the PFD fault tree for the 1oo2 architecture. The system can fail dangerously if both units fail dangerously due to a common-cause failure, detected or undetected. Other than common cause, it can fail dangerously only if both A and B fail dangerously. A first-order approximation for PFD and PFS can be derived from the fault tree. The equation for PFD is

$$\text{PFD}_{1oo2} = \lambda^{DUC} \times TI + \lambda^{DDC} \times RT + (\lambda^{DDN} \times RT + \lambda^{DUN} \times TI)^2 \quad (14\text{-}4)$$

It should be noted that Figure 14-8 and Figure 12-11 are equivalent. Although they appear different, the logic is equivalent. Likewise, Equation 12-29 will give the same results as Equation 14-4.

System Architectures

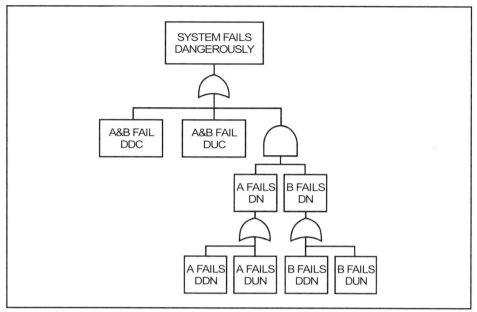

Figure 14-8. 1oo2 PFD Fault Tree.

EXAMPLE 14-12

Problem: Two SBCs are used in a 1oo2 architecture system. The failure rates are obtained from an FMEDA (Example 14-3). The average repair time is 8 hr. A periodic inspection is performed every year (8,760 hr). What is the approximate PFD?

Solution: Using Equation 14-4, $PFD_{1oo2} = 0.00147$. This is considerably better than the 1oo1 single-board controller PFD, which was 0.0267.

EXAMPLE 14-13

Problem: Two small safety controllers are wired into a 1oo2 architecture system. The failure rates are obtained from an FMEDA (Example 14-4). The average repair time is 8 hr. A periodic inspection is done every year (8,760 hr). The average time to restart the process after a shutdown is 24 hr. What is the approximate PFD?

Solution: Using Equation 14-4,

$$PFD_{1oo2} = 0.000032$$

PFS Fault Tree for 1oo2

Figure 14-9 shows the PFS fault tree for the 1oo2 architecture. This shows the trade-off in the architecture: Any safe failure from either unit will cause a false trip. An approximation for PFS can be quickly derived from the fault tree.

$$\text{PFS}_{1oo2} = (\lambda^{SDC} + \lambda^{SUC} + 2\lambda^{SDN} + 2\lambda^{SUN}) \times SD \tag{14-5}$$

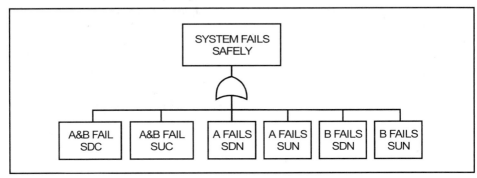

Figure 14-9. 1oo2 PFS Fault Tree.

EXAMPLE 14-14

Problem: Two SBCs are used in a 1oo2 architecture system. The failure rates are obtained from an FMEDA (Example 14-3). The average repair time is 8 hr. A periodic inspection is performed once a year (8,760 hr). The average time to restart the process after a shutdown is 24 hr. What is the approximate PFS?

Solution: Using Equation 14-5,

$$\text{PFS}_{1oo2} = 0.00032$$

EXAMPLE 14-15

Problem: Two small safety controllers are wired into a 1oo2 architecture system. The failure rates are obtained from an FMEDA (Example 14-4). The average repair time is 8 hr. A periodic inspection is performed every year (8,760 hr). The average time to restart the process after a shutdown is 24 hr. What is the approximate PFS?

Solution: Using Equation 14-5,

$$\text{PFS}_{1oo2} = 0.00048$$

Markov Model for 1oo2

The Markov model for a 1oo2 single-board system is shown in Figure 14-10. (Note: This is the same as Figure 12-12, with the same assumptions.) Three system success states exist. In state 0, both controllers operate. In state 1 and state 2, one controller has failed with outputs energized. The system is successful because the other controller can still de-energize as required. Since failures in state 1 are detected, an on-line repair rate returns from state 1 to state 0. States 3, 4, and 5 are the system failure states. In state 3, the system has failed with its outputs de-energized. In state 4, the system has failed detected with its outputs energized. An undetected failure with outputs energized has occurred in state 5. Note that only a dangerous common-cause failure will move the system from state 0 to states 4 or 5.

The on-line repair rate from state 4 to state 0 assumes that the repair technician will inspect the system and repair all failures when making a service call. If that assumption is not valid, state 4 must be split into two states: one with both units failed detected and the other with one detected and one undetected. The state with both detected will repair to state 0. The state with only one detected will repair to state 2. The assumption does simplify the model and is not significant unless coverage factors are low.

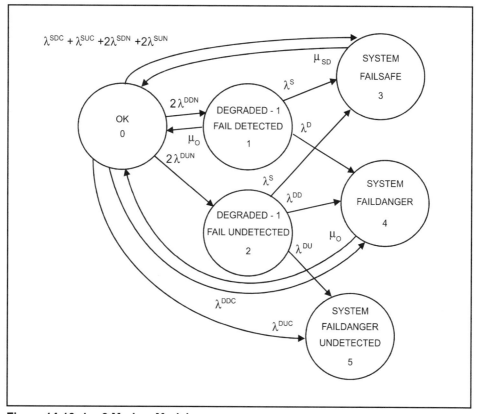

Figure 14-10. 1oo2 Markov Model.

The transition matrix, **P**, for the 1oo2 system is

$$P = \begin{bmatrix} 1-(\lambda^{DC}+2\lambda^{DN}+\lambda^{SC}+2\lambda^{SN}) & 2\lambda^{DDN} & 2\lambda^{DUN} & \lambda^{SC}+2\lambda^{SN} & \lambda^{DDC} & \lambda^{DUC} \\ \mu_O & 1-(\lambda^S+\lambda^D+\mu_O) & 0 & \lambda^S & \lambda^D & 0 \\ 0 & 0 & 1-(\lambda^S+\lambda^D) & \lambda^S & \lambda^{DD} & \lambda^{DU} \\ \mu_{SD} & 0 & 0 & 1-\mu_{SD} & 0 & 0 \\ \mu_O & 0 & 0 & 0 & 1-\mu_O & 0 \\ 0 & 0 & 0 & 0 & 0 & 1 \end{bmatrix}$$

Numeric solutions for PFD, PFS, MTTF, and other reliability metrics can be obtained from this matrix with a spreadsheet.

EXAMPLE 14-16

Problem: Two small safety controllers are used in a 1oo2 architecture. Using the failure rates from Example 14-4, calculate the system PFS and PFD at 8,760 hr and the system MTTF. The average repair time from a detectable failure is 8 hr. The average startup time from a shutdown is 24 hr.

Solution: MTTF is numerically calculated by following the method detailed in Chapter 8. First, the failure rate data are substituted into the transition matrix:

	0	1	2	3	4	5
0	0.9999682	0.0000116	0.0000002	0.0000199	0.0000002	0.0000000
1	0.1250000	0.8749838	0.0000000	0.0000101	0.0000061	0.0000000
2	0.0000000	0.0000000	0.9999838	0.0000101	0.0000060	0.0000001
3	0.0416667	0.0000000	0.0000000	0.9583333	0.0000000	0.0000000
4	0.1250000	0.0000000	0.0000000	0.0000000	0.8750000	0.0000000
5	0.0000000	0.0000000	0.0000000	0.0000000	0.0000000	1.0000000

with **P** = label on the left.

The **P** matrix is truncated to obtain the **Q** matrix. The **Q** matrix is subtracted from the identity matrix:

$$I - Q = \begin{bmatrix} 0.0000318 & -0.0000116 & -0.0000002 \\ -0.1250000 & 0.1250162 & 0.0000000 \\ 0.0000000 & 0.0000000 & 0.0000162 \end{bmatrix}$$

The **I – Q** matrix is inverted to obtain the **N** matrix.

$$N = \begin{bmatrix} 49295.10 & 4.56 & 662.72 \\ 49288.72 & 12.55 & 662.63 \\ 0.00 & 0.00 & 61873.53 \end{bmatrix}$$

System Architectures

> **EXAMPLE 14-16 (continued)**
>
> The MTTF is obtained by adding the row elements of the starting state. Assuming that the system starts in state 0,
>
> $$\text{MTTF} = 49295.1 + 4.6 + 662.7$$
> $$= 49{,}962 \text{ hr}$$
>
> The PFD and PFS at 8,760 hr are calculated by multiplying a starting row matrix (starting in state 0) times the **P** matrix repeatedly.
>
> $$\text{PFD} = 0.000032$$
> $$\text{PFS} = 0.000477$$

2oo2: Dual-Channel System

Another dual controller configuration was developed for situations where it is undesirable to fail with outputs de-energized. This system is used in energize-to-trip protection systems. The outputs of two controllers are wired in parallel (Figure 14-11). If one controller fails with its output de-energized, the other is still capable of energizing the load.

A disadvantage of the configuration is its susceptibility to failures in which the output is energized. If either controller fails with its output energized, the system has failed with output energized. This configuration is not suitable for de-energize-to-trip protection systems unless each unit is of an inherently fail-safe design.

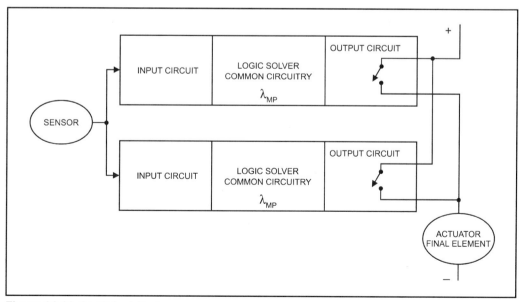

Figure 14-11. 2oo2 Architecture.

PFD Fault Tree for 2oo2

Since the controllers are wired in parallel, any short-circuit (dangerous) failure of the components results in a dangerous (outputs energized) failure of the system. This is shown in Figure 14-12. The first-order approximation equation to solve for PFD is

$$\text{PFD}_{2oo2} = \lambda^{DDC} \times RT + \lambda^{DUC} \times TI + 2\lambda^{DDN} \times RT + 2\lambda^{DUN} \times TI \qquad (14\text{-}6)$$

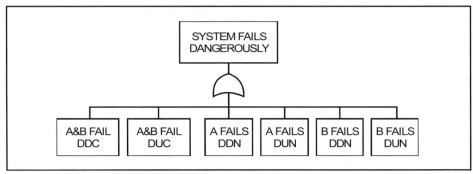

Figure 14-12. 2oo2 PFD Fault Tree.

EXAMPLE 14-17

Problem: Two SBCs are used in a 2oo2 architecture system. The failure rates are obtained from an FMEDA (Example 14-3). The average repair time is 8 hr. A periodic inspection is performed every year (8,760 hr). The average time to restart the process after a shutdown is 24 hr. What is the approximate PFD?

Solution: Using Equation 14-6,

$$\text{PFD}_{2oo2} = 0.053$$

EXAMPLE 14-18

Problem: Two small safety controllers are wired into a 2oo2 architecture system. The failure rates are obtained from an FMEDA (Example 14-4). The average repair time is 8 hr. A periodic inspection is performed every year (8,760 hr). The average time to restart the process after a shutdown is 24 hr. What is the approximate PFD?

Solution: Using Equation 14-6,

$$\text{PFD}_{2oo2} = 0.002$$

PFS Fault Tree for 2oo2

This architecture is designed to tolerate an open-circuit failure. The fault tree in Figure 14-13 shows this. The system will fail open circuit (de-energized, safe) if there is a safe common-cause failure. Other than common cause, an open-circuit failure on both A and B must occur. The first-order approximation equation to solve for PFS is

$$\text{PFS}_{2oo2} = \lambda^{SDC} \times SD + \lambda^{SUC} \times SD + (\lambda^{SDN} \times RT + \lambda^{SUN} \times TI)^2 \qquad (14\text{-}7)$$

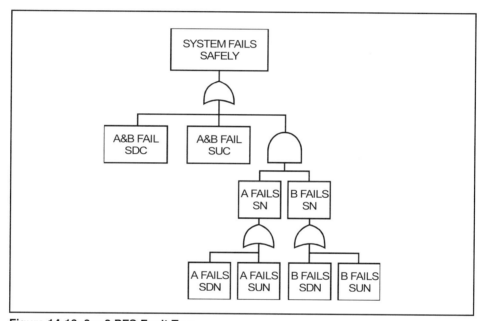

Figure 14-13. 2oo2 PFS Fault Tree.

EXAMPLE 14-19

Problem: Two SBCs are used in a 2oo2 architecture system. The failure rates are obtained from an FMEDA (Example 14-3). The average repair time is 8 hr. A periodic inspection is performed every year (8,760 hr). The average time to restart the process after a shutdown is 24 hr. What is the approximate PFS?

Solution: Using Equation 14-7,

$$\text{PFS}_{2oo2} = 0.0006$$

> **EXAMPLE 14-20**
>
> **Problem:** Two small safety controllers are wired into a 2oo2 architecture system. The failure rates are obtained from an FMEDA (Example 14-4). The average repair time is 8 hr. A periodic inspection is performed every year (8,760 hr). The average time to restart the process after a shutdown is 24 hr. What is the approximate PFS?
>
> **Solution:** Using Equation 14-7,
>
> $$PFS_{2oo2} = 0.000008$$

Markov Model for 2oo2

The single-board controller Markov model for the 2oo2 architecture is shown in Figure 14-14. The system is successful in three states: 0, 1, and 2. The system has failed with outputs de-energized in state 3. The system has failed with outputs energized in states 4 and 5. A comparison inspection of this Markov to that of the 1oo2 architecture will show a certain symmetry that makes sense given the architectures. The **P** matrix for this model is

$$\mathbf{P} = \begin{bmatrix} 1-(\lambda^{SC}+2\lambda^{SN}+\lambda^{DC}+2\lambda^{DN}) & 2\lambda^{SDN} & 2\lambda^{SUN} & \lambda^{SC} & \lambda^{DDC}+2\lambda^{DDC} & \lambda^{DUC}+2\lambda^{DUN} \\ \mu_O & 1-(\lambda^{S}+\lambda^{D}+\mu_O) & 0 & \lambda^{S} & \lambda^{D} & 0 \\ 0 & 0 & 1-(\lambda^{S}+\lambda^{D}) & \lambda^{S} & \lambda^{DD} & \lambda^{DU} \\ \mu_{SD} & 0 & 0 & 1-\mu_{SD} & 0 & 0 \\ \mu_O & 0 & 0 & 0 & 1-\mu_O & 0 \\ 0 & 0 & 0 & 0 & 0 & 1 \end{bmatrix}$$

Numeric solutions for the various reliability and safety metrics are practical and precise.

System Architectures

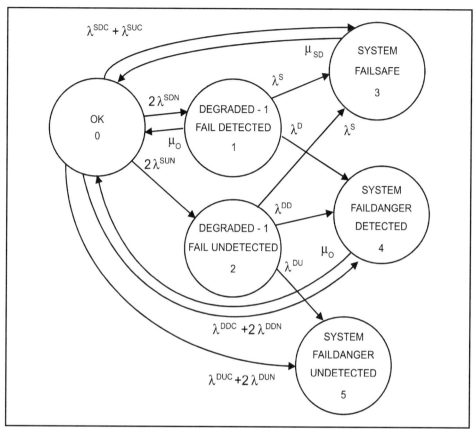

Figure 14-14. 2oo2 Markov Model.

EXAMPLE 14-21

Problem: Two small safety controllers are used in a 2oo2 architecture. Using the failure rates from Example 14-4, calculate the MTTF, PFS, and PFD for a time interval of 8,760 hr.

Solution: Substituting numerical values into the **P** matrix yields:

$$P = \begin{array}{c|cccccc} & 0 & 1 & 2 & 3 & 4 & 5 \\ \hline 0 & 0.9999682 & 1.94\text{E}{-}05 & 1.9\text{E}{-}07 & 3.03\text{E}{-}07 & 1.17\text{E}{-}05 & 2.21\text{E}{-}07 \\ 1 & 0.125 & 0.874984 & 0 & 1.01\text{E}{-}05 & 6.07\text{E}{-}06 & 0 \\ 2 & 0 & 0 & 0.999984 & 1.01\text{E}{-}05 & 5.96\text{E}{-}06 & 1.12\text{E}{-}07 \\ 3 & 0.0416667 & 0 & 0 & 0.958333 & 0 & 0 \\ 4 & 0.125 & 0 & 0 & 0 & 0.875 & 0 \\ 5 & 0 & 0 & 0 & 0 & 0 & 1 \end{array}$$

> **EXAMPLE 14-21 (continued)**
>
> Truncating the rows and columns of the failure states will provide the **Q** matrix. This matrix is subtracted from the identity matrix to obtain:
>
> $$\mathbf{I} - \mathbf{Q} = \begin{bmatrix} 3.184\text{E}{-}05 & -1.9\text{E}{-}05 & -1.9\text{E}{-}07 \\ -0.125 & 0.125016 & 0 \\ 0 & 0 & 1.62\text{E}{-}05 \end{bmatrix}$$
>
> The **I − Q** matrix is inverted to obtain the **N** matrix:
>
> $$\mathbf{N} = \begin{bmatrix} 80325.111 & 12.45986 & 944.8961 \\ 80314.727 & 20.45721 & 944.7739 \\ 0 & 0 & 61873.53 \end{bmatrix}$$
>
> Adding the row elements from row 0 provides the answer:
>
> $$\text{MTTF} = 81{,}282 \text{ hr}$$
>
> The PFS and PFD at 8,760 hr are calculated by multiplying the **P** matrix by a row matrix **S** starting with the system in state 0.
>
> $$\text{PFD} = 0.002$$
>
> $$\text{PFS} = 0.00000765$$
>
> Note that the PFS is very low. This is expected for the 2oo2 architecture.

1oo1D: Dual-Channel System

Figure 14-15 shows an architecture that uses a single controller channel with diagnostic capability and a second diagnostic channel wired in series to utilize the diagnostic signal to de-energize the output. This system represents an enhancement used for safety applications. A 1oo1D architecture can be built using an SBC and an external watchdog timer if the timer has an output that can be wired in series with the controller output. This is shown in Figure 14-16.

In more advanced systems, built-in diagnostics control an independent series output that will force the system to a de-energized state when a failure within the unit is detected. Diagnostics allow a detected dangerous failure to be converted into a safe failure. In general, additional failure rates must be included in quantitative analysis to account for the extra diagnostic channel. In systems using external diagnostic control devices (such as watchdog timers), additional failure rates for these external devices must be added to the single-board rates.

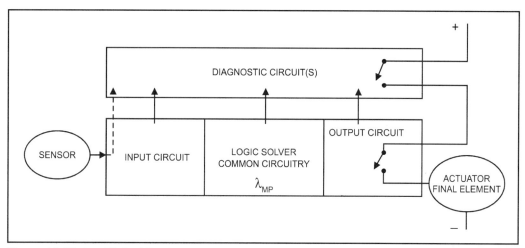

Figure 14-15. 1oo1D Architecture.

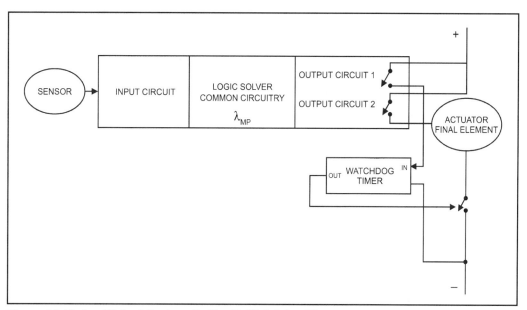

Figure 14-16. 1oo1D Architecture Built with Watchdog Timer.

PFD Fault Tree for 1oo1D

The 1oo1D architecture has a second diagnostic channel that will de-energize when failures are detected by the diagnostics. Therefore, the only failures that cause system failure with outputs energized are dangerous undetected failures. The fault tree has only one failure group, DU, as shown in Figure 14-17. The approximation for PFD is

$$\text{PFD}_{1oo1D} = \lambda^{DU} \times TI \tag{14-8}$$

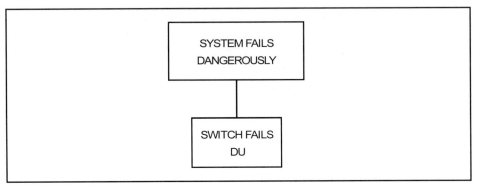

Figure 14-17. 1oo1D PFD Fault Tree.

EXAMPLE 14-22

Problem: A small safety controller is used in a 1oo1D architecture system. The failure rates are obtained from an FMEDA (Example 14-2); however, extra diagnostic channel components are added. The FMEDA results for the diagnostic channel are

$$\lambda^{SD} = 420 \text{ FITS}$$
$$\lambda^{SU} = 30 \text{ FITS}$$
$$\lambda^{DD} = 48 \text{ FITS}$$
$$\lambda^{DU} = 2 \text{ FITS}$$

The average repair time is 8 hr. A periodic inspection is done every year (8,760 hr). What is the approximate PFD?

Solution: First, the total failure rates are added since we assume that the failure of either the controller channel or the diagnostic channel will cause system failure. The total failure rates are

$$SD_{1oo1D} = 0.000010416$$
$$SU_{1oo1D} = 0.000000128$$
$$DD_{1oo1D} = 0.000006004$$
$$DU_{1oo1D} = 0.000000114$$

Using Equation 14-8,

$$\text{PFD}_{1oo1D} = 0.000000114 \times 8760 = 0.000999$$

PFS Fault Tree for 1oo1D

Figure 14-18 shows that a 1oo1D architecture will fail safely if the unit fails with SD, SU, or DD failures. The approximation techniques can be used to generate a formula for probability of failing safely from this fault tree:

$$\text{PFS}_{1oo1D} = (\lambda^{SD} + \lambda^{SU} + \lambda^{DD}) \times SD \qquad (14\text{-}9)$$

where SD is the time required to restart the process after a shutdown.

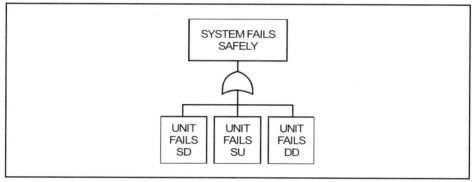

Figure 14-18. 1oo1D PFS Fault Tree.

EXAMPLE 14-23

Problem: A system consists of a 1oo1D small safety controller with failure rates from Example 14-22. The average time to restart the process after a shutdown is 24 hr. What is the approximate PFS?

Solution: Using Equation 14-9,

$$\text{PFS}_{1oo1D} = (0.1042 \times 10^{-4} + 0.13 \times 10^{-6} + 0.6 \times 10^{-5}) \times 24$$

$$= 0.000397$$

Markov Model for 1oo1D

The 1oo1D architecture can also be modeled using a Markov model (Figure 14-19). In the Markov model for this configuration, state 0 represents the condition where there are no failures. From this state, the controller can reach two other states. State 1 represents the fail-safe condition. In this state, the controller has failed with its outputs de-energized. The system has failed dangerously in state 3, and the failure is not detected by on-line diagnostics. The Markov model for the 1oo1D is similar to the 1oo1, except that the dangerous detected failures automatically trip the system (go to state 1).

The transition matrix, **P**, for the 1oo1D system is

$$\mathbf{P} = \begin{bmatrix} 1 - (\lambda^{SD} + \lambda^{SU} + \lambda^{DD} + \lambda^{DU}) & (\lambda^{SD} + \lambda^{SU} + \lambda^{DD}) & \lambda^{DU} \\ \mu_{SD} & 1 - \mu_{SD} & 0 \\ 0 & 0 & 1 \end{bmatrix}$$

The equation for 1oo1D MTTF is obtained in a manner similar to the 1oo1. After this process, it is discovered that this equation is identical to Equation 14-3. This makes sense. The 1oo1D architecture merely converts dangerous failures to safe failures; it does not provide any fault tolerance.

$$\text{MTTF} = \frac{1}{\lambda^S + \lambda^D}$$

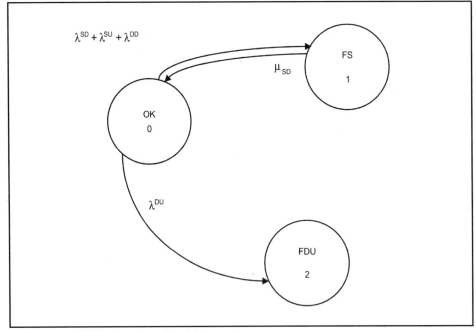

Figure 14-19. 1oo1D Markov Model.

System Architectures

EXAMPLE 14-24

Problem: Using failure rates given in Example 14-22, calculate the system MTTF for a 1oo1D architecture.

$$SD_{1oo1D} = 0.000010416$$
$$SU_{1oo1D} = 0.000000128$$
$$DD_{1oo1D} = 0.000006004$$
$$DU_{1oo1D} = 0.000000114$$

Solution: Using Equation 14-3,

$$MTTF = \frac{1}{0.1666 \times 10^{-4}}$$
$$= 60{,}016 \text{ hr}$$

EXAMPLE 14-25

Problem: Using failure rates given in Example 14-22, calculate the system PFS and PFD for a 1oo1D architecture. The average repair time is 8 hr. A periodic inspection is performed once a year (8,760 hr). The average time to restart the process after a shutdown is 24 hr.

Solution: Substituting numeric values into the **P** matrix:

	0	1	2
0	0.999983	1.65E–05	1.14E–07
1	0.041667	0.958333	0
2	0	0	1

Multiplication by the **S** matrix is used to calculate time-dependent state probabilities. For a time interval of 8,760 hr:

$$PFD = 0.000997$$

$$PFS = 0.000396$$

2oo3: Triple Controller System

It is difficult, if not impossible, to choose between failure modes in some applications of control systems. The 1oo2 configuration reduces dangerous (output energized) failures. The 2oo2 configuration reduces safe failures (output de-energized). More sophisticated architectures are required when both failure modes must be protected against. An architecture designed to tolerate both "safe" and "dangerous" failures is the 2oo3 (two units out of three are required for the system to operate). This architecture provides both safety and high availability with three controller units.

Two outputs from each controller unit are required for each output channel. The two outputs from the three controllers are wired in a "voting" circuit, which determines the actual output (Figure 14-20). The output will equal the "majority." When two sets of outputs conduct, the load is energized. When two sets of outputs are off, the load is de-energized.

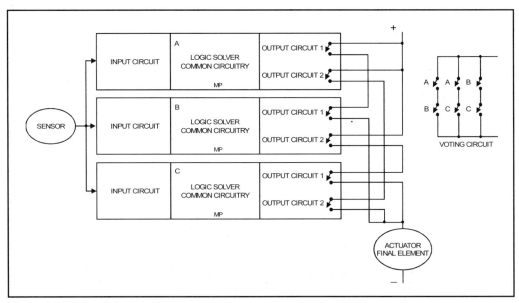

Figure 14-20. 2oo3 Architecture.

A closer examination of the voting circuit shows that it will tolerate a failure of either failure mode: dangerous (short circuit) or safe (open circuit). Figure 14-21 shows that when one unit fails open circuit, the system effectively degrades to a 1oo2 configuration. If one unit fails short circuit, the system effectively degrades to a 2oo2 configuration. In both cases, the system remains in successful operation.

System Architectures 365

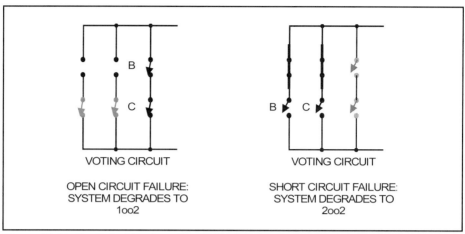

Figure 14-21. 2oo3 Architecture Single Fault Degradation Modes.

PFD Fault Tree for 2oo3

The 2oo3 architecture will fail dangerously only if two units fail dangerously (Figure 14-22). There are three ways in which this can happen: The AB leg can fail short circuit, the AC leg can fail short circuit, and the BC leg can fail short circuit. These are shown in the top-level events of the PFD fault tree of Figure 14-23. Each leg consists of two switches wired in series, as in a 1oo2 configuration. The subtree for each leg is developed for the 1oo2 configuration, and each looks like Figure 14-8 (the 1oo2 PFD fault tree). Note that the system will also fail if all three legs fail dangerously. This can happen due to common cause or a combination of three independent failures. Since this is a second-order effect, it can be assumed to be negligible for first-order approximation purposes. This is indicated in the fault tree with the "incomplete event" symbol. An approximation equation for PFD can be derived from the fault tree. The first-order approximate equation for PFD is

$$\text{PFD}_{2oo3} = 3\lambda^{DUC} \times TI + 3\lambda^{DDC} \times RT + 3(\lambda^{DDN} \times RT + \lambda^{DUN} \times TI)^2 \qquad (14\text{-}10)$$

Note that the simplified equations derived by approximation techniques are valid only for low failure rates.

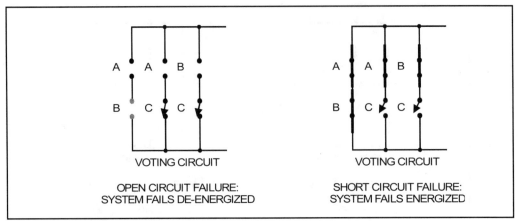

Figure 14-22. 2oo3 Architecture Dual Fault Failure Modes.

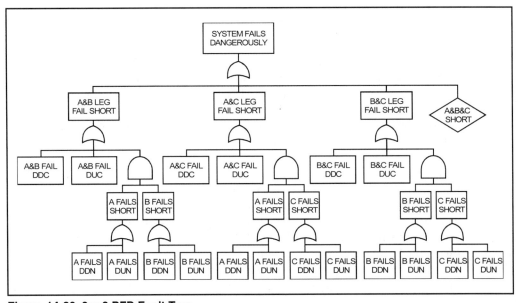

Figure 14-23. 2oo3 PFD Fault Tree.

System Architectures 367

EXAMPLE 14-26

Problem: Three SBCs are used in a 2oo3 architecture system. The failure rates are obtained from an FMEDA (Example 14-1). The system requires eight inputs and four outputs; therefore, each SBC will require eight input channels and eight output channels. The average repair time is 8 hr. A periodic inspection is performed every year (8,760 hr). The average time to restart the process after a shutdown is 24 hr. Common cause must be considered, and a beta factor of 0.03 is estimated. Calculate the total failure rate per unit and the approximate PFD using those failure rates.

Solution: The total failure rate per unit equals

$$\lambda^{SD} = 8 \times 0 + 3790 + 8 \times 26 = 3,998 \text{ FITS}$$

$$\lambda^{SU} = 8 \times 70 + 2210 + 8 \times 27 = 2,986 \text{ FITS}$$

$$\lambda^{DD} = 8 \times 0 + 2890 + 8 \times 4 = 2,922 \text{ FITS}$$

$$\lambda^{DU} = 8 \times 24 + 2720 + 8 \times 34 = 3,184 \text{ FITS}$$

The total failure rate equals

$$3998 + 2986 + 2922 + 3184 = 13,090 \text{ FITS}$$

Accounting for common cause:

$\lambda^{SDC} = 0.03 \times 0.000003998 \quad = 0.000000120$ failures per hr $= 120$ FITS

$\lambda^{SUC} = 0.03 \times 0.000002986 \quad = 0.000000090$ failures per hr $= 90$ FITS

$\lambda^{SDN} = (1 - 0.03) \times 0.000003998 = 0.000003878$ failures per hr $= 3878$ FITS

$\lambda^{SUN} = (1 - 0.03) \times 0.000002986 = 0.000002896$ failures per hr $= 2896$ FITS

$\lambda^{DDC} = 0.03 \times 0.000002922 \quad = 0.000000088$ failures per hr $= 88$ FITS

$\lambda^{DUC} = 0.03 \times 0.000003184 \quad = 0.000000096$ failures per hr $= 96$ FITS

$\lambda^{DDN} = (1 - 0.03) \times 0.000002922 = 0.000002834$ failures per hr $= 2834$ FITS

$\lambda^{DUN} = (1 - 0.03) \times 0.000003184 = 0.000003088$ failures per hr $= 3088$ FITS

Using these failure rates in Equation 14-10,

$$PFD_{2oo3} = 0.0047$$

This result should be compared with those of Example 14-12 (1oo2 SBC) and Example 14-17 (2oo2 SBC). The 1oo2 architecture PFD is 0.001474. The 2oo2 architecture PFD is 0.053. The 1oo2 PFD is about three times better than the 2oo3, but the 2oo3 is more than an order of magnitude better than the 2oo2.

EXAMPLE 14-27

Problem: Three small safety controllers are wired into a 2oo3 architecture system. The failure rates are obtained from an FMEDA (Example 14-2). The system requires eight inputs and four outputs; therefore, each SBC will require eight input channels and eight output channels. The average repair time is 8 hr. A periodic inspection is performed every year (8,760 hr). The average time to restart the process after a shutdown is 24 hr. The common-cause beta factor is estimated to be 0.03. Calculate the unit failure rates and the approximate PFD.

Solution: The unit failure rates are

$$\lambda^{SD} = 8 \times 87 + 8960 + 8 \times 85 = 10{,}336 \text{ FITS}$$

$$\lambda^{SU} = 8 \times 1 + 50 + 8 \times 10 = 138 \text{ FITS}$$

$$\lambda^{DD} = 8 \times 22 + 5580 + 8 \times 50 = 6{,}156 \text{ FITS}$$

$$\lambda^{DU} = 8 \times 1 + 100 + 8 \times 1 = 116 \text{ FITS}$$

The total failure rate equals

$$10{,}336 + 138 + 6156 + 116 = 16{,}746 \text{ FITS}$$

Accounting for common cause:

$\lambda^{SDC} = 0.03 \times 0.000010336 = 0.000000310$ failures per hr = 310 FITS

$\lambda^{SUC} = 0.03 \times 0.000000138 = 0.000000004$ failures per hr = 4 FITS

$\lambda^{SDN} = (1 - 0.03) \times 0.000010336 = 0.000010026$ failures per hr = 10026 FITS

$\lambda^{SUN} = (1 - 0.03) \times 0.000000138 = 0.000000134$ failures per hr = 134 FITS

$\lambda^{DDC} = 0.03 \times 0.000006156 = 0.000000185$ failures per hr = 185 FITS

$\lambda^{DUC} = 0.03 \times 0.000000116 = 0.000000003$ failures per hr = 3 FITS

$\lambda^{DDN} = (1 - 0.03) \times 0.000006156 = 0.000005971$ failures per hr = 5971 FITS

$\lambda^{DUN} = (1 - 0.03) \times 0.000000116 = 0.000000113$ failures per hr = 113 FITS

Substituting these failure rates into Equation 14-10,

$$PFD_{2oo3} = 0.000099$$

PFS Fault Tree for 2oo3

The 2oo3 is a symmetrical architecture that successfully tolerates a short-circuit or an open-circuit failure. It will fail with outputs de-energized only when two failures occur, as shown in Figure 14-22. The fault tree for safe failures is shown in Figure 14-24. It looks like the fault tree for dangerous failures, except that the failure modes are different. This is the result of the symmetrical nature of the architecture. Note that each major event in the top level of the fault tree is equivalent to the 2oo2 fault tree of Figure 14-13. The approximate equation for PFS derived from the fault tree is

$$\text{PFS}_{2oo3} = 3\lambda^{SUC} \times SD + 3\lambda^{SDC} \times SD + 3(\lambda^{SDN} \times RT + \lambda^{SUN} \times TI)^2 \quad (14\text{-}11)$$

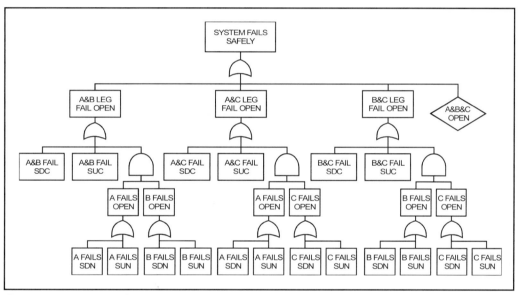

Figure 14-24. 2oo3 PFS Fault Tree.

EXAMPLE 14-28

Problem: Three SBCs are used in a 2oo3 architecture system. The average repair time is 8 hr. A periodic inspection is performed every year (8,760 hr). The average time to restart the process after a shutdown is 24 hr. Using failure rates from Example 14-26, what is the approximate PFS?

Solution: Using Equation 14-11,

$$\text{PFS}_{2oo3} = 0.00195$$

370 System Architectures

> **EXAMPLE 14-29**
>
> **Problem:** Three small safety controllers are wired into a 2oo3 architecture system. The average repair time is 8 hr. A periodic inspection is performed every year (8,760 hr). The average time to restart the process after a shutdown is 24 hr. Using failure rates from Example 14-27, what is the approximate PFS?
>
> **Solution:** Using Equation 14-11,
>
> $$PFS_{2oo3} = 0.000027$$

Markov Model for 2oo3

A Markov model can be created for the 2oo3 architecture. The model construction begins with all three units fully operational in state 0. All four failure modes of the three units must be placed in the model as exit rates from state 0. In addition, the four failure modes of common-cause failures must be included in the model. Since there are three combinations of two units — AB, AC, and BC — three sets of common-cause failures are included (much like the three sets in the fault tree diagrams). When all these failure rates are placed, the partial model looks like Figure 14-25. In state 1, one controller has failed in a safe detected manner. In state 2, one controller has failed in a safe undetected manner. In states 1 and 2, the system has degraded to a 1oo2. In state 3, one controller has failed in a dangerous detected manner. In state 4, one controller has failed in a dangerous undetected manner. The system has degraded to a 2oo2 in states 3 and 4. The system has not failed in any of these states.

From states 1 and 2, the system is operating. Further safe failures will fail the system with outputs de-energized (safe). Further dangerous failures lead to secondary degradation. From states 3 and 4, the system is operating in 2oo2 mode. Additional dangerous failures fail the system with outputs energized. Further safe failures degrade the system again. Repair rates are added to the diagram. It is assumed that the system is inspected and that all failed units are repaired if a service call is made. Therefore, all repair rates transition to state 0.

System Architectures

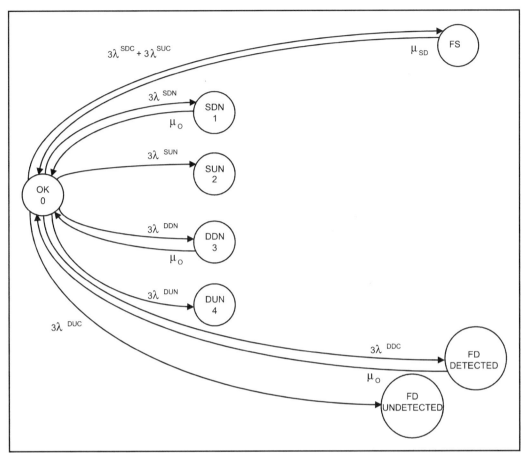

Figure 14-25. Partially Developed 2oo3 Markov Model.

The completed 2oo3 architecture Markov model is shown in Figure 14-26. Several interesting patterns are formed. The 2oo2 degradation mode when one dangerous failure occurs is shown in states 3, 5, 7, 9, 10, and 11. These are redrawn in Figure 14-27. Comparison of the failure rates in the 2oo2 Markov model of Figure 14-14 shows them to be nearly the same as in Figure 14-27. The transition matrix for Figure 14-26 is

$$\mathbf{P} = \begin{bmatrix} 1-\Sigma & 3\lambda^{SDN} & 3\lambda^{SUN} & 3\lambda^{DDN} & 3\lambda^{DUN} & 0 & 0 & 0 & 0 & 3\lambda^{SC} & 3\lambda^{DDC} & 3\lambda^{DUC} \\ \mu_O & 1-\Sigma & 0 & 0 & 0 & 2\lambda^{DDN} & 2\lambda^{DUN} & 0 & 0 & \lambda^{SC}+2\lambda^{SN} & \lambda^{DDC} & \lambda^{DUC} \\ 0 & 0 & 1-\Sigma & 0 & 0 & 0 & 0 & 2\lambda^{DDN} & 2\lambda^{DUN} & \lambda^{SC}+2\lambda^{SN} & \lambda^{DDC} & \lambda^{DUC} \\ \mu_O & 0 & 0 & 1-\Sigma & 0 & 2\lambda^{SDN} & 0 & 2\lambda^{SUN} & 0 & \lambda^{SC} & \lambda^{DC}+2\lambda^{DN} & 0 \\ 0 & 0 & 0 & 0 & 1-\Sigma & 0 & 2\lambda^{SDN} & 0 & 2\lambda^{SUN} & \lambda^{SC} & \lambda^{DDC}+2\lambda^{DDN} & \lambda^{DUC}+2\lambda^{DUN} \\ \mu_O & 0 & 0 & 0 & 0 & 1-\Sigma & 0 & 0 & 0 & \lambda^S & \lambda^D & 0 \\ \mu_O & 0 & 0 & 0 & 0 & 0 & 1-\Sigma & 0 & 0 & \lambda^S & \lambda^{DD} & \lambda^{DU} \\ \mu_O & 0 & 0 & 0 & 0 & 0 & 0 & 1-\Sigma & 0 & \lambda^S & \lambda^D & 0 \\ 0 & 0 & 0 & 0 & 0 & 0 & 0 & 0 & 1-\Sigma & \lambda^S & \lambda^{DD} & \lambda^{DU} \\ \mu_{SD} & 0 & 0 & 0 & 0 & 0 & 0 & 0 & 0 & 1-\Sigma & 0 & 0 \\ \mu_O & 0 & 0 & 0 & 0 & 0 & 0 & 0 & 0 & 0 & 1-\Sigma & 0 \\ 0 & 0 & 0 & 0 & 0 & 0 & 0 & 0 & 0 & 0 & 0 & 1 \end{bmatrix}$$

where $1 - \Sigma$ indicates one minus the sum of all other row elements. Reliability and safety metrics can be calculated using numerical techniques from this **P** matrix.

System Architectures

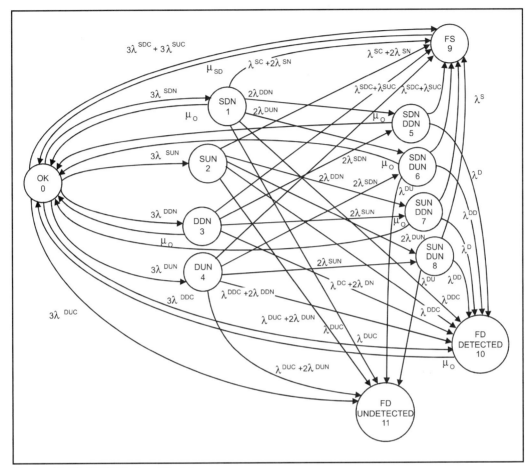

Figure 14-26. 2003 Markov Model.

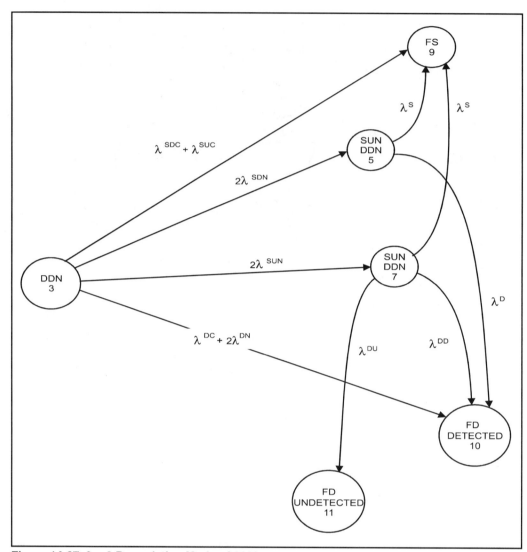

Figure 14-27. 2oo2 Degradation Mode of 2oo3.

EXAMPLE 14-30

Problem: Three small safety controllers are used in a 2oo3 architecture. Average repair time is 8 hr. Average startup time after a shutdown is 24 hr. Using the failure rates from Example 14-27 to account for two outputs per unit per channel, calculate the MTTF, PFS, and PFD for a time interval of 8,760 hr.

Solution: Substituting numerical values into the **P** matrix yields:

	0	1	2	3	4	5	6	7	8	9	10	11
0	0.9999498	3.008E–05	4.016E–07	1.791E–05	3.376E–07	0	0	0	0	9.4266E–07	5.5E–07	1E–08
1	0.125	0.874967	0	0	0	1.19E–05	2.25E–07	0	0	2.0634E–05	1.8E–07	3.5E–09
2	0	0	0.999967	0	0	0	0	1.194E–05	2.25E–07	2.0634E–05	1.8E–07	3.5E–09
P = 3	0.125	0	0	0.874967	0	2.005E–05	0	2.677E–07	0	3.1422E–07	1.2E–05	0
4	0	0	0	0	0.999967	0	2.005E–05	0	2.677E–07	3.1422E–07	1.2E–05	2.3E–07
5	0.125	0	0	0	0	0.8749833	0	0	0	1.0474E–05	6.3E–06	0
6	0.125	0	0	0	0	0	0.8749833	0	0	1.0474E–05	6.2E–06	1.2E–07
7	0.125	0	0	0	0	0	0	0.8749833	0	1.0474E–05	6.3E–06	0
8	0	0	0	0	0	0	0	0	0.9999833	1.0474E–05	6.2E–06	1.2E–07
9	0.0416667	0	0	0	0	0	0	0	0	0.95833333	0	0
10	0.125	0	0	0	0	0	0	0	0	0	0.875	0
11	0	0	0	0	0	0	0	0	0	0	0	1

Truncating the rows and columns of the failure states will provide the **Q** matrix. This matrix is subtracted from the identity matrix to obtain:

	0	1	2	3	4	5	6	7	8	9
0	5.024E–05	–3.008E–05	–4.016E–07	–1.791E–05	–3.376E–07	0	0	0	0	–9.427E–07
1	–0.125	0.125033	0	0	0	–1.194E–05	–2.25E–07	0	0	–2.063E–05
2	0	0	3.299E–05	0	0	0	0	–1.194E–05	–2.25E–07	–2.063E–05
I – Q = 3	–0.125	0	0	0.125033	0	–2.005E–05	0	–2.677E–07	0	–3.142E–07
4	0	0	0	0	3.299E–05	0	–2.005E–05	0	–2.677E–07	–3.142E–07
5	–0.125	0	0	0	0	0.1250167	0	0	0	–1.047E–05
6	–0.125	0	0	0	0	0	0.1250167	0	0	–1.047E–05
7	–0.125	0	0	0	0	0	0	0.1250167	0	–1.047E–05
8	0	0	0	0	0	0	0	0	1.675E–05	–1.047E–05
9	–0.0416667	0	0	0	0	0	0	0	0	0.04166667

The **I – Q** matrix is inverted to obtain the **N** matrix.

	0	1	2	3	4	5	6	7	8	9
0	1434546	345.09236	17462.613	205.53295	14678.718	0.0659322	2.3549883	1.6686143	469.34032	41.5047609
1	1434543.8	353.08973	17462.587	205.53264	14678.696	0.0666961	2.3549991	1.6686118	469.33961	41.5086591
2	1422672.7	342.23614	47630.646	203.83182	14557.227	0.0653865	2.3354968	4.5505122	872.80911	56.2754726
N = 3	1434404.2	345.05826	17460.888	213.51053	14677.268	0.0672085	2.3547556	1.6684665	469.29394	41.5007197
4	892851.14	214.78301	10868.605	127.92224	39448.494	0.0410357	6.3276581	1.0385336	776.7241	26.1839026
5	1434474	345.07505	17461.737	205.52264	14677.982	8.0648573	2.3548701	1.6685306	469.31677	41.5046894
6	1434474	345.07505	17461.737	205.52264	14677.982	0.0659289	10.353799	1.6685306	469.31677	41.5046894
7	1434474	345.07505	17461.737	205.52264	14677.982	0.0659289	2.3548701	9.667459	469.31677	41.5046894
8	897255.15	215.84244	10922.215	128.55322	9180.9923	0.0412381	1.4729576	1.0436562	60009.308	40.970791
9	1434546	345.09236	17462.613	205.53295	14678.718	0.0659322	2.3549883	1.6686143	469.34032	65.5047609

The MTTF is the sum of the state 0 row.

$$\text{MTTF} = 1{,}467{,}752 \text{ hr}$$

The PFS and PFD at 8,760 hr are calculated by multiplying the **P** matrix by a row matrix **S** starting with the system in state 0.

$$\text{PFD} = 0.0000986$$

$$\text{PFS} = 0.000024$$

2oo2D Architecture

The 2oo2D is a four-channel architecture that consists of two 1oo1D controllers arranged in a 2oo2 style (Figure 14-28). Since the 1oo1D protects against dangerous failures when diagnostics detect the failure, two units can be wired in parallel to protect against shutdowns. Effective diagnostics are important to this architecture, as an undetected dangerous failure on either unit will fail the system dangerously.

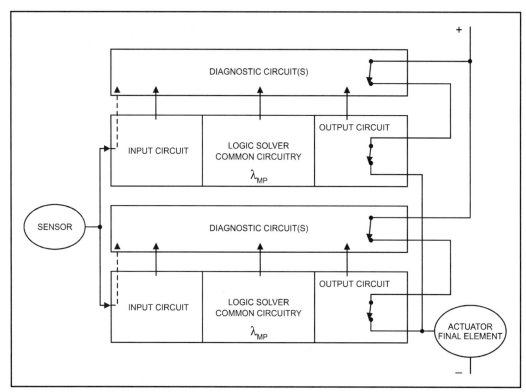

Figure 14-28. 2oo2D Architecture.

PFD Fault Tree for 2oo2D

The 2oo2D architecture will fail with outputs energized if either unit has a dangerous undetected failure or if the system experiences a dangerous undetected common-cause failure. This is shown in the fault tree of Figure 14-29. The approximate equation for PFD is

$$\text{PFD}_{2oo2D} = \lambda^{DUC} \times TI + 2\lambda^{DUN} \times TI \qquad (14\text{-}12)$$

System Architectures

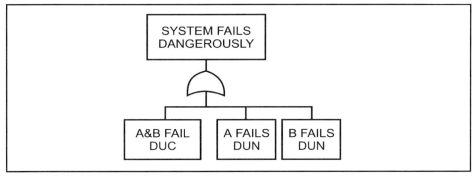

Figure 14-29. PFD Fault Tree for 2oo2D.

EXAMPLE 14-31

Problem: Two 1oo1D small safety controllers are used in a 2oo2D architecture system. The failure rates for each 1oo1D unit are (Example 14-22):

$$SD_{1oo1D} = 0.000010416$$
$$SU_{1oo1D} = 0.000000128$$
$$DD_{1oo1D} = 0.000006004$$
$$DU_{1oo1D} = 0.000000114$$

Using a beta factor of 0.03, what are the failure rates accounting for common cause? A periodic inspection is performed every year (8,760 hr). What is the approximate PFD?

Solution: The failure rates are divided using Equations 12-21 to 12-28:

$$SDC_{1oo1D} = 0.000000312$$
$$SUC_{1oo1D} = 0.000000004$$
$$SDN_{1oo1D} = 0.000010104$$
$$SUN_{1oo1D} = 0.000000124$$
$$DDC_{1oo1} = 0.000000180$$
$$DUC_{1oo1} = 0.000000003$$
$$DDN_{1oo1} = 0.000005824$$
$$DUN_{1oo1} = 0.000000111$$

Using Equation 14-12,

$$PFD_{2oo2D} = 0 \times 8760 + 2 \times 0.11 \times 10^{-6} \times 8760$$
$$= 0.00197$$

PFS Fault Tree for 2oo2D

Figure 14-30 shows that a 2oo2D architecture will fail safely only if both units fail safely. This can happen due to common-cause failures SDC, SUC, or DDC, or if A and B fail safely. The approximation techniques can be used to generate a formula for probability of failing safely from this fault tree:

$$\text{PFS}_{2oo2D} = (\lambda^{SDC} + \lambda^{SUC} + \lambda^{DDC}) \times SD + (\lambda^{SDN} \times RT + \lambda^{DDN} \times RT + \lambda^{SUN} \times TI)^2 \qquad (14\text{-}13)$$

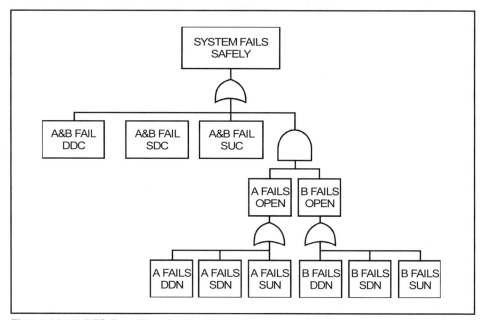

Figure 14-30. PFS Fault Tree for 2oo2D.

EXAMPLE 14-32

Problem: A system consists of two 1oo1D small safety controllers wired in a 2oo2D configuration with failure rates from Example 14-31. The average time to restart the process after a shutdown is 24 hr. Average repair time for a detected failure that does not shut down the process is 8 hr. The system is inspected every year (8,760 hr). What is the approximate PFS?

Solution: Using Equation 14-13,

$$\text{PFS}_{2oo2D} = 0.000013$$

Markov Model for 2oo2D

The Markov model for a 2oo2D system is presented in Figure 14-31, which shows three system success states that are similar to the other dual systems previously developed. State 1 is an interesting case. It represents a safe detected failure or a dangerous detected failure. The result of both failures is the same since the diagnostic cutoff switch de-energizes the output whenever a dangerous failure is detected. State 2, the only other system success state, represents the situation where one controller has failed in a safe undetected manner. The system operates because the other controller manages the load.

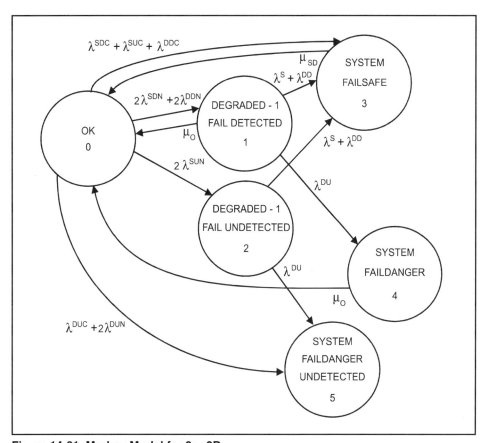

Figure 14-31. Markov Model for 2oo2D.

The 2oo2D architecture shows good tolerance to both "safe" and "dangerous" failures. However, since coverage is utilized to convert dangerous failures into safe failures, this tolerance depends in great part on the diagnostic coverage. The transition matrix for the single-board Markov model of Figure 14-31 is

$$\mathbf{P} = \begin{bmatrix} 1-\Sigma \ (2\lambda^{SDN}+2\lambda^{DDN}) & 2\lambda^{SUN} & (\lambda^{SC}+\lambda^{DDC}) & 0 & (\lambda^{DUC}+2\lambda^{DUN}) \\ \mu_O & 1-\Sigma & 0 & (\lambda^{S}+\lambda^{DD}) & (\lambda^{DU}) & 0 \\ 0 & 0 & 1-\Sigma & (\lambda^{S}+\lambda^{DD}) & 0 & (\lambda^{DU}) \\ \mu_{SD} & 0 & 0 & 1-\mu_{SD} & 0 & 0 \\ \mu_O & 0 & 0 & 0 & 1-\mu_O & 0 \\ 0 & 0 & 0 & 0 & 0 & 1 \end{bmatrix}$$

where Σ represents the sum of all other row elements.

EXAMPLE 14-33

Problem: Using failure rate values from Example 14-31, calculate the MTTF, PFS, and PFD of a 2oo2D system.

Solution: Substituting numerical values into the **P** matrix yields:

	0	1	2	3	4	5
0	0.99996718	0.00003185	0.00000025	0.00000050	0.00000000	0.00000022
1	0.12500000	0.87498334	0.00000000	0.00001655	0.00000011	0.00000000
2	0.00000000	0.00000000	0.99998334	0.00001655	0.00000000	0.00000011
3	0.04166667	0.00000000	0.00000000	0.95833333	0.00000000	0.00000000
4	0.12500000	0.00000000	0.00000000	0.00000000	0.87500000	0.00000000
5	0.00000000	0.00000000	0.00000000	0.00000000	0.00000000	1.00000000

Truncating and subtracting provides the numerical **I − Q** matrix:

$$\mathbf{I-Q} = \begin{bmatrix} 0.00003282 & -0.00003185 & -0.00000025 \\ -0.12500000 & 0.12501666 & 0.00000000 \\ 0.00000000 & 0.00000000 & 0.00001666 \end{bmatrix}$$

System Architectures

> **EXAMPLE 14-33 (continued)**
>
> The **N** matrix is obtained by inverting the **I – Q** matrix:
>
> $$N = \begin{bmatrix} 1027131.1 & 261.717561 & 15307.7179 \\ 1026994.21 & 269.681613 & 15305.6777 \\ 0 & 0 & 60016.8047 \end{bmatrix}$$
>
> Adding the row elements of state 0,
>
> MTTF = 1,042,700 hr
>
> PFD = 0.0019638 at 8760 hr
>
> PFS = 0.00001276 at 8760 hr

1oo2D Architecture

The 1oo2D architecture is similar to the 2oo2D, but has extra control lines to provide 1oo2 safety functionality. Figure 14-32 shows the 1oo2D; comparison with Figure 14-28 reveals added control lines. The 1oo2D is designed to tolerate both safe and dangerous failures.

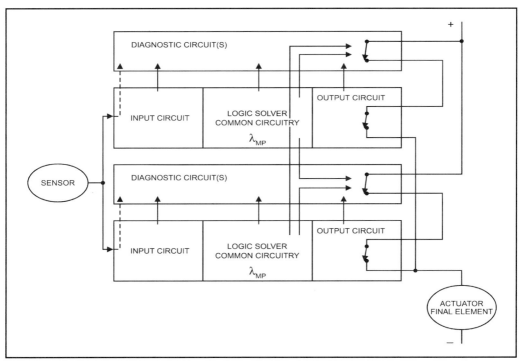

Figure 14-32. 1oo2D Architecture.

The primary difference between 2oo2D and 1oo2D can be seen when a dangerous undetected failure occurs in one unit. This is shown in Figure 14-33. The upper unit has a failure that causes one switch to be failed short circuit. The failure is not detected by the self-diagnostics in that unit, so the diagnostic switch is not opened by its control electronics. However, when the failure is detected by the other unit, the diagnostic switch is opened via the additional control line.

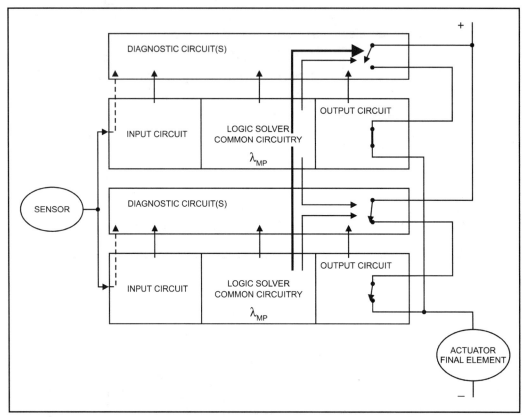

Figure 14-33. 1oo2D with Dangerous Undetected Failure in One Unit.

PFD Fault Tree for 1oo2D

The 1oo2D fails dangerously only if both units fail dangerously and if that failure is not detected by the diagnostics in either unit. The fault tree is shown in Figure 14-34. An approximate PFD equation developed from the fault tree is

$$\text{PFD}_{1oo2D} = \lambda^{DUC} \times TI + (\lambda^{DUN} \times TI)^2 \qquad (14\text{-}14)$$

Compare this with Equation 14-4 for the 1oo2 architecture. The 1oo2D should offer better safety performance than the 1oo2, because only undetected failures are included in the PFD.

System Architectures

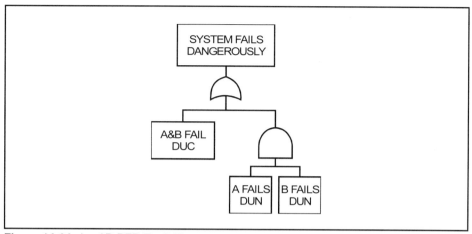

Figure 14-34. 1oo2D PFD Fault Tree.

EXAMPLE 14-34

Problem: Two small safety controllers with diagnostic channels are used in a 1oo2D architecture system. The failure rates for each unit are

$$SDC_{1oo1D} = 0.000000312$$
$$SUC_{1oo1D} = 0.000000004$$
$$SDN_{1oo1D} = 0.000010104$$
$$SUN_{1oo1D} = 0.000000124$$
$$DDC_{1oo1} = 0.000000180$$
$$DUC_{1oo1} = 0.000000003$$
$$DDN_{1oo1} = 0.000005824$$
$$DUN_{1oo1} = 0.000000111$$

What is the approximate PFD?

Solution: Using Equation 14-14,

$$PFD_{1oo2D} = 0.00003$$

This is slightly better than the 1oo2.

PFS Fault Tree for 1oo2D

Figure 14-35 shows that a 1oo2D architecture will fail safely if there is a common-cause safe failure, a common-cause dangerous detected failure, if either unit fails in a safe undetected manner, or if both units fail safely. The approximation techniques can be used to generate a formula for probability of failing safely from this fault tree:

$$\text{PFS}_{1oo2D} = (\lambda^{SDC} + \lambda^{SUC} + \lambda^{DDC}) \times SD + (\lambda^{SDN} \times RT + \lambda^{DDN} \times RT)^2 + (\lambda^{SUN} \times TI)^2 \quad (14\text{-}15)$$

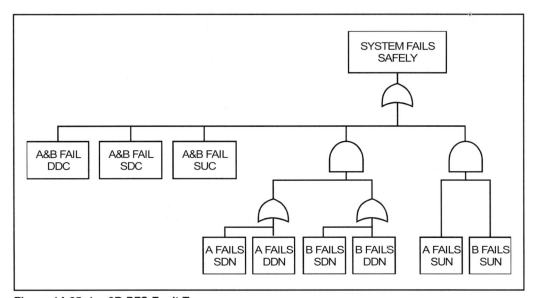

Figure 14-35. 1oo2D PFS Fault Tree.

EXAMPLE 14-35

Problem: A system consists of two small safety controllers with diagnostic channels wired in a 1oo2D configuration with failure rates from Example 14-34. The average time to restart the process after a shutdown is 24 hr. Average repair time for a detected failure that does not shut down the process is 8 hr. The system is inspected every year (8,760 hr). What is the approximate PFS?

Solution: Using Equation 14-15,

$$\text{PFS}_{1oo2D} = 0.000013$$

Markov Model for 1oo2D

In the Markov model for a 1oo2D system in Figure 14-36, four system success states are shown. State 1 represents a safe detected failure or a dangerous detected failure. Like the 2oo2D, the result of both failures is the same since the diagnostic switch de-energizes the output whenever a dangerous failure is detected. Another system success state, state 2, represents the situation where one controller has failed in a dangerous undetected manner. The system will operate correctly given a process demand because the other unit will detect the failure and de-energize the load via its 1oo2 control lines (Figure 14-33). The third system success state is shown in Figure 14-37. One unit has failed with its output de-energized. The system load is maintained by the other unit, which will still respond properly to a process demand.

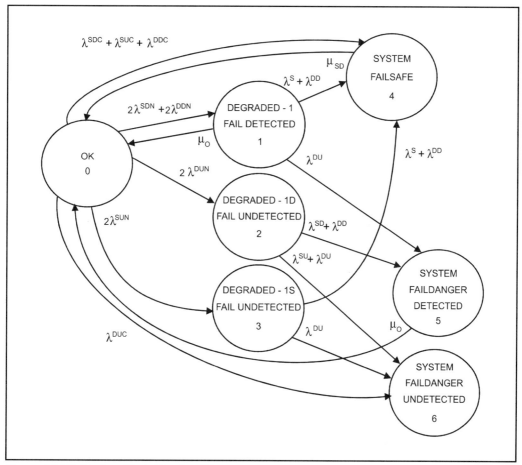

Figure 14-36. Markov Model for 1oo2D.

386   System Architectures

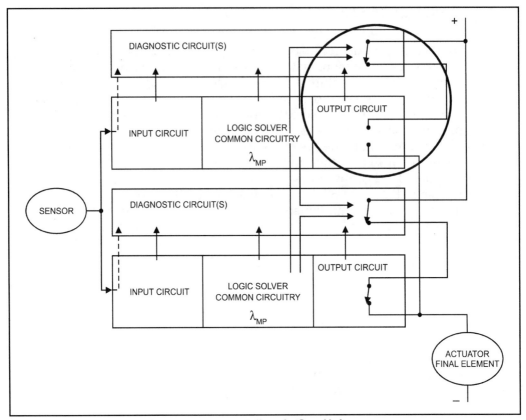

Figure 14-37. 1oo2D with Safe Undetected Failure in One Unit.

In state 1, the system has degraded to 1oo1D operation. A second safe failure or a dangerous detected failure will fail the system safely. Like the 1oo1D, a dangerous undetected failure will fail the system dangerously. In this model, though, the system fails to state 5, where one unit has failed detected and one unit has failed undetected. An assumption is that all units will be inspected and tested during a service call, so an on-line repair rate exits from state 5 to state 0.

One unit has failed with a dangerous undetected failure in state 2. The system is still successful since it will respond to a demand as described above. From this state, any other component failure will fail the system dangerously. In Figure 14-38, a dangerous undetected failure occurred first in the top unit and a safe detected failure has occurred in the lower unit. Since it is assumed that any component failure in a unit causes the entire unit to fail, it must be assumed that the control line from the lower unit to the upper unit will not work. Under those circumstances, the system will not respond to a process demand and is considered failed dangerously. Since the lower unit failed in a safe detected manner, the Markov model goes to state 5.

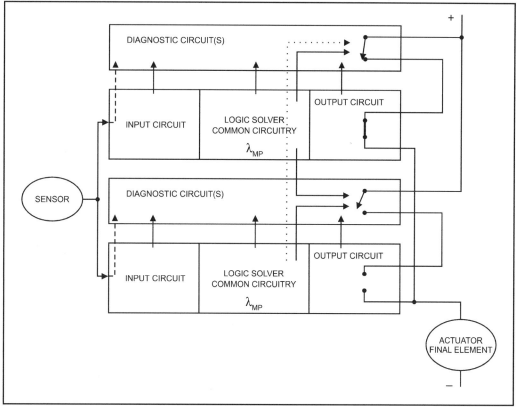

Figure 14-38. 1oo2D with Two Failures, State 5.

In state 3, one unit has failed in a safe undetected manner. In this condition, the system has also degraded to 1oo1D operation. Additional safe failures or dangerous detected failures will cause the system to fail safely. An additional dangerous undetected failure will fail the system dangerously — taking the Markov model to state 6, where both units have an undetected failure. Failures from this state are not detected until there is a maintenance inspection. The transition matrix for the Markov model of Figure 14-35 is

$$P = \begin{bmatrix} 1-\Sigma\,(2\lambda^{SDN}+2\lambda^{DDN}) & 2\lambda^{DUN} & 2\lambda^{SUN} & \lambda^{SC}+\lambda^{DDC} & 0 & \lambda^{DUC} \\ \mu_O & 1-\Sigma & 0 & 0 & \lambda^{S}+\lambda^{DD} & \lambda^{DU} & 0 \\ 0 & 0 & 1-\Sigma & 0 & 0 & \lambda^{SD}+\lambda^{DD} & \lambda^{SU}+\lambda^{DU} \\ 0 & 0 & 0 & 1-\Sigma & \lambda^{S}+\lambda^{DD} & 0 & \lambda^{DU} \\ \mu_{SD} & 0 & 0 & 0 & 1-\Sigma & 0 & 0 \\ \mu_O & 0 & 0 & 0 & 0 & 1-\Sigma & 0 \\ 0 & 0 & 0 & 0 & 0 & 0 & 1 \end{bmatrix}$$

where Σ represents the sum of all other row elements.

EXAMPLE 14-36

Problem: Using failure rate values from Example 14-31, calculate the MTTF, PFS, and PFD of a 1oo2D system.

Solution: Substituting numerical values into the **P** matrix yields:

$$P = \begin{array}{c|ccccccc} & 0 & 1 & 2 & 3 & 4 & 5 & 6 \\ \hline 0 & 0.999967672 & 0.000031855 & 0.000000221 & 0.000000248 & 0.000000496 & 0.000000000 & 0.000000003 \\ 1 & 0.125000000 & 0.874983338 & 0.000000000 & 0.000000000 & 0.000016548 & 0.000000114 & 0.000000000 \\ 2 & 0.000000000 & 0.000000000 & 0.999983338 & 0.000000000 & 0.000000000 & 0.000016420 & 0.000000242 \\ 3 & 0.000000000 & 0.000000000 & 0.000000000 & 1.000000000 & 0.000016548 & 0.000000000 & 0.000000114 \\ 4 & 0.041666667 & 0.000000000 & 0.000000000 & 0.000000000 & 0.958333333 & 0.000000000 & 0.000000000 \\ 5 & 0.125000000 & 0.000000000 & 0.000000000 & 0.000000000 & 0.000000000 & 0.875000000 & 0.000000000 \\ 6 & 0.000000000 & 0.000000000 & 0.000000000 & 0.000000000 & 0.000000000 & 0.000000000 & 1.000000000 \end{array}$$

Truncating and subtracting provides the numerical **I – Q** matrix:

$$I - Q = \begin{bmatrix} 0.3282 \times 10^{-4} & -0.3185 \times 10^{-4} & -0.22 \times 10^{-6} & -0.25 \times 10^{-6} \\ -0.12500000 & 0.12501666 & 0.00000000 & 0.00000000 \\ 0.00000000 & 0.00000000 & 0.00001666 & 0.00000000 \\ 0.00000000 & 0.00000000 & 0.00000000 & 0.00001666 \end{bmatrix}$$

The **N** matrix is obtained by inverting the **I – Q** matrix:

$$N = \begin{bmatrix} 1027131.102 & 261.7175606 & 13633.43623 & 15307.71787 \\ 1026994.207 & 269.6816131 & 13631.61919 & 15305.67768 \\ 0 & 0 & 60016.80471 & 0 \\ 0 & 0 & 0 & 60016.80471 \end{bmatrix}$$

Adding the row elements of state 0,

$$MTTF = 1{,}056{,}334 \text{ hr}$$

Using successive matrix multiplication, at 8,760 hr:

$$PFD = 0.000033$$

$$PFS = 0.00001277$$

1oo2D Architecture with Comparison

A comparison diagnostic technique can be added to any architecture with two or more logic solvers. This is frequently done in actual implementations. When the comparison detects a mismatch, this indicates a fault within the system. Figure 14-39 shows an interprocessor communication path added to the 1oo2D architecture in order to provide comparison capability between logic solvers. In systems with two logic solvers like the 1oo2D, if the self-diagnostics do not detect the fault both units will de-energize in order to ensure safety. The comparison provides another layer of diagnostics. The notation C_2 will be used to indicate this additional diagnostic coverage factor.

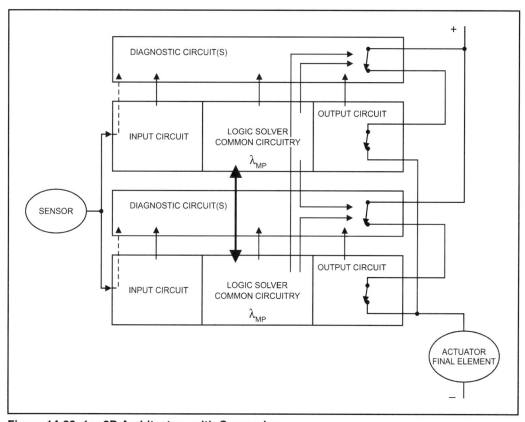

Figure 14-39. 1oo2D Architecture with Comparison.

PFD Fault Tree for 1oo2D with Comparison

The 1oo2D with comparison fails dangerously only if both units fail dangerously and that failure is not detected by the self-diagnostics in either unit or the comparison. The fault tree for this variation is shown in Figure 14-40. A first-order approximate PFD equation developed from the fault tree is

$$\text{PFD}_{1oo2D} = \lambda^{DUC} \times TI + [(1 - C_2)\lambda^{DUN} \times TI)]^2 \qquad (14\text{-}16)$$

Compare this with Equation 14-14 for the 1oo2D architecture. The 1oo2D with comparison should offer better safety performance because only failures undetected by both self-diagnostics and the comparison are included in the PFD.

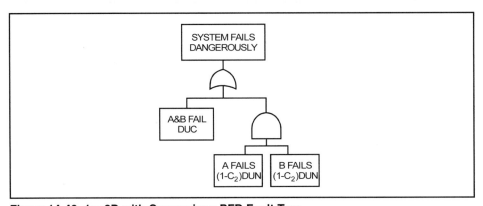

Figure 14-40. 1oo2D with Comparison PFD Fault Tree.

EXAMPLE 14-37

Problem: Two small safety controllers with diagnostic channels are used in a 1oo2D architecture system. The failure rates for each unit are

$$SDC_{1oo1D} = 0.000000312$$
$$SUC_{1oo1D} = 0.000000004$$
$$SDN_{1oo1D} = 0.000010104$$
$$SUN_{1oo1D} = 0.000000124$$
$$DDC_{1oo1} = 0.000000180$$
$$DUC_{1oo1} = 0.000000003$$
$$DDN_{1oo1} = 0.000005824$$
$$DUN_{1oo1} = 0.000000111$$

The coverage factor for the comparison is 99.9%. What is the approximate PFD?

Solution: Using Equation 14-16,

$$\text{PFD}_{1oo2D} = 0 \times 8760 + [(1 - 0.999) \times 0.11 \times 10^{-6} \times 8760]$$
$$= 0.00003$$

PFS Fault Tree for 1oo2D

Figure 14-41 shows that a 1oo2D architecture will fail safely if there is a common-cause safe failure, a common-cause dangerous detected failure, if either unit fails in a safe undetected manner, if both units fail safely, or if there is an undetected failure that causes a comparison mismatch. Note that the fault tree contains an incomplete symbol, which shows that the failures not detected by self-diagnostics or by the comparison are not included in the tree since they are considered insignificant. First-order approximation techniques can be used to generate a formula for probability of failing safely from this fault tree:

$$PFS_{1oo2D} = (\lambda^{SDC} + \lambda^{SUC} + \lambda^{DDC} + 2C_2\lambda^{SUN} + 2C_2\lambda^{DUN}) \times SD + (\lambda^{SDN} \times RT + \lambda^{DDN} \times RT)^2 \quad (14\text{-}17)$$

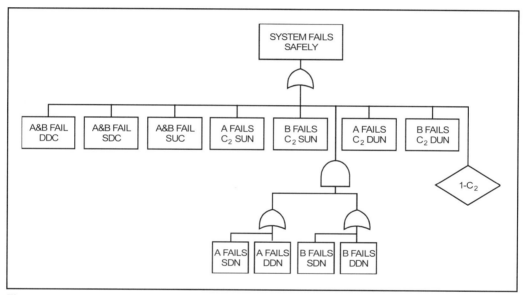

Figure 14-41. 1oo2D with Comparison PFS Fault Tree.

EXAMPLE 14-38

Problem: A system consists of two small safety controllers with diagnostic channels wired in a 1oo2D configuration with failure rates from Example 14-34. The average time to restart the process after a shutdown is 24 hr. Average repair time for a detected failure that does not shut down the process is 8 hr. The system is inspected every year (8,760 hr). What is the approximate PFS?

Solution: Using Equation 14-17,

$$PFS_{1oo2D} = 0.0000232$$

Markov Model for 1oo2D with Comparison

In the Markov model for a 1oo2D system with comparison in Figure 14-42, four system success states are shown. The states are much the same as the 1oo2D. The major difference is the failures detected by the comparison. Previously undetected safe and dangerous failures detected by the comparison are added to the arc from state 0 to state 4. Only failures undetected by the comparison cause a transition to states 2 and 3.

The transition matrix for the Markov model of Figure 14-42 is

$$P = \begin{bmatrix} 1-\Sigma & 2\lambda^{SDN}+2\lambda^{DDN} & 2(1-C_2)\lambda^{DUN} & 2(1-C_2)\lambda^{SUN} & \lambda^{SC}+\lambda^{DDC}+2C_2(\lambda^{SUN}+\lambda^{DUN}) & 0 & \lambda^{DUC} \\ \mu_O & 1-\Sigma & 0 & 0 & \lambda^S+\lambda^{DD} & \lambda^{DU} & 0 \\ 0 & 0 & 1-\Sigma & 0 & 0 & \lambda^{SD}+\lambda^{DD} & \lambda^{SU}+\lambda^{DU} \\ 0 & 0 & 0 & 1-\Sigma & \lambda^S+\lambda^{DD} & 0 & \lambda^{DU} \\ \mu_{SD} & 0 & 0 & 0 & 1-\Sigma & 0 & 0 \\ \mu_O & 0 & 0 & 0 & 0 & 1-\Sigma & 0 \\ 0 & 0 & 0 & 0 & 0 & 0 & 1 \end{bmatrix}$$

where Σ represents the sum of all other row elements.

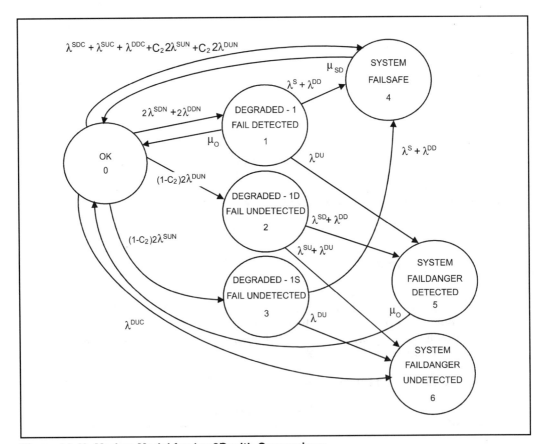

Figure 14-42. Markov Model for 1oo2D with Comparison.

EXAMPLE 14-39

Problem: Using failure rate values from Example 14-31, calculate the MTTF, PFS, and PFD of a 1oo2D system.

Solution: Substituting numerical values into the **P** matrix yields:

$$P = \begin{array}{c|ccccccc} & 0 & 1 & 2 & 3 & 4 & 5 & 6 \\ \hline 0 & 0.999967176 & 0.000031855 & 0.000000000 & 0.000000000 & 0.000000965 & 0.000000000 & 0.000000003 \\ 1 & 0.125000000 & 0.874983338 & 0.000000000 & 0.000000000 & 0.000016548 & 0.000000114 & 0.000000000 \\ 2 & 0.000000000 & 0.000000000 & 0.999983338 & 0.000000000 & 0.000000000 & 0.000016420 & 0.000000242 \\ 3 & 0.000000000 & 0.000000000 & 0.000000000 & 0.999983338 & 0.000016548 & 0.000000000 & 0.000000114 \\ 4 & 0.041666667 & 0.000000000 & 0.000000000 & 0.000000000 & 0.958333333 & 0.000000000 & 0.000000000 \\ 5 & 0.125000000 & 0.000000000 & 0.000000000 & 0.000000000 & 0.000000000 & 0.875000000 & 0.000000000 \\ 6 & 0.000000000 & 0.000000000 & 0.000000000 & 0.000000000 & 0.000000000 & 0.000000000 & 1.000000000 \end{array}$$

Truncating and subtracting provides the numerical **I – Q** matrix:

$$I - Q = \begin{bmatrix} 0.3282 \times 10^{-4} & -0.3185 \times 10^{-4} & 0.000000000 & 0.000000000 \\ -0.12500000 & 0.12501666 & 0.00000000 & 0.00000000 \\ 0.00000000 & 0.00000000 & 0.00001666 & 0.00000000 \\ 0.00000000 & 0.00000000 & 0.00000000 & 0.00001666 \end{bmatrix}$$

The **N** matrix is obtained by inverting the **I – Q** matrix:

$$N = \begin{bmatrix} 1027131.102 & 261.7175606 & 13.63343623 & 15.30771787 \\ 1026994.207 & 269.6816131 & 13.63161919 & 15.30567768 \\ 0 & 0 & 60016.80471 & 0 \\ 0 & 0 & 0 & 60016.80471 \end{bmatrix}$$

Adding the row elements of state 0,

$$\text{MTTF} = 1{,}027{,}422 \text{ hr}$$

At 8,760 hr,

$$\text{PFD} = 0.00002995$$

$$\text{PFS} = 0.00002327$$

Comparing Architectures

When the results of the analysis are examined, several of the main features of different architectures become apparent. The results of the fault tree analysis for the single-board controller architectures are compiled in Table 14-1. Of the four architectures, the highest safety rating is the 1oo2 architecture, with a PFD of 0.00147. The 2oo3 also does well, with a PFD of 0.00471. It should be noted that the results will be different for different parameter values.

	1oo1	1oo2	2oo2	2oo3
PFD_{ft}	0.02672	0.00147	0.05265	0.00471
RRF	37	679	19	212
PFS_{ft}	0.00016	0.00032	0.00060	0.00195

Table 14-1. SBC Model Results

Table 14-2 compares the results from the small safety controller models. It is interesting to note that the fault tree results and the Markov results are similar for this set of parameters. The failure rates used in the examples are sufficiently small to allow the first-order approximation to be reasonably accurate.

	1oo1	1oo2	2oo2	1oo1D	2oo3	2oo2D	1oo2D	$1oo2D_{Comp}$
PFD_{ft}	0.00102877	0.00003186	0.00202667	0.00099864	0.00009909	0.00196732	0.00003090	0.00002996
RRF	972.0364553	31388.448	493.41952	1001.3619	10091.7593	508.305509	32365.03357	33378.7274
PFS_{ft}	0.000242256	0.0004772	0.0000081	0.0003972	0.0000273	0.00001339	0.00001311	0.00002319
PFD_{mm}	0.00102794	0.00003179	0.00202329	0.00099775	0.00009856	0.00196386	0.00003312	0.00002995
PFS_{mm}	0.00024195	0.00047656	0.00000765	0.00039660	0.00002413	0.00001276	0.00001277	0.00002327
$MTTF_{mm}$	61,874	49,962	81,282	60,017	1,467,753	1,042,701	1,056,334	1,027,422

Table 14-2. Small Safety Controller Model Results

EXERCISES

14.1 An SBC has four analog input channels, two digital input channels, two analog output channels, and two digital output channels. The following failure rates are given:

> Analog input channel = 277 FITS
> Analog output channel = 212 FITS
> Digital input channel = 183 FITS
> Digital output channel = 290 FITS
> Main processing circuits = 2,555 FITS
> Software = 1,800 FITS

What is the total failure rate for the board?

14.2 We are given the following coverage factors for the SBC of Exercise 14.1:

> Analog input channel = 0.97
> Analog output channel = 0.99
> Digital input channel = 0.89
> Digital output channel = 0.95
> Main processing circuits = 0.99
> Software = 0.90

What are the failure rates of all failure categories when the board is used in a 1oo1 system configuration?

14.3 What is the system MTTF of a 1oo1 configuration using the failure rates of Exercise 14.2?

14.4 We are given a beta factor of 5% for hardware and 20% for software. The SBC of Exercise 14.2 will be used in a multichannel system. What are the normal and common cause failure rates?

14.5 What three system configurations are designed to avoid failures in which the outputs are energized (lowest PFD)?

14.6 What three system configurations have the highest MTTF?

14.7 Is there an advantage in having a repair policy that requires technicians to check all controllers when making a repair trip?

REFERENCES

1. Bukowski, J. V., and Lele, A., "The Case for Architecture-Specific Common Cause Failure Rates and How They Affect System Performance," *Proceedings of the Annual Reliability and Maintainability Symposium*, New York: IEEE, 1997.

2. Ramirez Valenzuela, C. E., and Delgadillo Valencia, M. A., "Fault Tolerance in a Distributed Control System for Combined Cycle Power Plants," *Proceedings from the 2nd Annual IFAC Symposium for Control on Power Plants and Power Systems*, Laxenburg, Austria: International Federation of Automatic Control, December 1995.

3. Simpson, K. "Design Selection Considerations for Safety System Hardware," *ISA TECH/EXPO Technology Update*, Vol. 1, Research Triangle Park, NC: ISA—The Instrumentation, Systems, and Automation Society, 1997.

4. Wilton, S. R., "System Safety Availability — 1oo2D and 2oo3," *ISA TECH/EXPO Technology Update*, Vol. 1, Research Triangle Park, NC: ISA—The Instrumentation, Systems, and Automation Society, 1997.

15
Safety Instrumented Systems

Introduction

A safety instrumented system (SIS) is a control system consisting of sensors, one or more controllers (frequently called logic solvers), and final elements. The purpose of an SIS is to monitor an industrial process for potentially dangerous conditions and to alarm or execute preprogrammed action to either prevent a hazardous event from occurring or mitigate the consequences of such an event should it occur. An SIS:

- Does not improve the yield of a process.

- Does not increase process efficiency.

- Saves money by loss reduction.

- Reduces *risk cost*.

Risk Cost

Risk is usually defined as the probability of a failure event times the consequences of the failure event. The consequence of the failure event is presented in terms of cost. The concept of risk cost is a statistical concept. An actual expense does not really occur each year, but occurs only when there is an event (an accident). The individual event cost can be quite high. Everyone works hard to keep the event costs low. If these event costs are averaged over many sites for many years, an "average" risk cost per year can be established. If actions are taken to reduce the chance of an event, average risk costs are lowered.

> **EXAMPLE 15-1**
>
> **Problem:** Records maintained over many sites for many years indicate that once every 15 years an industrial boiler will have an accident if no protection equipment such as an SIS is used. The average cost of each event is $1 million. What is the yearly risk cost with no protection equipment?
>
> **Solution:** The average yearly risk cost is
>
> $$\frac{\$1,000,000}{15} = \$66,667$$

Risk Reduction

There are risks in every activity of life. Admittedly, some activities involve more risk than others. According to Reference 1, the chance of dying during a 100-mile automobile trip in the midwestern United States is one in 588,000. The chance of dying from an earthquake or volcano is one in 11,000,000.

There is a certain risk inherent in the operation of an industrial process. Sometimes that risk is unacceptably high. A lower level of risk may be required by corporate rules, regulatory law, insurance companies, public opinion, or other interested parties. This leads to the concept of "acceptable risk." When inherent risk (perceived or actual) is higher than acceptable risk, then risk reduction is required (Figure 15-1).

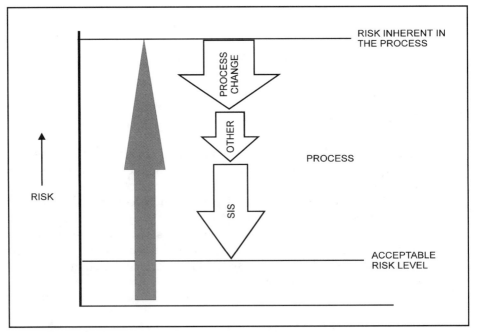

Figure 15-1. Risk Reduction.

While inherent risk and even acceptable risk are difficult to quantify, risk reduction is a little easier. Several methods have been proposed to determine the amount of needed risk reduction to at least to an order-of-magnitude level. The safety integrity levels (SIL) presented in Chapter 1 do this. Figure 15-2 shows the SILs of IEC61508 (Ref. 2), with examples of various industrial processes. Note that any particular industrial process can be assigned in different SILs depending on the actual or perceived effect of the event.

SAFETY INTEGRITY LEVEL	CORRESPONDING GERMAN APPL. CLASS (AK) - REFERENCE 5	AVERAGE PROBABILITY OF FAILURE ON DEMAND (PFDavg)	RISK REDUCTION FACTOR (RRF)	TYPICAL APPLICATIONS
4	7	< 0.0001	> 10,000	RAIL TRANSPORTATION / NUCLEAR POWER
3	5-6	0.001 - 0.0001	1,000 - 10,000	UTILITY BOILERS
2	4	0.01 - 0.001	100 - 1,000	INDUSTRIAL BOILERS / CHEMICAL PROCESSES
1	2-3	0.1 - 0.01	10 - 100	

Figure 15-2. Risk Reduction Categories.

Risk Reduction Factor

The risk reduction factor (RRF) is defined as:

$$\text{Risk reduction factor} = \frac{\text{Inherent risk}}{\text{Acceptable risk}}$$

An SIS provides risk reduction when it monitors a process looking for a dangerous condition (a process demand) and successfully performs its preprogrammed function to prevent an event. Assuming that the SIS has been properly programmed to prevent an event, the SIS reduces risk whenever it operates successfully in response to a process demand. It will not reduce risk if it fails to operate when there is a process demand. Therefore, an important measure of the risk reduction capability of an SIS is PFD, probability of failure on demand. In the case of a de-energize-to-trip system, this is the probability that the system will fail with its outputs energized. For these systems, RRF is defined as:

$$\text{RRF} = \frac{1}{\text{PFD}} \tag{15-1}$$

How Much RRF Is Needed?

It is hard to numerically estimate the needed RRF in an industrial process. That is why the standards provide an order-of-magnitude framework within which to work. This order-of-magnitude risk reduction framework is called "safety integrity level." Several methods to determine SIL are published in ISA's S84.01 standard (Ref. 3) and IEC61508 (Ref. 2). One such qualitative method developed to deal with personnel death and injury is shown in Figure 15-3. This method was derived from DIN V 19250 (now in IEC61508), and is called a *risk graph*. The user of the risk graph must determine four factors: the consequence (C), the frequency of exposure (F), the possibility of avoidance (P), and the probability of occurrence (W).

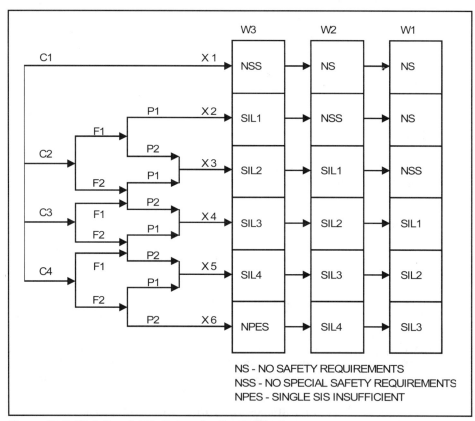

Figure 15-3. Risk Graph SIL Determination Method.

The probability of an event is characterized by three of the factors: exposure to a hazard, possibility of avoiding the hazard, and probability of event given external risk reduction measures. The fourth factor, consequence, completes the estimate of risk. Table 15-1 describes each factor.

	Consequence
C1	Minor injury
C2	Serious injury or single death
C3	Death to multiple persons
C4	Very many people killed
	Frequency and Exposure Time
F1	Rare to frequent
F2	Frequent to continuous exposure
	Possibility of Avoidance
P1	Avoidance possible
P2	Avoidance not likely, almost impossible
	Probability of Unwanted Occurrence
W1	Very slight probability
W2	Slight probability, few unwanted occurrences
W3	High probability

Table 15-1. Description of Risk Graph Factors

The consequences refer specifically to personnel injury or death. The frequency of exposure is a measure of the chances that personnel will be in the danger area when an accident occurs. If a process is only operated once a week and no operators are normally at the site, this is a relatively low risk compared to continuous operation with personnel always on duty. The possibility of avoidance considers such factors as warning signs, the speed at which the event develops, and other avoidance measures such as protective barriers. The probability of unwanted occurrence parameter allows consideration of other risk reduction devices.

The danger area is also an important consideration. If there is only risk of injury near the process unit, then perhaps only the operators and plant personnel need be considered in the analysis. But if a hazardous event could harm people in the neighborhood or over a wide geographic area, then that must be considered in the consequence estimation.

> **EXAMPLE 15-2**
>
> Problem: An industrial process may need an SIS to reduce risk. Use the risk graph of Figure 15-3 to determine this need.
>
> Solution: Experts estimate that an event may result in serious injury or a single death, C2. The process operates continuously and personnel are usually always present: therefore, exposure is rated as "frequent to continuous, F2." Since there is little or no warning of a dangerous condition, the probability of avoidance is rated "avoidance not likely, P2." Finally, it is judged that there is only a slight probability of occurrence, W2, because pressure relief valves are installed to provide another layer of protection. Using the risk graph, this process is SIL2. The calculated PFD for the SIS (including process connections, sensors, controllers, and valves) must be less than 0.01.

Many corporations have versions of risk graphs in corporate procedures. Frequently, these documents provide more detailed descriptions of how to classify the various parameters. Additional risk graphs are needed for equipment damage and environmental damage (Ref. 4). In some cases, three graphs are done for a process (personnel injury, equipment damage, and environmental damage), and the largest risk reduction need is the one specified for the SIS.

Another risk reduction determination method published in both ISA S84.01 and IEC61508 is called the hazardous event severity matrix. This method uses three parameters: the number of independent protection layers, event likelihood and event severity.

In the hazardous event severity matrix, the number of independent protection layers parameter allows consideration of other risk reduction mechanisms. The method assumes that there is only one SIS and that additional protection layers consist of alternative technology, such as mechanical or pneumatic. These alternative risk reduction mechanisms should have an RRF of at least 100. The other two parameters represent the primary variables of risk, chance of an event, and consequences of the event. Figure 15-4 shows a simplified hazardous event severity matrix.

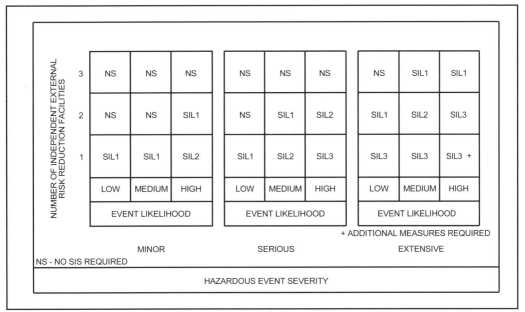

Figure 15-4. Hazard Event Severity Matrix.

EXAMPLE 15-3

Problem: An industrial process needs an SIS for additional risk reduction. There is an independent pneumatic protection system installed that has an estimated RRF of 120. Determine the needed integrity level via the hazardous event severity matrix method.

Solution: The process safety panel meets to review the situation. After a debate, the group decides that there is a "medium" event likelihood and "serious" event severity. The number of independent risk reduction facilities is two: the SIS and the pneumatic protection system. Using the matrix, an SIL1 risk reduction is required for the SIS. This means that the PFD of the SIS must be less than 0.1.

Quantitative Risk Reduction

In some cases, corporations and insurance companies determine RRFs based on quantitative risk analysis. Statistics are compiled on the frequency and severity of events. These statistics are used to determine RRFs based on risk cost goals or event probability goals. While these methods seem easy, it must be understood that the uncertainty of the statistical data and variability of the factors contributing to an event must be taken into account. This is typically done by increasing the required RRF by a certain safety margin.

> **EXAMPLE 15-4**
>
> **Problem:** The risk cost of operating an industrial boiler is estimated to be $66,667 per year. The insurance company that insures the operation of the boiler needs a risk cost of less than $5,000 per year in order to remain profitable without a significant increase in premiums. What is the needed RRF?
>
> **Solution:** The ratio of risk costs is 66,667/5,000, or 13.4. In order to account for the uncertainly of the data, the insurance company mandates that an SIS with a RRF of 100 be installed. Note: This is an SIL2 category system.

SIS Architectures

An SIS consists of sensors, controllers, and final elements that work together to detect and hopefully prevent or mitigate the effects of a hazardous event. It is very important to design a system with a high RRF. It is also important to maximize the production uptime. In order to achieve these goals, system designers use redundant controller architectures (Chapter 14). These architectural concepts extend to the field instruments as well as the controllers. Fault-tolerant configurations can be used for sensors and final elements.

Sensor Architectures

An SIS must have devices capable of sensing potentially dangerous conditions. Many types of sensors are used, including flame detectors (infrared or ultraviolet), gas detectors, pressure transmitters, thermocouples, resistance temperature detectors, and many types of discrete switches. These sensors can fail, and typically have more than one failure mode. Some of the sensor failures can be detected by on-line diagnostics in the sensor itself or in the controller to which it is connected. In general, all the reliability and safety modeling techniques can be used.

Safety Instrumented Systems

EXAMPLE 15-5

Problem: A two-wire pressure sensor transmits 4-20 mA to the analog input of an SIS. The manufacturer states that the device has an MTTF of 40 years. It is estimated that 20% of the failures are safe and 80% are potentially dangerous. The controller in the SIS has overcurrent and undercurrent detection for 4-20 mA analog signals. In addition, a "stuck signal" detection function block is available and will be used on this sensor. It is estimated that these diagnostic mechanisms will detect 75% of the failures in the sensor. What are the failure rates of this sensor?

Solution: The total failure rate is obtained using Equation 7-8, assuming that the sensor is a series system of its components, each of which has a constant failure rate:

$$\lambda = \frac{1}{(40 \times 8760)}$$
$$= 0.285 \times 10^{-5} \text{ failures per hour}$$

The total failure rate is divided into safe and dangerous according to the percentages given:

$$\lambda^S = 0.2 \times 0.285 \times 10^{-5}$$
$$= 0.57 \times 10^{-6}$$

$$\lambda^D = 0.8 \times 0.285 \times 10^{-5}$$
$$= 0.228 \times 10^{-5}$$

The safe and dangerous failure rates can be divided into detected and undetected using the coverage factor of 0.75 using Equations 12-15 to 12-18:

$$\lambda^{SD} = 0.75 \times 0.57 \times 10^{-6}$$
$$= 0.43 \times 10^{-6}$$

$$\lambda^{SU} = (1 - 0.75) \times 0.57 \times 10^{-6}$$
$$= 0.14 \times 10^{-6}$$

$$\lambda^{DD} = 0.75 \times 0.285 \times 10^{-5}$$
$$= 0.171 \times 10^{-5}$$

$$\lambda^{DU} = (1 - 0.75) \times 0.285 \times 10^{-5}$$
$$= 0.57 \times 10^{-6}$$

Figure 15-5 shows two discrete sensors measuring the same process variable. These two sensors can be configured in a 1oo2 architecture by simply adding logic to initiate a shutdown if either of the two sensors signals a dangerous condition. Like the 1oo2 controller architecture, this configuration will substantially reduce the chance of a dangerous failure but will almost double the chance of a safe failure. Note that common-cause failures apply when redundant configurations are used. The common-cause reduction rules (Chapter 10) apply. Avoid close physical installation, use high-strength sensors, use diverse design sensors, or apply some combination of all three techniques.

The 1oo2 sensor concept can be applied to analog sensors as well. Figure 15-6 shows two analog sensors measuring the same process variable. A high-select or low-select function block (depending on the fail-safe direction) is used to select which analog signal is used in the calculation. A comparison is made between the two signals to detect differences that indicate the failure of one or both sensors. If the comparison is not within the predetermined limits, a shutdown is initiated.

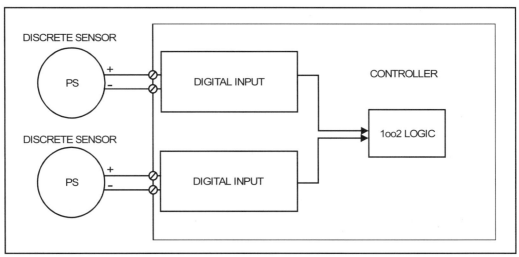

Figure 15-5. 1oo2 Digital Sensor.

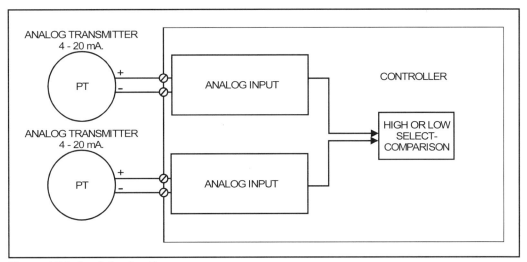

Figure 15-6. 1oo2 Analog Sensor.

EXAMPLE 15-6

Problem: Two analog pressure sensors are used in a 1oo2 configuration. Common-cause failures should be considered using a beta factor of 0.05. Sensor failures that are detected are repaired in an average of 8 hr. The system is restarted in an average of 24 hr after a shutdown. The system is operated for periods of one year between periodic inspections, where a complete test is performed on the system. If two sensors with failure rates from Example 15-5 are used, what is the sensor system PFS and PFD? How would it compare to a 1oo1 sensor system PFS and PFD?

Solution: The failure rates are first divided into normal and common cause using Equations 12-21 to 12-28. The failure rates (in failures per hour) are

$$\text{Beta} = 0.05$$
$$\lambda^{SDC} = 0.000000021$$
$$\lambda^{SDN} = 0.000000407$$
$$\lambda^{SUC} = 0.000000007$$
$$\lambda^{SUN} = 0.000000136$$
$$\lambda^{DDC} = 0.000000086$$
$$\lambda^{DDN} = 0.000001627$$
$$\lambda^{DUC} = 0.000000029$$
$$\lambda^{DUN} = 0.000000542$$
$$RT = 8$$
$$TI = 8,760$$
$$SD = 24$$

> **EXAMPLE 15-6 (continued)**
>
> Using the approximation technique, Equation 14-1 can be used to calculate the PFD for a 1oo1 architecture. Equation 14-2 can be used to calculate PFS for a 1oo1 architecture. For a 1oo2 configuration, note that the comparison will shut down the system if the two transmitters are different by more than a predetermined amount. Therefore, it could be argued that a difference would appear after any failure, safe or dangerous. In such a case, PFD should be modeled with only dangerous common-cause failure rates. It could also be argued that one sensor could fail dangerously without the comparison detecting the failure. In that case, Equation 14-4 can be used to calculate the PFD and Equation 14-5 can be used for PFS. Since Equation 14-4 will produce a more pessimistic result, that more conservative approach will be chosen. Substituting the failure rates:
>
> | | PFD | PFS |
> | ---- | ----------- | ----------- |
> | 1oo2 | 0.000273371 | 0.000026712 |
> | 1oo1 | 0.005013699 | 0.000013699 |
>
> As might be expected, the PFS of the 1oo2 is twice the PFS of the 1oo1. The PFD of the 1oo2 is much better than the 1oo1.

In most systems, a false shutdown is a serious problem — especially when process uptime is important. A 2oo3 sensor architecture as shown in Figure 15-7 can provide higher uptime just like the 2oo3 architecture for controllers, especially if common-cause avoidance is used in the design and installation. Three analog sensors measure the same process variable. A median selector is used to pick which signal is used for the calculation. Three sets of comparisons are used between the three combinations of two sensors. If one sensor is different from the other two, it is assumed that the deviant sensor is bad. Common-cause considerations are especially important in this architecture since three combinations of two sensors are susceptible to common-cause failures.

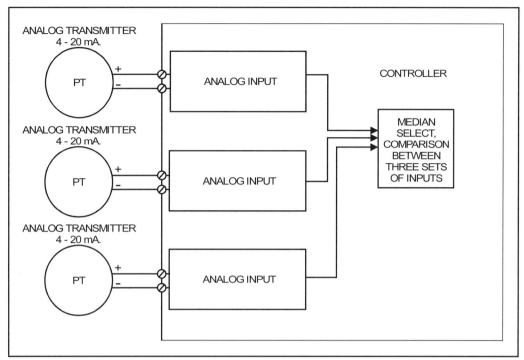

Figure 15-7. 2oo3 Analog Sensor.

EXAMPLE 15-7

Problem: Three analog pressure sensors are used in a 2oo3 configuration. Common-cause failures should be considered using a beta factor of 0.05. Sensor failures that are detected are repaired in an average of 8 hr. The system is restarted after a shutdown in an average of 24 hr. The system is operated for periods of one year between periodic inspections. If three sensors with failure rates from Example 15-6 are used, what is the 2oo3 sensor system PFS and PFD?

Solution: Using the approximation technique, Equation 14-10 can be used to calculate the PFD of a 2oo3 architecture and Equation 14-11 can be used to calculate the PFS. Substituting the failure rates:

$$PFS_{2oo3} = 0.0000063$$

$$PFD_{2oo3} = 0.00082$$

Like controller architectures, another configuration can provide high safety and high availability. A 1oo2D sensor architecture is shown in Figure 15-8. This architecture has a high level of safety similar to the 1oo2 via the same type of comparison algorithm. The 1oo2D will achieve higher availability depending on the ability of the diagnostics.

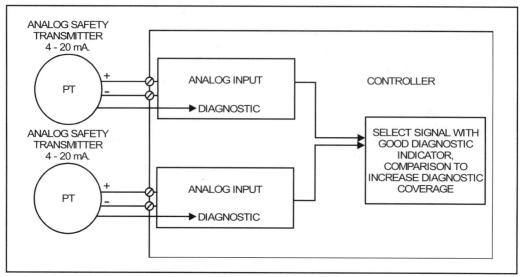

Figure 15-8. 1oo2D Analog Sensor.

Two signals are generated for each sensor: the process variable and a diagnostic signal. If a diagnostic signal from only one sensor is good, the selection algorithm in the controller will select the signal coming from the sensor with good diagnostic indication. If the diagnostic signals from both sensors are good, one of the signals is selected and a comparison is done (the 1oo2 function) to ensure high safety. The comparison will initiate a shutdown if the two units differ more than the predetermined amount. If both diagnostic signals indicate a problem, a shutdown is initiated. Depending on the coverage of the diagnostics, this architecture can provide effective reliability and safety.

> **EXAMPLE 15-8**
>
> **Problem:** Two analog pressure sensors are used in a 1oo2D configuration. Common-cause failures should be considered using a beta factor of 0.05. Sensor failures that are detected are repaired in an average of 8 hr. The system is restarted after a shutdown in an average of 24 hr. The system is operated for periods of one year between periodic inspections. If two sensors with failure rates from Example 15-6 are used, what is the 1oo2D sensor system PFS and PFD?
>
> **Solution:** Using the approximation technique, Equation 14-16 can be used to calculate the PFD of a 1oo2D sensor architecture with comparison and Equation 14-17 can be used for PFS. Substituting the failure rates:
>
> $$PFS_{1oo2D} = 0.000035$$
>
> $$PFD_{1oo2D} = 0.00025$$

Final Element Architectures

The final element, typically a valve and its actuator, are the third major category of devices in an SIS. Final elements have failure modes, failure rates, and potential diagnostics. The reliability and safety modeling techniques developed for other SIS devices are appropriate.

The simplest final element configuration is the 1oo1 (Figure 15-9). One design used for tight shutoff safety valves consists of a valve, its actuator, and a solenoid. The controller normally energizes a solenoid, which supplies air to the actuator. The actuator keeps the valve open (or closed if that is the desired position). When an unsafe condition is detected, the controller de-energizes the solenoid. The solenoid removes air from the actuator, and the spring return actuator moves the valve to its safe position. In such final element systems, failure rates must be obtained for the solenoid and the actuator/valve assembly.

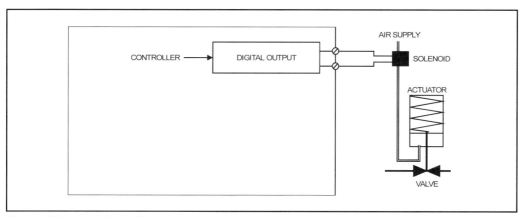

Figure 15-9. 1oo1 Valve System.

> **EXAMPLE 15-9**
>
> **Problem:** Solenoid failure rates are $\lambda^S = 600$ FITS and $\lambda^D = 400$ FITS. An actuator/valve assembly has failure rates of $\lambda^S = 1,800$ FITS and $\lambda^D = 1,500$ FITS. No diagnostics are available. When a shutdown occurs, the process is restarted in an average of 24 hr. The system is inspected once a year. What is the total safe and dangerous failure rate of the final assembly? What is the PFS? The PFD?
>
> **Solution:** The solenoid and the actuator/valve comprise a series system, so the failure rates are added. The total is $\lambda^S = 2,400$ FITS and $\lambda^D = 1,900$ FITS. Equation 14-1 can be used to approximate the PFD; all the λ^D failures are undetected. Equation 14-2 can be used to approximate the PFS. Substituting the failure rates:
>
> $$PFD = 0.0000019 \times 8760 = 0.0167$$
>
> $$PFS = 0.0000024 \times 24 = 0.000058$$

Figure 15-10 shows a 1oo2 configuration for the solenoid with a 1oo1 valve. This configuration can be effective if the PFD rating of the solenoid is much higher than that of the valve. This configuration is modeled as a series system comprised of a 1oo2 solenoid in series with a 1oo1 valve/actuator.

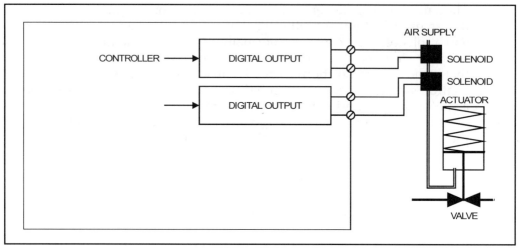

Figure 15-10. 1oo1 Valve System with 1oo2 Solenoids.

EXAMPLE 15-10

Problem: Solenoid failure rates are λ^S = 600 FITS and λ^D = 400 FITS. An actuator/valve assembly has failure rates of λ^S = 1,800 FITS and λ^D = 1,500 FITS. No diagnostics are available. The system is inspected once a year. Two solenoids are used in a 1oo2 configuration to supply air to a 1oo1 actuator/valve. Common-cause beta factor is 0.05 for the solenoids. What is the PFS? The PFD?

Solution: Failure rates (in failures per hour) are calculated for the solenoid using the beta model

$$\lambda_{Sol}^{DUC} = 0.000000020$$
$$\lambda_{Sol}^{DUN} = 0.000000380$$

for the dangerous failures and

$$\lambda_{Sol}^{SUC} = 0.000000030$$
$$\lambda_{Sol}^{SUN} = 0.000000570$$

for the safe failures.

Safety Instrumented Systems

> **EXAMPLE 15-10 (continued)**
>
> The solenoid is a 1oo2 configuration, and Equation 14-4 can be used to approximate PFD. Note that all failures are undetected as no diagnostics are available. Equation 14-1 is used for 1oo1 actuator/valve assembly. Substituting the failure rates,
>
> $$PFD_{system} = 0.01333$$
>
> This is an improvement over the system of Example 15-8, but not significant. The PFS for this system is calculated noting that any safe component failure results in a safe system failure. Substituting the failure rates:
>
> $$PFS = 0.00000003 \times 24 + 2 \times 0.00000057 \times 24 + 0.0000018 \times 24$$
>
> $$= 0.00007128$$
>
> | | PFD | PFS |
> |---|---|---|
> | 1oo1 | 0.016644000 | 0.000057600 |
> | S_{1oo2}/V_{1oo1} | 0.013326281 | 0.000071280 |

Figure 15-11 shows a common implementation of a full 1oo2 configuration for final elements. The valves close when a dangerous condition is detected. The system is considered operational if either valve successfully closes. This configuration is intended to provide higher safety and can be effective even if common-cause considerations are made.

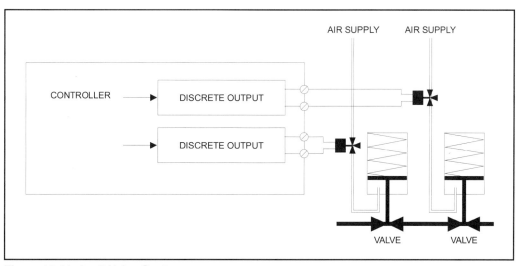

Figure 15-11. 1oo2 Valve System.

> **EXAMPLE 15-11**
>
> **Problem:** Using failure rates of Example 15-9, what is the PFS and PFD of a full 1oo2 final element assembly?
>
> **Solution:** The failure rates (in failures per hour) are
>
> $$\lambda_{Sol}^{SUC} = 0.000000030$$
>
> $$\lambda_{Sol}^{SUN} = 0.000000570$$
>
> $$\lambda_{Val}^{SUC} = 0.000000090$$
>
> $$\lambda_{Val}^{SUN} = 0.000001710$$
>
> $$\lambda_{Sol}^{DUC} = 0.000000020$$
>
> $$\lambda_{Sol}^{DUN} = 0.000000380$$
>
> $$\lambda_{Val}^{DUC} = 0.7 \times 10^{-7}$$
>
> $$\lambda_{Val}^{DUN} = 0.142 \times 10^{-5}$$
>
> Using a first-order approximation, Equation 14-4 (PFD_{1oo2}), the PFD of this system is
>
> $$PFD = (0.00000002 + 0.7 \times 10^{-7}) \times 8760 + [(0.00000038 + 0.142 \times 10^{-5}) \times 8760]^2$$
> $$= 0.0010822$$
>
> This is considerably better than the other final element architectures.
>
> $$PFD = (0.00000003 + 0.00000009) \times 24 + 2 \times (0.00000057 + 0.00000171) \times 24$$
> $$= 0.00011232$$

System Architectures

An SIS consists of process connection, sensor, power supplies, controller, and final element. The "process-to-process" approach must be used to ensure that the safety and reliability analysis includes all elements. Do not forget about the sensor "impulse line" and possibly a manifold. Some systems are modeled as series systems. Sometimes the system must be modeled as series/parallel components, depending on how the various components are connected.

Safety Instrumented Systems

EXAMPLE 15-12

Problem: A sensor is connected to the process via an impulse line. Records indicate that the impulse line clogs up on average once every four years. This condition is dangerous, as the protection system may not respond to a demand. An algorithm based on signal rate of change can detect 90% of the clogs. When detected, the clogs are repaired in an average of 8 hr. If the impulse line is manually inspected every year, how will the diagnostic algorithm affect the PFD?

Solution: Assuming a constant failure rate for the impulse line, the failure rate (in failures per hour) is calculated using Equation 4-18:

$$\lambda = \frac{1}{4 \times 8760}$$
$$= 0.00002854$$

Equation 14-1 can be used to approximate the PFD. Without the diagnostic algorithm, the entire failure rate must be classified as dangerous undetected. Therefore,

$$PFD_{impulse} = 0.00002854 \times 8760$$
$$= 0.25$$

Note that the approximation techniques are only valid for small failure rates. This example will likely have unacceptable error. Therefore, as a check, repeat the calculation using Equation 4-16:

$$PFD = 1 - e^{-\lambda t}$$
$$= 0.2212$$

With the diagnostic algorithm, the failure rate is divided into dangerous detected (0.000025685 failures per hour) and dangerous undetected (0.000002855 failures per hour). Using Equation 14-1,

$$PFD_{impulse} = 0.0252$$

The diagnostics provide a considerable improvement.

EXAMPLE 15-13

Problem: A sensor is connected to an impulse line with rate-of-change diagnostics. The subsystem is inspected every year. The sensor has the following failure rates (in failures per hour):

$$\lambda^{SD} = 0.000000428$$
$$\lambda^{SU} = 0.000000143$$
$$\lambda^{DD} = 0.000001712$$
$$\lambda^{DU} = 0.000000571$$

For a one-year inspection interval, the sensor has a PFD of 0.005 (Example 15-6). What is the PFD of the system consisting of the impulse line and the sensor?

Solution: If the impulse line or the sensor fails dangerously, the subsystem fails dangerously. In a fault tree this is represented by an OR gate. The one-year PFD for the sensor is 0.005 (Example 15-6), and the one-year PFD for the impulse line is 0.0252 (Example 15-12). Given these probabilities, Equation 6-3 can be used:

$$PFD_{subsystem} = 0.005 + 0.0252 - (0.005 \times 0.0252)$$
$$= 0.03008$$

In practice, this is frequently approximated by merely adding the failure probabilities:

$$PFD_{subsystem} = 0.005 + 0.0252$$
$$= 0.0302$$

Alternatively, Equation 14-1 could be used to obtain an approximation. The total failure rates are

$$\lambda^{SD} = 0.43 \times 10^{-6}$$
$$\lambda^{SU} = 0.14 \times 10^{-6}$$
$$\lambda^{DD} = 0.171 \times 10^{-5} + 0.2568 \times 10^{-4}$$
$$= 0.274 \times 10^{-4}$$
$$\lambda^{DU} = 0.57 \times 10^{-6} + 0.285 \times 10^{-5}$$
$$= 0.342 \times 10^{-5}$$
$$PFD = 0.274 \times 10^{-4} \times 8 + 0.342 \times 10^{-5} \times 8760$$
$$= 0.03022$$

Safety Instrumented Systems

EXAMPLE 15-14

Problem: A controller with a PFD of 0.0001 (RRF = 10,000) is purchased for an SIL2-rated SIS. Two sensor/impulse line subsystems from Example 15-13 are used to detect two different dangerous conditions. If either dangerous condition is detected, a valve closes. The valve failure rates are from Example 15-9. All system components, including both sensor subsystems and the valve, must operate for the system to be successful. The system is inspected and completely tested once a year. What is the system PFD?

Solution: The system will fail dangerously if any component fails dangerously. The PFD of each sensor/impulse line subsystem is 0.03. The PFD of the valve is 0.0167. The PFD of the system can be approximated by adding component PFDs. For this example,

$$PFD_{system} = 2 \times 0.03 + 0.0001 + 0.0167 = 0.0768$$

$$RRF = 13$$

This barely meets the requirements of SIL1 and is not acceptable for SIL2. This shows a common problem in SIS design. Frequently, the entire system is not considered in the PFD analysis, and a falsely high RRF is obtained.

EXAMPLE 15-15

Problem: The SIL2 system design does not meet requirements. It is proposed that two sets of two sensors, each with an independent impulse line, be used in 1oo2 configurations to decrease the system PFD. Will this configuration meet the requirements of SIL2?

Solution: First, the new PFD for a 1oo2 sensor/impulse line subsystem should be approximated. The sensor/impulse lines should be physically separated and mounted in different orientations to avoid common cause. The sensors in each 1oo2 set are electrically isolated from each other. Accounting for common cause, the beta factor is estimated to be 0.02.

EXAMPLE 15-15 (continued)

The failure rates (in failures per hour) for the sensor/impulse line subsystem are

$$\lambda^{SDC} = 0.43 \times 10^{-6} \times 0.02$$
$$= 0.9 \times 10^{-8}$$

$$\lambda^{SUC} = 0.14 \times 10^{-6} \times 0.02$$
$$= 0.3 \times 10^{-8}$$

$$\lambda^{DDC} = 0.274 \times 10^{-4} \times 0.02$$
$$= 0.55 \times 10^{-6}$$

$$\lambda^{DUC} = 0.342 \times 10^{-5} \times 0.02$$
$$= 0.7 \times 10^{-7}$$

$$\lambda^{SDN} = 0.43 \times 10^{-6} \times (1 - 0.02)$$
$$= 0.42 \times 10^{-6}$$

$$\lambda^{SUN} = 0.14 \times 10^{-6} \times (1 - 0.02)$$
$$= 0.000000140$$

$$\lambda^{DDN} = 0.274 \times 10^{-4} \times (1 - 0.02)$$
$$= 0.2685 \times 10^{-4}$$

$$\lambda^{DUN} = 0.342 \times 10^{-5} \times (1 - 0.02)$$
$$= 0.336 \times 10^{-5}$$

Using Equation 14-4,

$$PFD_{1oo2sensor} = 0.7 \times 10^{-7} \times 8760 + 0.55 \times 10^{-6} \times 8$$
$$+ (0.2685 \times 10^{-4} + 0.336 \times 10^{-5} \times 8760)^2$$
$$= 0.0015$$

Next, the system PFD can be approximated by adding the subsystem PFDs. For this example,

$$PFD_{system} = 2 \times 0.0015 + 0.0001 + 0.0167$$
$$= 0.0198$$

RRF = 50

This system does not meet SIL2.

Safety Instrumented Systems 419

> **EXAMPLE 15-16**
>
> **Problem:** The SIL2 system design still does not meet requirements. It is proposed that two valves in a 1oo2 configuration be used to further increase the system PFD. Will this configuration meet the requirements of SIL2?
>
> **Solution:** The 1oo2 configuration of two valves has a PFD of 0.001, as approximated in Example 15-11. The system PFD is approximated by adding the subsystem PFDs. For this example,
>
> $$PFD = 2 \times 0.0015 + 0.0001 + 0.001$$
> $$= 0.0041$$
>
> $$RRF = 243$$
>
> This meets SIL2.

It should be noted that the use of probability summation as an approximation is correct, but only if subsystems are independent. If a failure in one subsystem will affect the operation of another subsystem, then more sophisticated techniques are needed to evaluate system probabilities.

EXERCISES

15.1 A process is manned continuously and has no risk reduction mechanism. An accident could cause death to several persons. Dangerous conditions build slowly and alarm mechanisms should warn of dangerous conditions before an accident. Using a risk graph, determine how much risk reduction is needed.

15.2 A quantitative risk assessment indicates an inherent risk cost of $250,000 per year for an industrial process. Plant management would like to reduce the risk cost to less than $25,000 per year. What risk reduction factor is required? What SIL classification is this?

15.3 What components must be considered in the analysis of an SIS?

15.4 A process connection clogs every year on average. This is a dangerous condition. No diagnostics can detect this failure. Assuming a constant failure rate, what is the dangerous undetected failure rate?

15.5 Using the failure rate of Exercise 15.4, what is the approximate PFD for a six-month inspection interval?

15.6 Using the failure rate of Exercise 15.4, what is the PFD for a two-year inspection interval?

REFERENCES

1. "A Fistful of Risks," *Discover Magazine*, May 1996.

2. IEC1508, *Functional Safety — Safety Related Systems*, 1997.

3. S84.01, *Applications of Safety Instrumented Systems for the Process Industries*, 1996.

4. Korteweg, L., "The Importance of Field Instrumentation in Safety Instrumented Systems," *Proceedings of the Safety in the Process Industry Symposium*, September 1996.

5. Faller, R. I., "Aspects of TUV Type Certification and Safety-Related Application of Programmable Electronic Systems," *Proceedings of the Safety in the Process Industry Symposium*, September 1996.

BIBLIOGRAPHY

1. Brombacher A. C., Rouvroye J. L., and Spiker R. Th. E., "How Safe is 'Safe': Uncertainty in Safety," *Proceedings NIRIA Symposium on Process Safety*, October 1994.

2. DIN V 19250, *Basic Safety — Assessment for I&C Protection Devices*, May 1994.

3. Fisher, T. G., *Alarm and Interlock Systems*, Research Triangle Park, NC: ISA—The Instrumentation, Systems, and Automation Society, 1984.

4. Geddes, B., "Playing It Safe," *Control*, April 1997.

5. Green, D. L., and Dowell, A. M. III, "How to Design, Verify and Validate Emergency Shutdown Systems," *ISA Transactions*, Vol. 34, 1995.

6. Maggioli, V. J., "The Safety Matrix," *Fifth Process Computer Users Forum of the CMA*, Arlington, VA: Chemical Manufacturers Association, May 1986.

7. Stavrianidis, P., Bhimavarapu, K., and Moore, L., "Performance-based Safety Standards: An Integrated Risk Assessment Program," *Proceedings of TECH97*, Research Triangle Park, NC: ISA—The Instrumentation, Systems, and Automation Society, 1997.

8. Stavrianidis, P., and Karydas, D. M., "Reliability Certification of Electronic Systems Used for Protection", *Proceedings of the International Conference on Process Safety Management*, March 1994.

9. Stavrianidis, P., and Karydas, D. M., "Methodology for the Reliability Evaluation of an Electronic Gauging System Used for Safety in Power Boilers," *Proceedings of the International Conference on Reliability, Availability, Maintainability and Quality Control*, 1992.

10. Stavrianidis, P., and Rennie, J., "Improving Management of Industrial Risk: Reliability Certification of Safety Systems," *ISA Transactions*, Vol. 35, 1996.

16
Life Cycle Costing

The Language of Money

A reliability and safety analysis will provide one or more measures of system success or failure. These numbers develop meaning for an experienced engineer who has done many reliability analysis comparisons. To an inexperienced engineer, and especially to corporate management, the significance between a system with an availability of 0.99 and a system with an availability of 0.999 may be hard to grasp. This can be strikingly true when the difference in purchase price between two different systems will exceed thousands of dollars.

The language of money is readily understood, particularly by management. Since this common quantitative language exists, a conversion between reliability analysis output and the language of money is a useful communication tool. System reliability has a substantial effect on operating cost; ask anyone who has ever owned an automobile. The relationship between reliability and operating cost has been developed in an effective conversion technique known as "life-cycle cost" analysis.

Life-cycle costing (LCC) is a technique that encourages those who are responsible for system selection to consider all the costs incurred over the lifetime of a product rather than just the purchase costs. This comprehensive attitude is especially relevant to industrial systems, where the cost of equipment failure can be many times the initial cost of the equipment. One fundamental concept is that *all costs that will occur during the life of the equipment are essentially determined at purchase time*.

The cost factors that need to be included in a life-cycle cost vary from system to system. Specific cost factors may not be present in a specific system. The intent is to include all costs of procurement and ownership over the life span of a system. The various costs occur at different times during the product life. This is not a problem, since commonly used LCC techniques account for the time value of money.

The two primary categories of costs are procurement cost and operating cost. Procurement cost includes system design costs, capital costs, installation costs, initial training costs, and start-up costs. Operating costs include energy consumption costs, consumable costs, and maintenance costs. These costs are added to obtain total life cycle cost:

$$LCC = C_{PRO} + C_{OP} \tag{16-1}$$

Given a consistent level of technology, a relationship exists between reliability and life-cycle cost. At higher reliability levels, procurement costs are higher and operating costs are lower. This is the fundamental concept behind a portion of the quality movement, the primary slogan of which is "Quality is free" (Ref. 1). Experts contend that the slogan should have been "Quality is free for those who are willing to pay up-front." Since higher quality normally brings higher reliability, the point is that higher procurement costs bring lower operating costs. Therefore, quality is free. This works to a point. Depending on the exact relationships, a specific optimal point will exist in which minimum life cycle costs are achieved. This is illustrated in Figure 16-1.

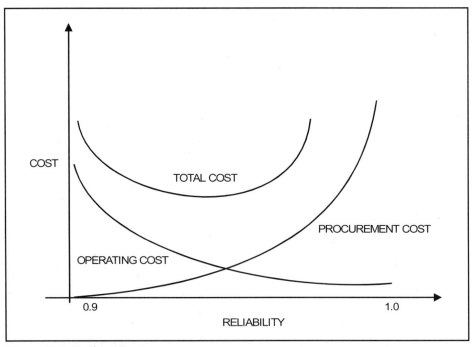

Figure 16-1. Optimal Cost.

Procurement Costs

Procurement costs include the system design costs, the purchase cost of the equipment, the installation cost, initial training, and the start-up/system commissioning cost. These costs occur only once. The total is obtained by summing:

$$C_{PRO} = C_{DESIGN} + C_{PURCHASE} + C_{INSTALLATION} + C_{STARTUP} \qquad (16\text{-}2)$$

System Design Costs

System design costs can be significant for complex control systems. Included in this category are detailed design, system FMEA, drawings, program management, system configuration, safety test verification, and custom software (if necessary). Much of the conceptual design work needs to be done in order to prepare a request for quotation and is often not considered in life-cycle cost.

The next step, detailed design work, usually represents the largest cost. The effort required depends to a great extent on the system tools and experience of the engineering staff. System tools that include a control point database, good report-generation facilities, a graphics language programming interface, and other computer-aided design assistance can really cut costs. Drawings and other documentation costs are also affected by tools; computer-aided graphics tools save time and reduce systematic errors.

Good training can reduce engineering costs. This is especially true when the engineers have no experience with the control system. A good training program can jump-start the design effort. Configuration libraries are collections of control system configurations. These libraries reduce detailed design time. Libraries of proven system configurations can often be used "as is," thus saving tremendous time and reducing system-level design error. Even when a standard configuration cannot be used, it is easier to modify an existing library design than to create a new one from scratch.

System design costs are initially estimated. Engineering estimates are, by nature, uncertain. Many factors that affect the estimate are generally undefined when the estimate must be made. Given this uncertainty, it is best to estimate upper and lower bounds on these numbers. The life-cycle cost can then be done with similar upper and lower bounds.

Purchase Costs

Purchase costs always include the cost of equipment. Sometimes neglected are the costs of cabinetry, environmental protection (air conditioning, etc.), wire termination panels, factory acceptance tests (if required), initial service contracts, and shipping costs. Purchase costs normally represent the focal point of any system comparison. This is quite understandable, since capital expenditures often require a tedious approval procedure and many factors compete for a limited capital budget. These costs are also easy to obtain and have a high level of certainty.

Installation Costs

Installation costs must account for delivery of the equipment, siting and mounting, any weatherization required, piping, and wiring. These costs are affected by the nature of the equipment. Small modular cabinets are easier to move and mount. Wiring is simplified when wire termination panels are easy to access. Some systems offer field-mountable termination panels to reduce wiring costs. "Remote I/O" and "distributed I/O" are also concepts that are designed to reduce wiring costs by placing the I/O near the signals.

Start-up Costs

Start-up costs must account for the inevitable system test, debug process, and safety functional verification. Configuration tools that help manage system change and automatic change documentation can cut costs in this area. Testing and full functional verification is a big portion of start-up cost. Testing is expedited when the control system allows detail monitoring and forcing of I/O points. On-line display of system variables, ideally within the context of the system design drawings, can also speed testing and debugging of complex systems.

Operating Costs

Operating costs can overshadow procurement costs in many systems. Depending on the consequences of lost production, operating costs may dominate any life-cycle cost study. One experienced system designer stated, "Our plant manager simply cannot tolerate a shutdown due to control system failure!" There are reasons for this attitude; the cost of a shutdown can be extreme. The cost of a shutdown, however, is not the only operating cost. Other operating costs include the cost of system engineering changes, consumption costs, and fixed maintenance costs, and are generally incurred each year. The cost of direct repair labor and the cost of lost production are incurred whenever a failure occurs. These costs vary directly with system failure probability.

In the case of a safety instrumented system (SIS), the "risk cost" is increased when the system probability of failure on demand goes up. While safety ratings in terms of the required safety integrity level are usually dictated by a safety study, reduced risk cost may justify even higher levels of safety integrity.

Engineering Changes

Engineering changes are part of operating costs. Everything cannot be anticipated by the original designers, and all systems are inevitably changed. As a system is used, it becomes obvious that changes will improve system operation. System-level design faults must be repaired, thus increasing the strength of the system design.

The cost of making a change can vary widely. Upper-bound and lower-bound estimates should be made. Factors to be considered when making the estimate include the procedures required to change the system, the ease of changing the documentation, and the ease of testing and verifying the changes. In the case of SISs, a review of initial hazard analysis and factors such as an update of the periodic maintenance test procedures must be considered. System capabilities affect these estimates. Systems that have automatic updating of system design documentation will be changed with much less effort.

Consumption Costs

System operation requires energy consumption and parts consumption. These costs are typically estimated on an annual basis. Energy consumption includes both the energy required to operate the control equipment and the energy required to maintain the equipment environment. A control system may not have sufficient internal strength to withstand specific environmental stress factors. Environmental protection is added to the system. Special environments actively created will cost money to maintain.

Fixed Maintenance Costs

Certain maintenance costs are incurred each year regardless of the number of system failures. Fixed maintenance costs include periodic inspection, the cost of repair inventory, the cost of continued maintenance training, annual costs of full-time staff, and amortization of test equipment. There is a relationship between these costs and system failure probability. Low failure probability means low fixed maintenance costs.

Cost of System Failure

One of the most significant operating costs can occur when a system fails. In some systems, the exact failure mode is not important. A single failure state is all that is required for a reliability and safety model. If an oven fails energized while baking a soufflé, the dish is burned. If an oven fails de-energized, the soufflé falls and is ruined. In both cases, the cost of lost product as well as repair costs must be paid.

Other systems have multiple failure modes that must be distinguished. If the stove fails de-energized while heating some soup, we fix the stove and then the soup. No product is lost. If the stove fails energized, the soup is burned. The cost of that failure includes both the cost of the lost product and the cost of repair. (Under certain circumstances, we may even have to throw away the pot, the stove, or even the house!)

Timing of the failure may affect cost. If the braking system of an automobile fails just as another vehicle pulls in front of it, the costs are expected to be high. Brake failure on a long, straight, empty road is likely to cost much less. This illustrates the concept of failure on demand. The concept is particularly important to an SIS. A catastrophic event can be extremely expensive.

All these factors complicate the task of estimating the expected cost of system failure. Though complicated, general risk analysis techniques apply. Reliability parameters can be used to calculate the expected costs. Costs of failure are multiplied by the probability of failure. Reliability analysis techniques provide the expected probability of failure information. Costs are estimated in two ways: time based and event based. Time-based failure costs are those that are directly related to failure time. A prime example is the cost of lost production in a repairable system. The longer the system is down, the higher the cost. Event-based failure costs are those related to an event. Examples of this include replacement parts, fixed repair call charges, or cost of lost batch production.

Time-based Failure Costs

In many systems, the most significant failure costs are time based. A majority of systems are fully repairable with one or two failure states. For those systems, the cost of repair and the cost of lost production vary directly with system downtime. Repair labor and lost production should be estimated with consistent units (typically on a per hour basis). For a fully repairable, single-failure state system:

$$C_{Failure} = [(C_R + C_{LP}) \times U] \times \text{operating hours} \tag{16-3}$$

where:

C_R = repair cost per hour

C_{LP} = cost of lost production per hour

U = unavailability (average probability)

By convention, the cost of failure is calculated on a yearly basis. Therefore, the number of operating hours in one year is used in Equation 16-3.

For fully repairable, multiple failure state systems:

$$C_{Failure} = [(C_{R1} + C_{LP1}) \times U_1 + (C_{R2} + C_{LP2}) \times U_2] \times \text{operating hours} \quad (16\text{-}4)$$

where:

C_{R1} = repair cost per hour for failure state 1
C_{LP1} = cost of lost production per hour for failure state 1
C_{R2} = repair cost per hour for failure state 2
C_{LP2} = cost of lost production per hour for failure state 2
U_1 = unavailability (average probability of failure) for failure state 1
U_2 = unavailability (average probability of failure) for failure state 2

The number of operating hours used in Equation 16-4 is typically the number of hours in one year (8,760).

Event-based Failure Costs

In a fully repairable system, the average probability of an event can be used to calculate yearly failure costs as a function of each event.

$$C_{Failure} = C_E \times U_E \quad (16\text{-}5)$$

where

C_E = cost per event
U_E = probability of an event during the year

The total operating cost is the sum of the failure costs (both event and time based), the engineering change costs (C_{EC}), the fixed maintenance costs (C_{FM}), and the consumable costs (C_{CC}). If engineering change costs, fixed maintenance costs, and consumable costs are estimated on a yearly basis, they must be multiplied by the system lifetime (in years). Since the cost of failure calculation accounts for operating hours in one year, that cost must also be multiplied by the system lifetime. The total operating costs are

$$C_{OP} = (C_{EC} + C_{FM} + C_{CC} + C_{Failure}) \times \text{years life} \quad (16\text{-}6)$$

A life-cycle cost analysis requires a reliability analysis, costs estimates, and a life-cycle analysis. The hardest part is sometimes getting started. What factors must be accounted for in your system? Table 16-1 shows a checklist of life-cycle cost items that should be considered.

Procurement Cost	
System Design Cost	Specification Preparation
	Request for Quote
	Detailed Design
	Drawings
	Periodic Test Procedures
	System FMEA
	Design Training
	System Configuration
Purchase Cost	Program Management
	Control Hardware
	Wire Termination Panels
	Service Contract
	Environmental Protection
	Acceptance Test
	Initial Spare Parts
	Initial Maintenance Training
Installation Cost	Weatherization
	Mounting
	Piping
	Wiring
Startup Cost	System Debugging
	Safety Verification
	Engineering Changes
	Test Equipment
Operating Cost	
	Lost Production Cost
	Risk Cost
Maintenance Cost	Repair Labor
Consumable Cost	Spare Parts
	Energy

Table 16-1. Life Cycle Cost Checklist

EXAMPLE 16-1

Problem: For a 10-year operating life, calculate the approximate life-cycle cost of the forging machine of Figure 8-5. Do not account for the time value of money.

Solution: Our estimating department gives us the following data:

Design Cost

52 hr @ $75/hr engineering time
22 hr @ $45/hr drawing/documentation time
16 hr @ $75/hr safety review

Purchase Cost: $120,000

Installation Cost

Truck rental: $300
32 hr @ $75/hr

Start-up Cost

Training course fee: $1,500
80 hr @ $75/hr training time
10 hr @ $75/hr equipment assembly

Engineering Change Cost: $500/year

Fixed Maintenance Costs: $5,000/Year

One repair technician for 10 machines

Consumption Costs

Electricity: $1,200/year
Lubricating oil: $200/year
Filters: $100/year

Direct Repair Costs

Labor rate: $100/hr

Lost Production Cost: $2,000/hr

To calculate the procurement costs, we add all the expenses:

52 × $75 =	$3,900
22 × $45 =	$990
16 × $75 =	$1,200
Purchase =	$120,000
Truck rental =	$300
32 × $75 =	$2,400
Course =	$1,500
80 × $75 =	$6,000
10 × $75 =	$750
Total =	$137,040

EXAMPLE 16-1 (continued)

The reliability analysis has determined that the forging machine has a limiting state availability of 98.04% and an unavailability of 1.96%. Using the unavailability, the yearly failure costs are calculated using Equation 16-3:

$$\$2100 \times 0.0196 \times 8760 \text{ hr/year} = \$360,561.60$$

To calculate total operating costs, we first add yearly expenses:

Engineering changes =	$500
Fixed maintenance =	$5000
Electricity =	$1200
Oil =	$200
Filters =	$100
Failure cost =	$360,561.60
Total yearly =	**$367,561.60**

These are multiplied by the expected life:

$$\$367,561.60 \times 10 = \$3,675,616$$

Life-cycle costs are the sum of procurement and operating costs:

$$\$3,675,616 + \$137,040 = \$3,812,656$$

EXAMPLE 16-2

Problem: A different forging machine is available. It has a purchase cost of $200,000. The availability improves from 98.04% to 98.9%. What is the difference in life-cycle cost?

Solution: The difference in purchase cost is +$80,000. The new failure cost is found using Equation 16-3:

$$\$2,100 \times (1 - 0.989) \times 8760 \text{ hr/year} \times 10 \text{ years} = \$2,023,560$$

The difference is life cycle costs are:

$$\$2,023,560 - \$3,675,616 + \$80,000 = -\$1,572,056$$

We have discovered that an increase in purchase cost of $80,000 results in a savings of $1,572,056 over 10 years.

> **EXAMPLE 16-3**
>
> **Problem:** An extruding machine has a system to control material feed temperature. The control system has a procurement cost of $270,000. Operating costs include engineering change costs of $2,000 per year, maintenance costs of $5,000 per year, and consumption costs of $10,000 per year. The machine and its control system are inspected and overhauled each year.
>
> There are two failure modes in the temperature-control system. If the control system fails with its outputs energized, the excess heat will destroy the entire extruding machine. Including the cost of lost production, this event would cost $1,200,000. If the temperature control system fails with its outputs de-energized, the heat is removed and the material in the extruder will harden. The extruding machine must be rebuilt if this occurs. The cost of a machine rebuild, including the cost of lost production, is $200,000 per event.
>
> The reliability analysis shows an average probability of failure in the energized state to be 0.0017 for the one-year interval. The probability of failure in the de-energized state is 0.06 for the one-year interval. Find the life-cycle cost for a 10-year life.
>
> **Solution:** Procurement costs are given. Operating costs, including the cost of failure, must be determined for each year. The cost of failure is obtained by multiplying the cost of an event by the probability of that event (Equation 16-5). In this case, the cost of failure is
>
> $$(0.0017 \times \$1,200,000) + (0.06 \times \$200,000) = \$14,040/\text{year}$$
>
> Total operating costs for a 10-year life are
>
> $$(\$2,000 + \$5,000 + \$10,000 + \$14,040) \times 10 = \$310,400$$
>
> This is added to the procurement cost to obtain a 10-year life-cycle cost of $580,400.

Time Value of Money

More precision can be added to life-cycle cost estimates if the time value of money is included in the analysis. This occurs because cash in one year cannot be compared to cash from an earlier or later period of time. The value may be changed. This is especially true when systems are used for longer lifetimes. Money spent early in the life of a product is weighted more, because interest swells its value.

Discount Rate

Almost everyone is familiar with the concept of compound interest. An amount of money (or principal) is invested at a particular interest rate. The interest earned is reinvested so that it too earns interest. Inflation is another familiar concept. When considering the time value of money, both interest rates and inflation rates must be taken into account. A term that combines both is called the *discount rate*. The concept of discount rate also includes some judgement regarding the uncertainty of numbers estimated for the future. Risky projects are often given a higher discount rate. The discount rate is expressed as a percentage. The best source for the discount rate estimate is the corporate financial officer.

If a principal of M dollars is invested at a percentage of $100R$ (5% means $R = 0.05$), then the compound amount of money at the beginning of the second year is

$$M + MR$$

This amount is considered the principal for the second year. At the beginning of the third year, the compounded amount of money is

$$(M + MR) + (M + MR)R \tag{16-7}$$

Factoring out the M, we obtain:

$$M[(1 + R) \times (1 + R)] = M(1 + R)^2 \tag{16-8}$$

This formula can be generalized for any number of years because the pattern continues. For any quantity of years, N, the future value of money, FV, after N years equals:

$$FV = M(1 + R)^N \tag{16-9}$$

where:

FV = future value of the money

M = principal

R = discount rate

N = number of time periods

EXAMPLE 16-4

Problem: A purchase cost of $250,000 is required for a control system. What is the equivalent future value of this investment for a 10-year period at a 6% per year rate?

Solution: Using Equation 16-9,

$$FV = \$250,000(1+0.06)^{10} = \$250,000 \times 1.7908$$
$$= \$447,700$$

Present Value

Another way to account for the time value of money is to calculate the present value of some future expense. This is the equivalent of asking how much money to invest now in order to pay for some future expense. The equation for present value is obtained from Equation 16-9. Look at Example 16-4. The compound amount $447,700 is the future value. The principal of $250,000 is the present value of $447,700. To directly calculate a present value when a future value is known, Equation 16-9 must be solved for PV. This results in:

$$PV = \frac{M}{(1+R)^N} \qquad (16\text{-}10)$$

where:

PV = present value of money
M = amount of future money
R = discount rate
N = number of time periods

EXAMPLE 16-5

Problem: An operating expense of $5,000 will be needed five years from now. What is the present value of this expense if the discount rate is 5%?

Solution: Using Equation 16-10,

$$PV = \frac{\$5000}{(1+0.05)^5} = \$3917.64$$

EXAMPLE 16-6

Problem: Reevaluate the life-cycle cost analysis of Example 16-1, accounting for the time value of money. Assume that the average rate is 5%, compounded yearly. Calculate the cost at the end of a 10-year operating life.

Solution: Costs must be calculated on a yearly basis. Those costs are listed for each year in which they are incurred. Equation 16-9 is then used to calculate added finance cost for each year. First, given cost data are summarized:

Procurement costs = $137,000.00
Yearly fixed costs = $7,000.00

EXAMPLE 16-6 (continued)

The failure costs on a yearly basis must be calculated. This is done by multiplying the probable number of downtime hours in a year by the cost of each hour:

$$C_{Failure} = \$2100 \times 0.0196 \times 8760 \text{ hr/year}$$
$$= \$360,561.60/\text{year}$$

Total yearly costs are the sum of the yearly fixed costs and the yearly failure costs, or $367,561.60.

For each cost, calculate the cost with discount. This is done using Equation 16-9. Initial costs must pay finance charges for the entire 10-year period. The calculation is as follows:

$$FV = \$137,000 \times (1.05)^{10} = \$233,158.56$$

For yearly costs, a similar calculation for the appropriate period is done. For the first yearly cost:

$$FV = \$367,561.60 \times (1.05)^{9} = \$570,208.68$$

The calculation is repeated for each year and the results are added. A personal computer spreadsheet program is highly recommended for this type of problem. Such programs are quick to set up, reduce mistakes, and provide great flexibility. Discount rates and costs can be varied each year. A listing obtained from a spreadsheet shows yearly costs for the problem.

Costs per Year:	
Initial Costs	$223,158.56
Year 1 Costs	$570,208.68
Year 2 Costs	$543,055.89
Year 3 Costs	$517,196.08
Year 4 Costs	$492,567.70
Year 5 Costs	$469,112.09
Year 6 Costs	$446,773.42
Year 7 Costs	$425,498.50
Year 8 Costs	$405,236.66
Year 9 Costs	$385,939.68
Year 10 Costs	$367,561.60

Life-cycle costs (in terms of future value) = $4,846,308.87.

Life Cycle Costing

EXAMPLE 16-7

Problem: The accounting department has advised that life-cycle costs are normally done in terms of present value, not future value. Recalculate Example 16-1, accounting for the time value of money in terms of present value.

Solution: For each yearly expense, Equation 16-10 can be used to convert future amounts into present value. For the first year:

$$PV = \frac{\$367,561.60}{(1.05)^1}$$
$$= \$350,058.67$$

Using a spreadsheet, the calculations are repeated for each year.

Year 1 Present Value	350,058.67
Year 2 Present Value	333,389.21
Year 3 Present Value	317,513.53
Year 4 Present Value	302,393.84
Year 5 Present Value	287,994.13
Year 6 Present Value	274,280.13
Year 7 Present Value	261,219.17
Year 8 Present Value	248,780.16
Year 9 Present Value	236,933.48
Year 10 Present Value	225,650.94
Initial costs	137,000.00

Life-cycle costs (in terms of present value) = $2,975,213.25.

The result can be checked by calculating the future value of this amount. Using Equation 16-9,

$$FV = 2,975,213.25 \times (1.05)^{10}$$
$$= \$4,846,308.87$$

This result agrees with Example 16-6.

Annuities

An annuity is a sequence of payments made at fixed periods of time over a given time interval. (It is assumed that payments are made at the end of the period.) Yearly life-cycle costs can be modeled as an annuity. Both future value costs and present value costs can be calculated.

The future value of an annuity can be obtained by using Equation 16-9 for each year:

$$FV = M + M(1 + R)^1 + M(1 + R)^2 + \ldots + M(1 + R)^{N-1}$$

If this equation is multiplied by $(1 + R)$, the result is

$$FV(1 + R) = M(1 + R) + M(1 + R)^2 + \ldots + M(1 + R)^N$$

When these two equations are subtracted:

$$FV - FV(1 + R) = M + M(1 + R) - M(1 + R) + \ldots - \ldots - M(1 + R)^N$$

Therefore,

$$FV(1 - 1 - R) = M - M(1 + R)^N$$

This can be simplified into:

$$FV_A = M \frac{[(1 + R)^N - 1]}{R} \qquad (16\text{-}11)$$

EXAMPLE 16-8

Problem: No spreadsheet program is available to calculate life-cycle cost. Solve Example 16-6 (life-cycle cost in future value) without a spreadsheet.

Solution: The yearly expenses can be modeled as an annuity. Using Equation 16-11,

$$FV = 367{,}561.60 \times \frac{(1.05)^{10} - 1}{0.05}$$

$$= \$4{,}623{,}150.30$$

This is added to the future value of the initial expense, $223,158.56, to obtain a total of $4,846,308.87.

Life Cycle Costing

In many cases, it is customary to calculate life-cycle costs in terms of present value rather than future value (as in Example 16-8). The present value of an annuity is the sum of the present values of all payments. It represents the amount of money that must invested now in order to make the required future payments.

The present value of an annuity can be obtained using Equation 16-10. Assuming payments due at the end of a period, for N payments of M (dollars) at an interest rate of R:

$$PV_A = M(1 + R)^{-1} + M(1 + R)^{-2} + \ldots + M(1 + R)^{-N}$$

This can be algebraically arranged as:

$$PV_A = M \frac{1 - (1 + R)^{-N}}{R} \tag{16-12}$$

EXAMPLE 16-9

Problem: Evaluate the life-cycle cost analysis of Example 16-1 accounting for the time value of money in terms of present value without using a spreadsheet. Assume that the discount rate is 5%.

Solution: Total yearly costs of $367,561.60 are the equivalent of an annuity. Thus, Equation 16-12 can be used to determine the present value for yearly costs.

$$PV = 367{,}561{,}60 \frac{[1 - (1.05)^{-10}]}{0.05}$$

$$= \$2{,}838{,}213.25$$

This is added to the initial costs of $137,000 to obtain a result of $2,975,213.25. This result agrees with that of Example 16-7.

SIS Life-Cycle Cost

An SIS represents a special case, as failure costs consist primarily of risk costs. Risk costs are not necessarily paid each year unless they are paid as insurance premiums. However, risk costs can and should be considered in a life-cycle cost analysis.

Risk costs are calculated by multiplying the cost of an event times the probability of an event. If an SIS is used, the probability of an event equals the probability of an event without the SIS times the PFD of the SIS.

$$P_{\text{EVENT with SIS}} = P_{\text{EVENT without SIS}} \times PFD_{SIS} \tag{16-13}$$

Risk costs are included as yearly costs and may be treated as an annuity.

EXAMPLE 16-10

Problem: With no risk reduction, the probability of an event is 0.01. If an SIS is added with a PFD of 0.001, what is the probability of an event?

Solution: Using Equation 16-13,

$$P_{\text{EVENT with SIS}} = 0.01 \times 0.001$$
$$= 0.00001$$

Other cost factors that must be considered when adding an SIS to a process include the cost of a false trip. An SIS may fail safely, falsely tripping the system. The decreased risk cost must be greater than the increased production downtime cost. Other life-cycle cost factors are much the same for an SIS as for other process control systems.

EXAMPLE 16-11

Problem: A decision must be made to purchase an SIS. The purchase and installation cost of the SIS is $50,000. The SIS is periodically inspected every year and has a PFD of 0.001 and a PFS of 0.01 for the one-year interval. Yearly operational costs for the SIS are $600. The cost of a false trip is $1,000 per trip. The probability of an event without the SIS is 0.01. The cost of an event is $2,000,000. The system will be used for longer than five years. The discount rate is 5%. Should the SIS be added to the system?

Solution: The risk cost without the SIS equals 0.01 × $2,000,000 = $20,000/year. The risk cost with the SIS equals 0.01 × 0.001 × $2,000,000 = $20/year. With the SIS added to the system, incremental trip costs equal 0.01 × $1,000 = $10/year. The cost comparisons are

	No SIS	SIS
Procurement Costs	$0	$50,000
Yearly Risk/Failure Cost	$20,000	$30
Yearly Operational Cost	$0	$600

Life Cycle Costing

> **EXAMPLE 16-11 (continued)**
>
> Converting yearly expenses to present value and adding the totals for each year:
>
> | | No SIS | Discount | 5% |
> |---|---|---|---|
> | Total Yearly | $20,000 | Cumulative | $0 |
> | PV year 1 | $19,048 | Year 1 | $19,048 |
> | PV year 2 | $18,141 | Year 2 | $37,188 |
> | PV year 3 | $17,277 | Year 3 | $54,465 |
> | PV year 4 | $16,454 | Year 4 | $70,919 |
> | PV year 5 | $15,671 | Year 5 | $86,590 |
>
> Total life-cycle costs for five years = $86.590
>
> | | SIS | Discount | 5% |
> |---|---|---|---|
> | Total Yearly | $630 | Cumulative | $50,000 |
> | PV year 1 | $600 | Year 1 | $50,600 |
> | PV year 2 | $571 | Year 2 | $51,171 |
> | PV year 3 | $544 | Year 3 | $51,716 |
> | PV year 4 | $518 | Year 4 | $52,234 |
> | PV year 5 | $494 | Year 5 | $52,728 |
>
> Total life-cycle costs for five years = $52,728
>
> Since the process is to be operated for five years or more, the SIS is less expensive and thus should be added.

EXERCISES

16.1 A control system has a procurement cost of $100,000. Fixed yearly costs are $5,000. The system has an availability of 99%. The failure costs are $1,000 per hour. The system will be used for five years. During this time period, inflation and interest are identical so the discount rate will be zero (no need to account for the time value of money). Calculate the life cycle cost.

16.2 The control system of Exercise 16.1 is fully repairable, with availability calculated using MTTF equals 9,900 hr and MTTR equals 100 hr. An expert system can be purchased at a cost of $10,000. This expert system will diagnose error messages and identify failed components. The MTTR is reduced to 50 hr. What is the new availability? What is the new life cycle cost? Should the expert system be purchased?

16.3 The control system of Exercise 16.1 is available with dual redundant modules. The extra modules will increase fixed yearly costs to $6,000. The MTTF will increase to 100,000 hr. The MTTR remains at 50 hr when the expert system is used. The procurement cost increases to $210,000. What is the new availability? What is the new life-cycle cost? Should the dual redundant modules be purchased?

16.4 An expenditure of $100,000 must be made. The discount rate is 4%. What is the future value of this expenditure after five years?

16.5 An expenditure of $150,000 in today's cost must be made in five years. What is the amount that must be invested now (present value) in order to purchase the item in five years? Assume that the discount rate is 5%.

16.6 Repeat Exercise 16.1 assuming that the discount rate is 5%. How much did the life-cycle cost change?

REFERENCES AND BIBLIOGRAPHY

1. Crosby, P. H., *Quality Is Free*, New York: McGraw-Hill, 1979.

2. Carrubba, E. R., "Integrating Life-Cycle Cost and Cost-of-Ownership in the Commercial Sector," *Annual Reliability and Maintainability Symposium Proceedings*, New York: IEEE, 1992.

3. Dacko, L. M., and Darlington, R. F., "Life Cycle Cost Procedure for Commercial Aircraft Subsystems," *Annual Reliability and Maintainability Symposium Proceedings*, New York: IEEE, 1988.

4. Feiler, A. M., "Incorporating Risk Analysis into Life Cycle Costing," *Annual Reliability and Maintainability Symposium Proceedings*, New York: IEEE, 1986.

5. Shelton, K. W., Skweres, P. C., and Jones, R. B., "Total Cost of Ownership — True Process Instrument Evaluation," *Proceedings of the 56th Annual Symposium on Instrumentation for the Process Industries*, Research Triangle Park, NC: ISA—The Instrumentation, Systems, and Automation Society, 1991.

APPENDIX A

Standard Normal Distribution Table

	0.00	0.01	0.02	0.03	0.04
0.0	0.500000	0.503989	0.507978	0.511967	0.515954
0.1	0.539828	0.543796	0.547759	0.551717	0.555670
0.2	0.579260	0.583167	0.587065	0.590955	0.594835
0.3	0.617912	0.621720	0.625516	0.629301	0.633072
0.4	0.655422	0.659098	0.662758	0.666403	0.670032
0.5	0.691463	0.694975	0.698469	0.701945	0.705402
0.6	0.725748	0.729070	0.732372	0.735654	0.738915
0.7	0.758037	0.761149	0.764239	0.767306	0.770351
0.8	0.788146	0.791031	0.793893	0.796732	0.799547
0.9	0.815941	0.818590	0.821215	0.823816	0.826392
1.0	0.841346	0.843754	0.846137	0.848496	0.850831
1.1	0.864335	0.866502	0.868644	0.870763	0.872858
1.2	0.884931	0.886862	0.888769	0.890653	0.892513
1.3	0.903201	0.904903	0.906584	0.908242	0.909878
1.4	0.919244	0.920731	0.922197	0.923643	0.925067
1.5	0.933194	0.934479	0.935745	0.936993	0.938221
1.6	0.945202	0.946302	0.947385	0.948450	0.949498
1.7	0.955435	0.956368	0.957285	0.958186	0.959071
1.8	0.964070	0.964853	0.965621	0.966376	0.967117
1.9	0.971284	0.971934	0.972572	0.973197	0.973811
2.0	0.977251	0.977785	0.978309	0.978822	0.979325
2.1	0.982136	0.982571	0.982998	0.983415	0.983823
2.2	0.986097	0.986448	0.986791	0.987127	0.987455
2.3	0.989276	0.989556	0.989830	0.990097	0.990359
2.4	0.991803	0.992024	0.992240	0.992451	0.992657
2.5	0.993791	0.993964	0.994133	0.994297	0.994458

	0.00	0.01	0.02	0.03	0.04
2.6	0.995339	0.995473	0.995604	0.995731	0.995855
2.7	0.996533	0.996636	0.996736	0.996834	0.996928
2.8	0.997445	0.997523	0.997599	0.997673	0.997745
2.9	0.998134	0.998193	0.998250	0.998305	0.998359
3.0	0.998650	0.998694	0.998736	0.998777	0.998817
3.1	0.999033	0.999065	0.999096	0.999126	0.999156
3.2	0.999313	0.999337	0.999359	0.999381	0.999403
3.3	0.999517	0.999534	0.999550	0.999566	0.999581
3.4	0.999663	0.999675	0.999687	0.999698	0.999709
3.5	0.999768	0.999776	0.999784	0.999792	0.999800
3.6	0.999841	0.999847	0.999853	0.999859	0.999864
3.7	0.999892	0.999897	0.999901	0.999904	0.999908
3.8	0.999928	0.999931	0.999933	0.999936	0.999939
3.9	0.999952	0.999954	0.999956	0.999958	0.999959
4.0	0.999969	0.999970	0.999971	0.999972	0.999973
4.1	0.999980	0.999980	0.999981	0.999982	0.999983
4.2	0.999987	0.999987	0.999988	0.999989	0.999989
4.3	0.999992	0.999992	0.999992	0.999993	0.999993
4.4	0.999995	0.999995	0.999995	0.999995	0.999996
4.5	0.999997	0.999997	0.999997	0.999997	0.999997
4.6	0.999998	0.999998	0.999998	0.999998	0.999998

	0.05	0.06	0.07	0.08	0.09
0.0	0.519939	0.523922	0.527903	0.531882	0.535857
0.1	0.559618	0.563560	0.567495	0.571424	0.575346
0.2	0.598707	0.602569	0.606420	0.610262	0.614092
0.3	0.636831	0.640577	0.644309	0.648028	0.651732
0.4	0.673646	0.677243	0.680823	0.684387	0.687934
0.5	0.708841	0.712261	0.715662	0.719044	0.722406
0.6	0.742155	0.745374	0.748572	0.751749	0.754904
0.7	0.773374	0.776374	0.779351	0.782306	0.785237
0.8	0.802339	0.805107	0.807851	0.810571	0.813268
0.9	0.828945	0.831474	0.833978	0.836458	0.838914
1.0	0.853142	0.855429	0.857692	0.859930	0.862145
1.1	0.874929	0.876977	0.879001	0.881001	0.882978
1.2	0.894351	0.896166	0.897959	0.899729	0.901476
1.3	0.911493	0.913086	0.914658	0.916208	0.917737
1.4	0.926472	0.927856	0.929220	0.930564	0.931889
1.5	0.939430	0.940621	0.941793	0.942948	0.944084
1.6	0.950529	0.951544	0.952541	0.953522	0.954487
1.7	0.959942	0.960797	0.961637	0.962463	0.963274
1.8	0.967844	0.968558	0.969259	0.969947	0.970622
1.9	0.974413	0.975003	0.975581	0.976149	0.976705

	0.05	0.06	0.07	0.08	0.09
2.0	0.979818	0.980301	0.980774	0.981238	0.981692
2.1	0.984223	0.984614	0.984997	0.985372	0.985738
2.2	0.987776	0.988090	0.988397	0.988697	0.988990
2.3	0.990614	0.990863	0.991106	0.991344	0.991576
2.4	0.992858	0.993054	0.993245	0.993431	0.993613
2.5	0.994614	0.994767	0.994915	0.995060	0.995202
2.6	0.995976	0.996093	0.996208	0.996319	0.996428
2.7	0.997021	0.997110	0.997197	0.997282	0.997365
2.8	0.997814	0.997882	0.997948	0.998012	0.998074
2.9	0.998411	0.998462	0.998511	0.998559	0.998605
3.0	0.998856	0.998894	0.998930	0.998965	0.998999
3.1	0.999184	0.999211	0.999238	0.999264	0.999289
3.2	0.999423	0.999443	0.999462	0.999481	0.999499
3.3	0.999596	0.999611	0.999624	0.999638	0.999651
3.4	0.999720	0.999730	0.999740	0.999750	0.999759
3.5	0.999808	0.999815	0.999822	0.999828	0.999835
3.6	0.999869	0.999874	0.999879	0.999884	0.999888
3.7	0.999912	0.999915	0.999919	0.999922	0.999925
3.8	0.999941	0.999944	0.999946	0.999948	0.999950
3.9	0.999961	0.999963	0.999964	0.999966	0.999967
4.0	0.999975	0.999976	0.999977	0.999978	0.999979
4.1	0.999984	0.999984	0.999985	0.999986	0.999986
4.2	0.999990	0.999990	0.999990	0.999991	0.999991
4.3	0.999993	0.999994	0.999994	0.999994	0.999995
4.4	0.999996	0.999996	0.999996	0.999996	0.999997
4.5	0.999998	0.999998	0.999998	0.999998	0.999998
4.6	0.999999	0.999999	0.999999	0.999999	0.999999

APPENDIX B

Matrix Math

The Matrix

A matrix is an array of numeric values or algebraic variables. The matrix is written with rows and columns enclosed in brackets. There is no single numerical value. An example 3×3 matrix (third order) is shown below.

$$\mathbf{P} = \begin{bmatrix} a_{11} & a_{12} & a_{13} \\ a_{21} & a_{22} & a_{23} \\ a_{31} & a_{32} & a_{33} \end{bmatrix}$$

Column Matrix

If a matrix has only one column, it is known as a *column matrix* or *column vector*. An example is shown below.

$$\mathbf{C} = \begin{bmatrix} 3 \\ 1 \\ 0.6 \\ 2 \end{bmatrix}$$

Row Matrix

If a matrix has only one row, it is known as a *row matrix* or *row vector*. An example is shown below.

$$S = \begin{bmatrix} 1 & 0 & 0 & 0 & 0 \end{bmatrix}$$

Identity Matrix

The identity matrix is a square matrix in which all the elements are zero except those on a diagonal from upper left to lower right. The diagonal elements are unity.

$$\mathbf{I} = \begin{bmatrix} 1 & 0 & 0 & 0 & 0 \\ 0 & 1 & 0 & 0 & 0 \\ 0 & 0 & 1 & 0 & 0 \\ 0 & 0 & 0 & 1 & 0 \\ 0 & 0 & 0 & 0 & 1 \end{bmatrix}$$

Matrix Addition

Two matrices of the same size can be added. The elements in corresponding positions are summed.

$$\begin{bmatrix} a & b & c \\ d & e & f \end{bmatrix} + \begin{bmatrix} g & h & i \\ j & k & l \end{bmatrix} = \begin{bmatrix} a+g & b+h & c+i \\ d+j & e+k & f+l \end{bmatrix} \quad \text{(B-1)}$$

Matrix Subtraction

Two matrices of the same size can be subtracted. The elements in corresponding positions are subtracted.

$$\begin{bmatrix} a & b & c \\ d & e & f \end{bmatrix} - \begin{bmatrix} g & h & i \\ j & k & l \end{bmatrix} = \begin{bmatrix} a-g & b-h & c-i \\ d-j & e-k & f-l \end{bmatrix} \quad \text{(B-2)}$$

Matrix Multiplication

Two matrices may be multiplied if the number of columns in the first matrix equals the numbers of rows in the second matrix. The result will be a matrix that has a quantity of rows equal to that of the first matrix and a quantity of columns equal to that of the second matrix.

$$\begin{bmatrix} a & b & c \\ d & e & f \end{bmatrix} \times \begin{bmatrix} g & h \\ i & j \\ k & l \end{bmatrix} = \begin{bmatrix} a \times g + b \times i + c \times k & a \times h + b \times j + c \times l \\ d \times g + e \times i + f \times k & d \times h + e \times j + f \times l \end{bmatrix} \quad \text{(B-3)}$$

Matrix Inversion

There is no procedure that allows matrices to be divided. However, we can obtain the "reciprocal" or inverse of a square matrix. The inverse of one matrix can be multiplied by another matrix in a manner analogous to algebraic division.

The inverse (M^{-1}) of a square matrix (M) is another square matrix defined by the relation:

$$M \times M^{-1} = I \tag{B-4}$$

where I is the identity matrix. This is similar to stating that a number multiplied by its inverse equals one.

$$4 \times \frac{1}{4} = 1$$

EXAMPLE B-1

Problem: Multiply the matrix M by the matrix M^{-1}, where:

$$M = \begin{bmatrix} 4 & 0 & 5 \\ 0 & 1 & -6 \\ 3 & 0 & 4 \end{bmatrix}$$

and

$$M^{-1} = \begin{bmatrix} 4 & 0 & -5 \\ -18 & 1 & 24 \\ -3 & 0 & 4 \end{bmatrix}$$

Solution: When these two matrices are multiplied:

$$\begin{bmatrix} 4 & 0 & 5 \\ 0 & 1 & -6 \\ 3 & 0 & 4 \end{bmatrix} \times \begin{bmatrix} 4 & 0 & -5 \\ -18 & 1 & 24 \\ -3 & 0 & 4 \end{bmatrix} = \begin{bmatrix} 1 & 0 & 0 \\ 0 & 1 & 0 \\ 0 & 0 & 1 \end{bmatrix}$$

All square matrices are not invertible. In reliability evaluation, we are normally working with matrices derived from the transition matrix. These usually do not present a problem.

Matrix Inversion via Row Manipulation

One method of obtaining the inverse of a matrix requires a sequence of row manipulations applied to the target matrix in combination with an identity matrix of the same order. The two matrices are written side by side in the form:

$$\left[\begin{array}{ccc|ccc} a & b & c & 1 & 0 & 0 \\ d & e & f & 0 & 1 & 0 \\ g & h & i & 0 & 0 & 1 \end{array}\right]$$

Row operations are applied to this combination matrix until it appears in the form:

$$\left[\begin{array}{ccc|ccc} 1 & 0 & 0 & j & k & l \\ 0 & 1 & 0 & m & n & o \\ 0 & 0 & 1 & p & q & r \end{array}\right]$$

The matrix:

$$\mathbf{M}^{-1} = \begin{bmatrix} j & k & l \\ m & n & o \\ p & q & r \end{bmatrix}$$

is the inverse of the matrix:

$$\mathbf{M} = \begin{bmatrix} a & b & c \\ d & e & f \\ g & h & i \end{bmatrix}$$

The row operations are:

1. Any two rows can be interchanged.

2. Any row may be multiplied by a nonzero scalar.

3. Any row may be added to a multiple of another row.

EXAMPLE B-2

Problem: Obtain the inverse of the matrix:

$$M = \begin{bmatrix} 4 & 0 & 5 \\ 0 & 1 & -6 \\ 3 & 0 & 4 \end{bmatrix}$$

Solution: The matrix is written along with the identity matrix in a composite form:

$$\left[\begin{array}{ccc|ccc} 4 & 0 & 5 & 1 & 0 & 0 \\ 0 & 1 & -6 & 0 & 1 & 0 \\ 3 & 0 & 4 & 0 & 0 & 1 \end{array} \right]$$

The objective is to manipulate matrix rows until the left side of the composite equals the identity matrix. One good strategy is to put zeros into the left side. As a first step, manipulate rows in order to replace the five with a zero. Multiply row 3 by $\frac{5}{4}$. The result is

$$\left[\begin{array}{ccc|ccc} 4 & 0 & 5 & 1 & 0 & 0 \\ 0 & 1 & -6 & 0 & 1 & 0 \\ \frac{15}{4} & 0 & 5 & 0 & 0 & \frac{5}{4} \end{array} \right]$$

Replace row 1 with (row 1 − row 3).

$$\left[\begin{array}{ccc|ccc} \frac{1}{4} & 0 & 0 & 1 & 0 & -\frac{5}{4} \\ 0 & 1 & -6 & 0 & 1 & 0 \\ \frac{15}{4} & 0 & 5 & 0 & 0 & \frac{5}{4} \end{array} \right]$$

Next, replace the $\frac{15}{4}$ with a zero. To accomplish this, use rule 2: Any row may be multiplied by a nonzero scalar (row 1 = 15 × row 1).

$$\left[\begin{array}{ccc|ccc} \frac{15}{4} & 0 & 0 & 15 & 0 & -\frac{75}{4} \\ 0 & 1 & -6 & 0 & 1 & 0 \\ \frac{15}{4} & 0 & 5 & 0 & 0 & \frac{5}{4} \end{array} \right]$$

EXAMPLE B-2 (continued)

Next, use rule 3: Any row may be added to the multiple of another row (row 3 = row 3 − row 1).

$$\left[\begin{array}{ccc|ccc} \frac{15}{4} & 0 & 0 & 15 & 0 & -\frac{75}{4} \\ 0 & 1 & -6 & 0 & 1 & 0 \\ 0 & 0 & 5 & -15 & 0 & \frac{80}{4} \end{array}\right]$$

The −6 on the left side is the next target. Multiply row 3 by $\frac{6}{5}$.

$$\left[\begin{array}{ccc|ccc} \frac{15}{4} & 0 & 0 & 15 & 0 & -\frac{75}{4} \\ 0 & 1 & -6 & 0 & 1 & 0 \\ 0 & 0 & 6 & -18 & 0 & 24 \end{array}\right]$$

Using rule 3, row 2 is replaced by (row 2 + row 3).

$$\left[\begin{array}{ccc|ccc} \frac{15}{4} & 0 & 0 & 15 & 0 & -\frac{75}{4} \\ 0 & 1 & 0 & -18 & 1 & 24 \\ 0 & 0 & 6 & -18 & 0 & 24 \end{array}\right]$$

Row 3 is multiplied by $\frac{1}{6}$ and row 1 is multiplied by $\frac{4}{15}$.

$$\left[\begin{array}{ccc|ccc} 1 & 0 & 0 & 4 & 0 & -5 \\ 0 & 1 & 0 & -18 & 1 & 24 \\ 0 & 0 & 1 & -3 & 0 & 4 \end{array}\right]$$

The identity matrix is present in the left side of the composite matrix. The job is finished and the inverted matrix equals:

$$M^{-1} = \begin{bmatrix} 4 & 0 & -5 \\ -18 & 1 & 24 \\ -3 & 0 & 4 \end{bmatrix}$$

EXAMPLE B-3

Problem: Figure 9-1 shows the Markov model for an ideal dual-control system. The **I – Q** matrix for this model is

$$\mathbf{I-Q} = \begin{bmatrix} 2\lambda & -2\lambda \\ -\mu & \lambda+\mu \end{bmatrix}$$

The equation for MTTF (Equation 9-1) is found from the **N** matrix, the inverse of the **I – Q** matrix. Invert the **I – Q** matrix and find the equation for MTTF.

Solution: The composite matrix is written:

$$\left[\begin{array}{cc|cc} 2\lambda & -2\lambda & 1 & 0 \\ -\mu & \lambda+\mu & 0 & 1 \end{array} \right]$$

Multiply row 1:

$$R1 = R1 \times \frac{\lambda+\mu}{2\lambda}$$

$$\left[\begin{array}{cc|cc} \lambda+\mu & -\lambda+\mu & \frac{\lambda+\mu}{2\lambda} & 0 \\ -\mu & \lambda+\mu & 0 & 1 \end{array} \right]$$

Row 1 equals (row 1 + row 2):

$$\left[\begin{array}{cc|cc} \lambda & 0 & \frac{\lambda+\mu}{2\lambda} & 1 \\ -\mu & \lambda+\mu & 0 & 1 \end{array} \right]$$

Multiply row 1 and replace row 2 with the sum of row 1 and row 2.

$$R1 = R1 \times \frac{\mu}{\lambda}$$

$$\left[\begin{array}{cc|cc} \mu & 0 & \frac{(\lambda+\mu)\times\mu}{2\lambda^2} & \frac{\mu}{\lambda} \\ 0 & \lambda+\mu & \frac{(\lambda+\mu)\times\mu}{2\lambda^2} & \frac{\lambda+\mu}{\lambda} \end{array} \right]$$

EXAMPLE B-3 (continued)

Finally, rescale the results by multiplying both row 1 and row 2.

$$R1 = R1 \times \frac{1}{\mu}$$

$$R2 = R2 \times \frac{1}{\lambda + \mu}$$

$$\begin{bmatrix} 1 & 0 & \bigg| & \frac{\lambda + \mu}{2\lambda^2} & \frac{1}{\lambda} \\ 0 & 1 & \bigg| & \frac{\mu}{2\lambda^2} & \frac{1}{\lambda} \end{bmatrix}$$

The inverse of the **I – Q** matrix is the **N** matrix.

$$\mathbf{N} = \begin{bmatrix} \frac{\lambda + \mu}{2\lambda^2} & \frac{1}{\lambda} \\ \frac{\mu}{2\lambda^2} & \frac{1}{\lambda} \end{bmatrix}$$

MTTF is obtained by adding the row elements of the starting state. Assuming that the system starts in state 0, the upper row is added.

$$MTTF = \frac{\lambda + \mu}{2\lambda^2} + \frac{2\lambda}{2\lambda^2} = \frac{3\lambda + \mu}{2\lambda^2}$$

APPENDIX C

Probability Theory

Introduction

There are many events in the world for which insufficient knowledge exists to predict accurately an outcome. If a penny is flipped into the air, no one can predict with certainty that it will land "heads up." If a pair of dice is rolled, no one possesses enough knowledge to state which numbers will appear face up on each die. These events are called *random*.

It is conceivable, given such information as the precise weight distribution of a penny, the exact direction and magnitude of the launching force vector, and the air friction and spring coefficient of the landing surface, that one could calculate with great certainty that the penny would land heads up. But outcome calculation of the penny flip process, like many processes, is not practical or even possible. In many situations, the relationships that affect the process may not be understood. The input variables may not be measurable with sufficient precision. For these processes, one resorts to statistics.

Probability

Probability is a quantitative method of expressing chances. A probability is assigned a number between zero and one, inclusive. A probability assignment of zero means that the event is never expected. A probability assignment of one means that the event is always expected.

Probabilities are often assigned based on historical "frequency of occurrence." An experiment is repeated many times, say N. A quantity is tabulated for each possible outcome of the experiment. For any particular outcome, the probability is determined by dividing the number of occurrences, n, by the number of trials.

Probability Theory

$$P(E) = \frac{n}{N} = \frac{\text{Number of occurrences of event } E}{\text{Number of trials}} \tag{C-1}$$

The values become more certain as the number of trials is increased. A definition of probability based on this concept is

$$P(E) = \lim_{N \to \infty} \frac{n}{N} \tag{C-2}$$

Venn Diagrams

A convenient way to depict the outcomes of an experiment is through the use of the Venn diagram. These diagrams were created by John Venn (1834-1923), an English mathematician and cleric. They provide visual representation of data sets, including experimental outcomes. The diagrams are drawn by using the area of a rectangle to represent all possible outcomes; this area is known as the *sample space*. Any particular outcome is shown by using a portion of the area within the rectangle.

A "fair" coin is defined as a coin that is equally likely to give a heads result or a tails result. For a fair coin flip, the Venn diagram of possible outcomes is shown in Figure C-1. There are two expected outcomes: heads and tails. Each has a well-known probability of one-half. The diagram shows the outcomes, with each allocated area in proportion to its probability.

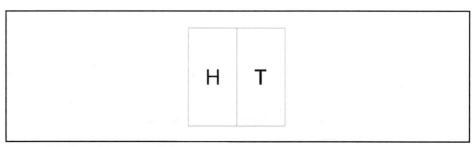

Figure C-1. Venn Diagram - Coin Toss.

For the toss of a fair pair of dice, the possible outcomes are shown in Figure C-2. The outcomes do not occupy the same area on the diagram. The probabilities of some outcomes are more likely than others; these occupy more area. For example, the area occupied by an outcome of "2" is $\frac{1}{36}$ of the total. The area occupied by the outcome "7" is $\frac{6}{36}$ of the total. Again, the area occupied by each outcome is proportional to its probability.

A Venn diagram is often used to identify the attributes of possible outcomes. Outcomes are grouped into sets that are based on some characteristic or combination of characteristics. The graphical nature of the Venn diagram is especially useful in showing these combinations of sets: unions, intersections, and complementary events.

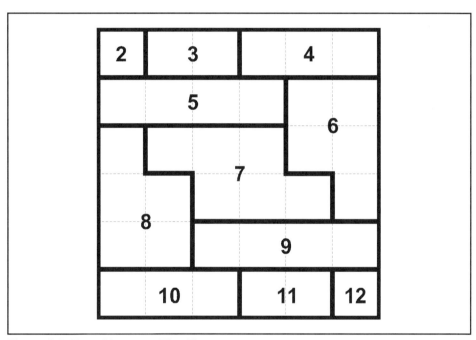

Figure C-2. Venn Diagram - Dice Throw.

A *union* of some number of sets (A, B, C) is defined as any event in either set A or set B or set C. This is represented in a Venn diagram as shown in Figure C-3. Vertical lines extend through three circles: the A circle, the B circle, and the C circle.

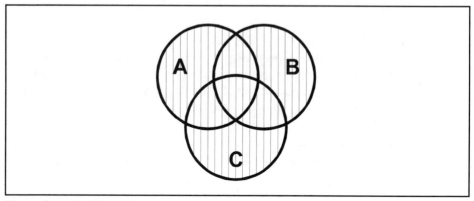

Figure C-3. P(AUBUC) Union.

An *intersection* of sets (A, B, C) is defined as any event in sets A and B and C. This is represented in a Venn diagram as shown in Figure C-4. Only one small area is in all three circles. That area is marked with lines.

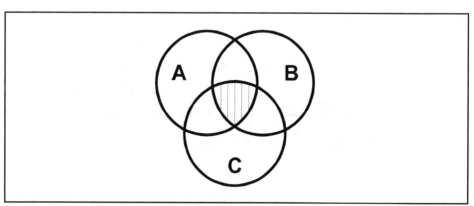

Figure C-4. Intersection of ABC.

Probability Theory

Complementary sets are easily shown on Venn diagrams. Since the diagram represents the entire sample space, all area not enclosed within an event is the complement of the event. In Figure C-5, the set A is represented by a circle. Its complement is set B, represented by the remainder of the diagram.

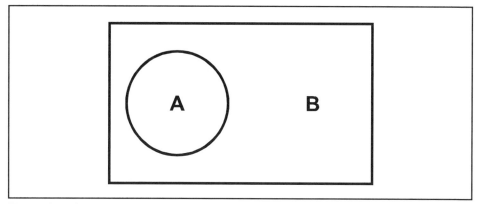

Figure C-5. Complementary Events.

Mutually exclusive sets are defined as sets that cannot happen at the same time. Mutually exclusive event sets are easily recognized on a Venn diagram. Figure C-6 shows the event sets A and B. There is no common space within the A circle and the B circle. There is no intersection between A and B. They cannot happen at the same time and are, therefore, mutually exclusive.

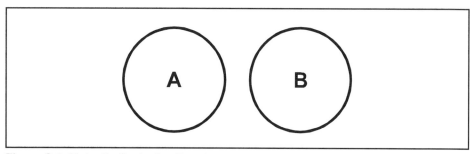

Figure C-6. Mutually Exclusive Events.

Venn diagrams can be useful in control systems reliability engineering because they can be used to represent the set of all possible failures. Failures might be categorized according to source; for instance, sources could include hardware failures, software failures, operation failures, and maintenance failures. A Venn diagram illustrating failure categories is shown in Figure C-7.

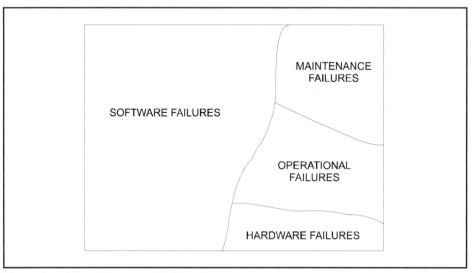

Figure C-7. Venn Diagram of Failure Sources.

Combining Probabilities

Certain rules help to combine probabilities. Combinations of events are common in the field of reliability evaluation. Often, system failures occur only when certain combinations of events happen during certain times.

Independent Events

If the occurrence of an event from set A does not affect the probability of events from set B, then sets A and B are defined to be *independent*; for example, the outcome of one coin toss does not affect the next toss. The outcome of one coin toss does not provide any information about the outcome of a subsequent independent coin toss. When two event sets are independent, the probability of getting an event from set A and set B (the intersection) is given by:

$$P(A \cap B) = P(A) \times P(B) \tag{C-3}$$

Probability Theory

Independent events are different from mutually exclusive events. Consider two events, A and B, that are mutually exclusive. Knowing that A has occurred tells us that B cannot occur. If events A and B are independent, knowing that A occurred tells us nothing about B. Two events, A and B, cannot be both mutually exclusive and independent.

EXAMPLE C-1

Problem: Two fair coins are flipped into the air. What is the probability that both coins will land with heads showing?

Solution: Each coin toss has only two possible outcomes: heads or tails. Each outcome has a probability of one-half. The coin tosses are independent. Therefore,

$$P(H_1 \cap H_2) = P(H_1) \times P(H_2)$$
$$= \frac{1}{2} \times \frac{1}{2}$$
$$= \frac{1}{4}$$

EXAMPLE C-2

Problem: A pair of fair (well-balanced) dice is rolled. What is the probability of getting "snake eyes" — one dot on each die?

Solution: The outcome of one die does not affect the outcome of the other die. Therefore, the events are independent. The probability of getting one dot can be obtained by noting that there are six sides on the die and that each side is equally likely. The probability of getting one dot is one-sixth ($\frac{1}{6}$). The probability of getting snake eyes is represented as:

$$P(1, 1) = \frac{1}{6} \times \frac{1}{6}$$
$$= \frac{1}{36}$$

Check the area occupied by the "2" result on Figure C-2. Is that area equal to $\frac{1}{36}$?

> **EXAMPLE C-3**
>
> **Problem:** A controller fails only if the input power fails and the controller battery fails. Assume that these factors are independent. The probability of input power failure is 0.0001. The probability of battery failure is 0.01. What is the probability of controller failure?
>
> **Solution:** Since input power and battery failure are independent, the probability of both events is given by Equation C-3:
>
> $$P(\text{Controller fail}) = P(\text{Input power fail}) \times P(\text{Battery fail})$$
> $$= 0.0001 \times 0.01$$
> $$= 0.000001$$

Probability Summation

If the probability of getting a result from set A equals 0.2 and the probability of getting a result from set B equals 0.3, what is the probability of getting a result from either set A or set B?

It would be natural to assume that the answer is 0.5, the sum of the above probabilities, but that answer is not always correct. Look at the Venn diagram in Figure C-8. If the area of set A ($\frac{6}{36}$) is added to the area of set B ($\frac{6}{36}$), the answer ($\frac{12}{36}$) is too large. (The answer should be $\frac{11}{36}$.) Since there is an intersection between sets A and B, the area of the intersection has been counted twice. When summing probabilities, the intersections must be subtracted. Thus, the probability of the union of event sets A and B is given by:

$$P(A \cup B) = P(A) + P(B) - P(A \cap B) \tag{C-4}$$

If set A and set B are mutually exclusive so that there is no intersection, then the following can be stated:

$$P(A \cup B) = P(A) + P(B) \tag{C-5}$$

> **EXAMPLE C-4**
>
> **Problem:** A pair of fair dice is rolled. What is the probability of getting a sum of seven?
>
> **Solution:** A sum of seven dots on the dice can be obtained in a number of different ways; these are described by the sets {1,6}, {2,5}, {3,4}, {4,3}, {5,2}, and {6,1}. Each specific combination has a probability of $\frac{1}{36}$. The combinations are mutually exclusive; therefore, Equation C-5 can be used.
>
> $$P(\Sigma \text{ of } 7) = \frac{1}{36} + \frac{1}{36} + \frac{1}{36} + \frac{1}{36} + \frac{1}{36} + \frac{1}{36} = \frac{1}{6}$$

EXAMPLE C-5

Problem: A control system fails if it experiences a hardware failure, a software failure, or a maintenance failure. Assume that the failures are mutually exclusive, since one failure stops the system such that another failure cannot occur. The probability of a hardware failure is 0.001. The probability of a software failure is 0.01. The probability of a maintenance failure is 0.00001. What is the probability of controller failure?

Solution: Using Equation C-5,

$$P(\text{Controller fail}) = P(HW) + P(SW) + P(\text{Maint.})$$
$$= 0.001 + 0.01 + 0.00001$$
$$= 0.01101$$

EXAMPLE C-6

Problem: A pair of fair dice is rolled. What is the probability of getting an even number on both dice?

Solution: On each die are six numbers. Three of the numbers are odd (1, 3, 5) and three of the numbers are even (2, 4, 6). All numbers are mutually exclusive. Equation C-5 gives the probability of getting an even number on one die.

$$P(\text{Even}) = P(2, 4, 6) = P(2) + P(4) + P(6)$$
$$= \frac{1}{6} + \frac{1}{6} + \frac{1}{6}$$
$$= \frac{1}{2}$$

The outcome of one die is independent of the other die. Therefore,

$$P(\text{Even, even}) = P(\text{Set A even}) \times P(\text{Set B even})$$
$$= \frac{1}{2} \times \frac{1}{2}$$
$$= \frac{1}{4}$$

EXAMPLE C-7

Problem: A pair of fair dice is rolled. What is the probability of getting two dots on either or both dice?

Solution: The probability of getting two dots on die A or B equals $\frac{1}{6}$. The probability of getting two dots on both dice, however, is $\frac{1}{36}$. We can use Equation C-4.

$$P(A \cup B) = P(A) + P(B) - P(A \cap B)$$
$$= \frac{1}{6} + \frac{1}{6} - \frac{1}{36}$$
$$= \frac{11}{36}$$

This is evident in Figure C-8, a Venn diagram of the problem.

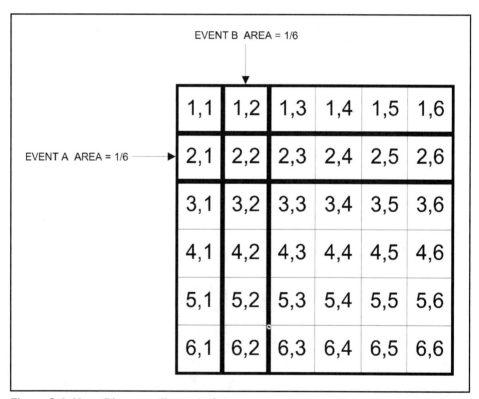

Figure C-8. Venn Diagram - Example C-8.

Conditional Probability

The probability of some event under specific circumstances often must be calculated. For example, the probability of event A, given that event B has occurred, may need to be calculated. Such a probability is called a *conditional probability*. The situation can be envisioned by examining the Venn diagram in Figure C-9.

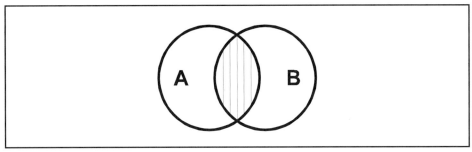

Figure C-9. Conditional Probability.

Normally, the probability of event A would be given by the area of circle A divided by the total area. Conditional probability is different. Event B has occurred. This means that only the state space within the area of circle B needs to be examined. This is a substantially reduced area! The desired probability is the area of circle A within circle B, divided by the area of circle B, expressed by:

$$P(A|B) = \frac{P(A \cap B)}{P(B)} \qquad (C\text{-}6)$$

which reads: The probability of A, given B, is equal to the probability of the intersection of A and B divided by the probability of B. The area of circle A within circle B represents the probability of the intersection of A and B. The area of circle B equals the probability of B.

EXAMPLE C-8

Problem: A pair of fair dice is rolled. What is the probability of getting a two on both dice given that one die has a two?

Solution: The probability of {2,2}, given that one die has a two, is given by Equation C-6:

$$P(2,2) = \frac{1/36}{1/6}$$
$$= \frac{1}{6}$$

In this case, the answer is intuitive since the outcome of each die is independent. The problem could have been solved by noting that

$$P(A \cap B) = P(A) \times P(B) = \frac{1}{6} \times \frac{1}{6}$$

as given in Equation C-3. Substituting this into Equation C-6:

$$P(A|B) = \frac{P(A) \times P(B)}{P(B)} \qquad (C\text{-}7)$$
$$= P(A)$$

In the example, the result could have calculated by merely knowing the probability of getting a two on the second die.

The equation for conditional probability can be rearranged as:

$$P(A \cap B) = P(A|B) \times P(B) \qquad (C\text{-}8)$$

This states that the intersection of events A and B can be obtained by multiplying the probability of A, given B, times the probability of B. When the statistics are kept in a conditional format, this equation can be useful.

EXAMPLE C-9

Problem: A pair of fair dice is rolled. What is the probability of getting a sum of seven, given that exactly one die shows a two?

Solution: There are only two ways to get a sum of seven, given that one die has a two. Those two combinations are {2,5} and {5,2}. There are 10 combinations that show a two on exactly one die. These sets are {2,1}, {2,3}, {2,4}, {2,5}, {2,6}, {1,2}, {3,2}, {4,2}, {5,2}, and {6,2}. Using Equation C-6,

$$P(A|B) = \frac{P(A \cap B)}{P(B)}$$
$$= \frac{2/36}{10/36}$$
$$= 2/10$$

This is shown graphically in Figure C-10.

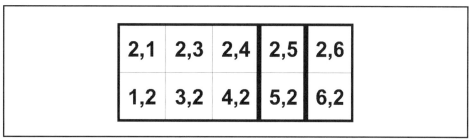

Figure C-10. Probability - Example C-9.

Bayes' Rule

Consider an event A. The state space in which it exists is divided into two mutually exclusive sections, B and B' (Figure C-11). Event A can be written as:

$$A = (A \cap B) \cup (A \cap B') \tag{C-9}$$

Since AB and AB' are mutually exclusive,

$$P(A) = P(A \cap B) + P(A \cap B') \tag{C-10}$$

Substituting Equation C-8,

$$P(A) = P(A|B) \times P(B) + P(A|B') \times P(B') \tag{C-11}$$

which states: The probability of event A equals the conditional probability of A, given that B has occurred, plus the conditional probability of A, given that B has not occurred. This is known as *Bayes' rule*. It is widely used in many aspects of reliability engineering.

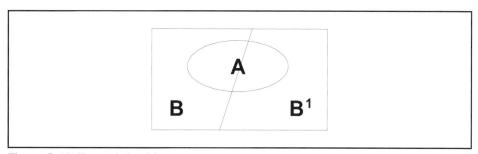

Figure C-11. Event A Partitioned.

EXAMPLE C-10

Problem: The workday is divided into three mutually exclusive time periods: day shift, evening shift, and night shift. Day shift lasts 10 hr. Evening shift is 8 hr. Night shift is 6 hr. Logs show that in the last year (8,760 hr), one failure occurred during the day shift (one failure in 3,650 hr), two failures occurred during the evening shift (two failures in 2,920 hr), and seven failures occurred during the night shift (seven failures in 2,190 hr). What is the overall probability of failure?

Solution: Define event A as failure. Define event B1 as the day shift, B2 as the evening shift, and B3 as the night shift. The probability of failure given, event B1 (day shift) is calculated knowing that one failure occurred in 3,650 hr (one-third of the hours in one year). A variation of Equation C-11 can be used where $P(B1)$ is day shift probability, $P(B2)$ is the evening shift probability, and $P(B3)$ is the night shift probability.

$$P(\text{fail}) = P(\text{fail}|B1) \times P(B1) \qquad \text{(C-12)}$$
$$+ P(\text{fail}|B2) \times P(B2)$$
$$+ P(\text{fail}|B3) \times P(B3)$$

The probabilities of failure for each shift are calculated by dividing the number of failures during each shift by the numbers of hours in each shift. Substituting the numbers into Equation C-12:

$$P(\text{fail}) = \left(\frac{1}{3650} \times \frac{10}{24}\right) + \left(\frac{2}{2920} \times \frac{8}{24}\right) + \left(\frac{7}{2190} \times \frac{6}{24}\right)$$

$$= 0.000114 + 0.000226 + 0.000799$$

$$= 0.001139$$

EXAMPLE C-11

Problem: A company manufactures controllers at two locations. Sixty percent are manufactured in plant X. Forty percent are manufactured in plant Y. Controllers manufactured in plant X have a 0.00016 probability of failure in a one-year period. Controllers manufactured in plant Y have a 0.00022 probability of failure in one year. A purchased controller can come randomly from either source. What is the probability of a controller failure?

Solution: Define controller failure as event A. Define event B1 as plant X manufacture. Define event B as plant Y manufacture. Using Equation C-11, substitute the values to obtain:

$$P(\text{fail}) = (0.00016 \times 0.6) + (0.00022 \times 0.4)$$
$$= 0.000096 + 0.000088 = 0.000184$$

Permutations and Combinations

Often, probabilities are determined by knowing how many different outcomes exist in an experiment. If all outcomes are equally probable, then the inverse of the number of different outcomes determines the probability. In many cases, the number of different outcomes can be determined by using counting rules.

Basic Counting Rule

If an event is created by x steps, each having a number of variations (nx), then the total number of different events equals the product of the variation count for each step ($n1 \times n2 \times n3 \times ... \times nx$).

EXAMPLE C-12

Problem: How many character strings of length two can be constructed using the letters A, B, and C if letters cannot be repeated?

Solution: A character string, two letters in length, takes two steps. In the first step, any of the three letters can be chosen. In the second step, either can be chosen from the remaining two letters. The number of different character strings is

$$3 \times 2 = 6$$

This can be verified by creating all the character strings. Starting with the letter A, they are {A,B}, {A,C}, {B,A}, {B,C}, {C,A}, and {C,B}.

EXAMPLE C-13

Problem: A vendor offers three different models of a controller. On each model are four communications options: two I/O options and two memory options. How many different kinds of controllers exist?

Solution: Four steps are required to build the controller. In the first step, one of three models is selected. One of four communications options is chosen in the second step. One of two I/O options is chosen in the third step. One of two memory options is chosen in the fourth step. Using the first counting rule, the number of variations are

$$3 \times 4 \times 2 \times 2 = 48$$

> **EXAMPLE C-14**
>
> **Problem:** Four letters, A, B, C, and D, are available to be arranged in a sequence that is four letters in length. How many different arrangements can be done without repeating a letter?
>
> **Solution:** Four steps are required to build a sequence. Using the basic counting rule, four letters are available in the first position. Three letters are available for the second position. Two letters are available in the third position. Only one letter is left for the last position. Accordingly, the number of different sequences is
>
> $$4 \times 3 \times 2 \times 1 = 24$$

Permutations

An ordered arrangement of objects without repetition is known as a *permutation*. The four-letter sequences from Example C-14 are permutations. The number of permutations of n objects is $n!$. ($n!$ is pronounced "n factorial" and is the mathematical notation for the product $1 \times 2 \times 3 \times \ldots \times n-1 \times n$.)

> **EXAMPLE C-15**
>
> **Problem:** The control system selection committee has six members. The company photographer is coming. How many different ways can the members be lined up for the committee portrait?
>
> **Solution:** Since committee members line up without repetition, they can be arranged in any of $n!$ different permutations:
>
> $$6! = 720$$

If a subset of the objects is arranged, the number of permutations is reduced. If two items are taken from a set of four, the basic counting rule tells that 12 permutations exist (4×3). In general, if only some members of the set are used (quantity r, where $r < n$), this is written:

$$P(n, r) = n(n-1)(n-2)\ldots(n-r+1) \tag{C-13}$$

If this product is multiplied by

$$\frac{(n-r)(n-r-1)\ldots 1}{(n-r)(n-r-1)\ldots 1}$$

then

$$P(n, r) = \frac{n(n-1)(n-2)\ldots(n-r+1)(n-r)(n-r-1)\ldots 1}{(n-r)(n-r-1)\ldots 1} \quad \text{(C-14)}$$

The numerator is $n!$. The denominator is $(n-r)!$. Therefore:

$$P(n, r) = \frac{n!}{(n-r)!} \quad \text{(C-15)}$$

Equation C-15 is used to determine permutations, the number of ways that r objects from a set of n objects can be arranged *in order*.

EXAMPLE C-16

Problem: Given the letters A, B, C, and D, how many permutations can be obtained with two letters?

Solution: Using Equation C-15, the number of permutations equals

$$P(4, 2) = \frac{4!}{(4-2)!}$$
$$= \frac{24}{2}$$
$$= 12$$

This can be verified by listing the permutations: AB, AC, AD, BA, BC, BD, CA, CB, CD, DA, DB, and DC.

EXAMPLE C-17

Problem: Given the letters A, B, C, D, E, and F, how many different ways can four letters be arranged where order counts?

Solution: Using Equation C-15, the number of permutations equals

$$P(6, 4) = \frac{6!}{(6-4)!}$$
$$= \frac{720}{2}$$
$$= 360$$

This can be verified by using the basic counting rule. The first step has six possibilities. The second step has five. The third step has four possibilities, and the fourth step has three variations. The basic counting rule tells us:

$$P(6, 4) = 6 \times 5 \times 4 \times 3$$
$$= 360$$

Combinations

Combinations are groupings of elements in which order does not count. Since order does not count, the number of combinations will always be lower than the number of permutations. Consider the number of permutations of three letters (3! = 6). They are: ABC, ACB, BAC, BCA, CAB, and CBA. If order does not count, all these arrangements are the same. There is only one combination! The number has been reduced by a factor of 3!.

When n objects are taken r at a time, the number of combinations is also less than the number of permutations. Consider Example C-16, where the 12 permutations of four letters taken two at time were AB, AC, AD, BA, BC, BD, CA, CB, CD, DA, DB, and DC. If order does not count, AB is not different from BA. AC is not different from CA. Eliminating duplicates from the list leaves AB, AC, AD, BC, BD, and CD. The number has been reduced by 2!. The formula for combinations of n objects taken r at a time is

$$C(n, r) = \frac{n!}{r!(n-r)!} \tag{C-16}$$

Comparing this formula with Equation C-15, note that the number of permutations is reduced by a factor $r!$ to obtain the number of combinations.

EXAMPLE C-18

Problem: Given the letters A, B, C, D, E, and F, how many combinations can be obtained using four letters at a time?

Solution: The numbers of different combinations can be found using Equation C-16.

$$\begin{aligned} C(6, 4) &= \frac{6!}{4!(6-4)!} \\ &= \frac{720}{24 \times 2} \\ &= 15 \end{aligned}$$

EXERCISES

C.1 Three fair coins are tossed into the air. What is the probability that all three will land heads up?

C.2 A control loop has a temperature transmitter, a controller, and a valve. All three devices must operate successfully for the loop to operate successfully. The loop will be operated for one year before the process is shut down and overhauled. The temperature transmitter has a probability of failure of 0.01 during the next year. The controller has a probability of failure of 0.005 during the next year. The valve has a probability of failure of 0.05 during the next year. Assume that temperature transmitter, controller, and valve failures are independent. What is the probability of failure for the control loop?

C.3 Three fair coins are tossed into the air. What is the probability that at least one coin will land heads up?

C.4 A pair of dice is rolled. If the result is a six or an eight, you hit Park Place or Boardwalk and go broke. If the result is nine or more, you pass GO and collect $200. What is the probability of going broke on the next turn? What is the probability of passing GO?

C.5 A control system has four temperature-control loops. The system operates if three or more loops operate. The system fails if fewer than three loops operate. How many combinations of loop failures cause system failure?

C.6 A vendor evaluation committee has three members. How many ways can they line up for a committee photograph?

C.7 Five letters, A, B, C, D, and E, are available to be arranged in a sequence. How many different sequences are possible without repeating a letter?

C.8 Given the letters A, B, and C, how many combinations can be obtained using two letters at a time?

C.9 A control system uses two controllers. The system is successful if one controller is successful. A controller is repairable and has a steady-state probability of success of 0.95. Therefore, each controller has a steady-state probability of failure that equals 0.05. Assume that controller failures are independent. What is the probability of having one failed controller and one successful controller?

C.10 A control system uses three controllers. The system is successful if two of the three controllers are successfully operating. The system fails if two controllers fail or if all three controllers fail. A controller is repairable and has a steady-state probability of success of 0.95. What is the probability that two controllers are failed and the other controller is operating? How many combinations of two failed controllers and one successful controller exist? What other combinations result in system failure? What is the overall probability of system failure?

C.11 Using the Venn diagram of Figure C-7, estimate the probabilities of the various failure sources.

C.12 A safety instrumented system (SIS) uses two pressure sensors, a logic solver, and a valve. The pressure sensors have a probability of failing dangerously of 0.000005 per hour. The logic solver has a probability of failing dangerously of 0.00001 per hour. The valve has a probability of failing dangerously of 0.00005 per hour. The safety system fails dangerously if any of the system components fail dangerously. What is the system probability of failing dangerously?

C.13 During a year, the probability of getting a dangerous condition in an industrial boiler is 0.00001. The probability of a safety instrumented protection system failing to respond to the dangerous demand is 0.000001. If there is a dangerous condition AND the protection system does not respond, there will be a boiler explosion. What is the probability of boiler explosion?

C.14 The probability of an explosion in an industrial process is 0.00002. The insurance underwriter wants an SIS designed that will reduce the probability of an explosion to 0.0000001. What is the maximum probability of failing dangerously allowed in the SIS?

BIBLIOGRAPHY

1. Johnsonbaugh, R., *Essential Discrete Mathematics*, New York: Macmillan, 1987.

2. Ross, S. M., *Introduction to Probability Models*, Orlando, FL.: Academic Press, 1987.

3. Paulos, J. A., *Innumeracy*, New York: Hill and Wang, 1988.

APPENDIX D

Reliability Parameters

Test Data

Reliability parameters can be calculated from life test data. Table 4-2 shows accelerated reliability life test data for a set of 50 modules. A number of variables are used to describe these data. The original number of modules in the test is denoted by the variable N_o. The number of modules surviving after each time period is denoted by the variable N_s. The cumulative number of modules that have failed is denoted by the variable N_f. The various reliability functions can be calculated as follows. The reliability function is

$$R(t) = \frac{N_s(t)}{N_o} \tag{D-1}$$

The unreliability function is

$$F(t) = \frac{N_f(t)}{N_o} \tag{D-2}$$

The probability density function (pdf) must be calculated in a piecewise linear manner. For the time period between samples, n, the formula is

$$f(t) = \frac{F(t_n) - F(t_{n-1})}{\Delta t}$$

This equals

$$f(t) = \frac{1}{N_o} \frac{N_f(t_n) - N_f(t_{n-1})}{\Delta t} \tag{D-3}$$

475

Since the failure rate equals the pdf divided by the reliability function,

$$\lambda(t) = \frac{f(t)}{R(t)} = \frac{1}{R(t)} \times f(t)$$

$$= \frac{N_o}{N_s(t_{n-1})} \times \frac{1}{N_o} \frac{N_f(t_n) - N_f(t_{n-1})}{\Delta t}$$

Canceling out the N_o terms yields:

$$\lambda(t) = \frac{1}{N_s(t_{n-1})} \frac{N_f(t_n) - N_f(t_{n-1})}{\Delta t} \tag{D-4}$$

Using the data from Table 4-2, at the end of the first week 41 modules survived and nine modules failed. The calculations for week one are

$$R(t_1) = \frac{41}{50} = 0.82$$

$$F(t_1) = \frac{9}{50} = 0.18$$

$$f(t_1) = \frac{1}{50} \frac{9-0}{24 \times 7} = 0.00107 \text{ failures/hr}$$

$$\lambda(t_1) = \frac{1}{50} \frac{9-0}{24 \times 7} = 0.00107 \text{ failures/hr}$$

The calculations for week two are

$$R(t_2) = \frac{36}{50} = 0.72$$

$$F(t_2) = \frac{14}{50} = 0.28$$

$$f(t_2) = \frac{1}{50} \frac{14-9}{24 \times 7} = 0.00059 \text{ failures/hr}$$

$$\lambda(t_2) = \frac{1}{50} \frac{14-9}{24 \times 7} = 0.00072 \text{ failures/hr}$$

Reliability Parameters

Table D-1 shows the calculations for the first 10 weeks.

Week (t_n)	Failures	R(t)	F(t)	f(t)	Failure Rate
1	9	0.82	0.18	0.00107	0.00107
2	5	0.72	0.28	0.00059	0.00072
3	3	0.66	0.34	0.00036	0.00050
4	2	0.62	0.38	0.00024	0.00036
5	2	0.58	0.42	0.00024	0.00038
6	1	0.56	0.44	0.00012	0.00021
7	2	0.52	0.48	0.00024	0.00043
8	1	0.50	0.50	0.00012	0.00023
9	1	0.48	0.52	0.00012	0.00024
10	0	0.48	0.52	0.00000	0.00000

Table D-1. Accelerated Life Test Reliability Calculations

Figures D-1 to D-4 show graphs of reliability, unreliability, probability of failure, and failure rate for the entire time period, respectively. Reliability starts at a value of one and decreases. Unreliability starts at a value of zero and increases. The probability of failure decreases, then remains relatively constant. The failure rate plot decreases, remains relatively constant, and then increases. This type of behavior is characteristic of many types of products.

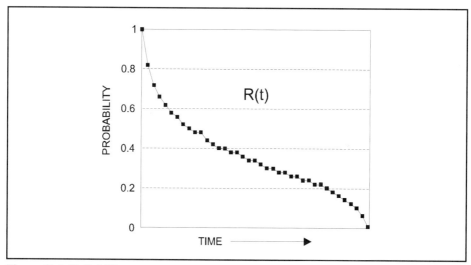

Figure D-1. Reliability.

478   Reliability Parameters

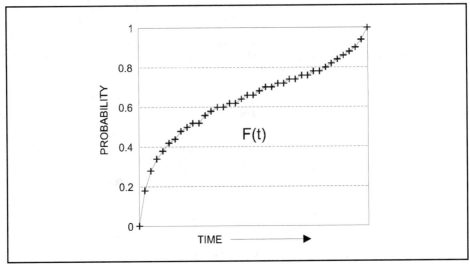

Figure D-2. Unreliability.

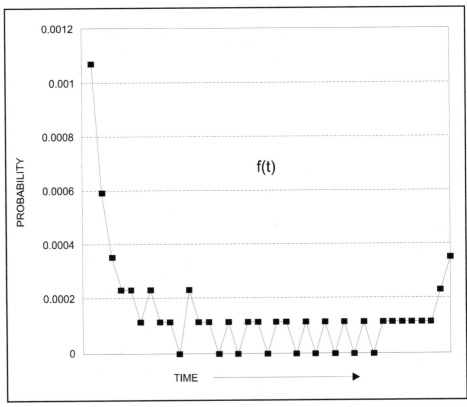

Figure D-3. Probability of Failure.

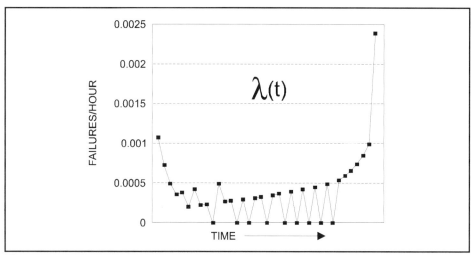

Figure D-4. Failure Rate.

Uncensored Data

It should be noted that the exact failure times of any module are not known from the data in Table 4-2. The failure times are known only within a one-week (168 hr) period. This "sampling error" causes noisy data and requires some interpretation. Table D-2 shows the exact module failure times for the same life test that was presented in Table 4-2. These data are called "uncensored."

The failure rate calculated from Table D-2 is shown in Figure D-5. Compare this with Figure D-4, which was calculated from summary (or censored) data. The censored data produce a "noisy" plot. In general, the more accurate the data, the better the analysis. Failure rate analysis from censored data must be interpreted in the presence of noise.

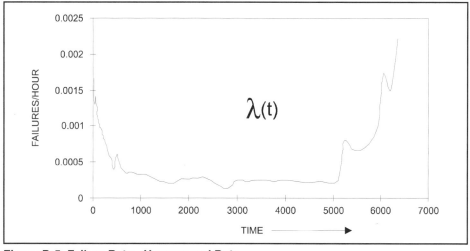

Figure D-5. Failure Rate - Uncensored Data.

Failure Number	Time, hr	Failure Number	Time, hr
1	12	26	1,488
2	26	27	1,681
3	42	28	1,842
4	57	29	2,010
5	75	30	2,181
6	95	31	2,365
7	116	32	2,772
8	139	33	2,992
9	163	34	3,234
10	189	35	3,486
11	219	36	3,771
12	251	37	4,050
13	289	38	4,384
14	333	39	4,741
15	383	40	5,100
16	454	41	5,225
17	502	42	5,387
18	566	43	5,577
19	650	44	5,764
20	740	45	5,933
21	832	46	6,049
22	937	47	6,215
23	1,045	48	6,365
24	1,164	49	6,381
25	1,312	50	6,384

Table D-2. Exact Failure Times for Modules

APPENDIX E

Continuous-Time Markov Modeling

Introduction

When the size of the time increment in a discrete Markov model is reduced, accuracy is increased. Taken to the limit, the time increment is near zero. The limit as the time increment (Δt) goes to zero is labeled dt. At the limit, we have achieved "continuous" time.

$$\lim_{\Delta t \to 0} \Delta t = dt$$

Using continuous-time Markov models, analytical solutions for time-dependent state probabilities can be obtained if it is assumed that only constant failure rates and constant repair rates exist. As time increments approach zero, transition arcs are labeled with failure rates (failure rate equals instantaneous failure probability). The notation $S_n(t)$ is used to indicate time-dependent probability for state n.

Single Nonrepairable Component

The Markov model for a single nonrepairable component is shown in Figure E-1. An analytical solution for state probabilities for this model can be developed by using a little logic and a little calculus.

Assume the model starts in state 0 at time t. The model will be in state 0 during the next instant (time = $t + \Delta t$) only if it stays in state 0. This can be expressed mathematically as:

$$S_0(t + \Delta t) = S_0(t)(1 - \lambda \Delta t)$$

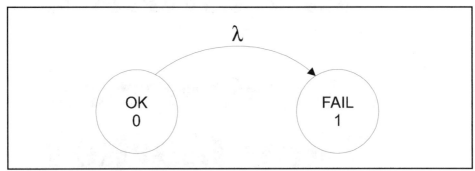

Figure E-1. Markov Model for Single Non-Repairable Component.

This can be rearranged as

$$S_0(t + \Delta t) - S_0(t) = -\lambda S_0(t)\Delta t$$

Dividing both sides by Δt:

$$\frac{S_0(t + \Delta t) - S_0(t)}{\Delta t} = -\lambda S_0(t) \tag{E-1}$$

The left side of Equation E-1 is the deviation with respect to time. Taking the limit as Δt goes to zero results in:

$$\frac{dS_0(t)}{dt} = -\lambda S_0(t) \tag{E-2}$$

Using a similar process:

$$\frac{dS_1(t)}{dt} = \lambda S_0(t) \tag{E-3}$$

Equations E-2 and E-3 are first-order differential equations with constant coefficients. One of the easiest ways to solve such equations is to use a Laplace transform to convert from the time domain (t) to the frequency domain (s). Taking the Laplace transforms:

$$sS_0(s) - S_0(0) = -\lambda S_0(s)$$

and

$$sS_1(s) - S_1(0) = \lambda S_0(s)$$

Continuous-Time Markov Modeling

Since the system starts in state 0, substitute $S_0(0) = 1$ and $S_1(0) = 0$. This results in:

$$sS_0(s) - 1 = -\lambda S_0(s) \tag{E-4}$$

and

$$sS_1(s) = \lambda S_0(s) \tag{E-5}$$

Rearranging Equation E-4:

$$(s + \lambda)S_0(s) = 1$$

Therefore:

$$S_0(s) = \frac{1}{s + \lambda} \tag{E-6}$$

Substituting Equation E-6 into Equation E-5 and solving:

$$sS_1(s) = \frac{\lambda}{s + \lambda}$$

A little algebra gives:

$$S_1(s) = \frac{\lambda}{s(s + \lambda)}$$

Taking the inverse transform results in:

$$S_0(t) = e^{-\lambda t} \tag{E-7}$$

and

$$S_1(t) = 1 - e^{-\lambda t} \tag{E-8}$$

Since state 0 is the success state, reliability is equal to $S_0(t)$ and is given by Equation E-7. Unreliability is equal to $S_1(t)$ and is given by Equation E-8. This result is identical to the result obtained in Chapter 4 (Equations 4-16 and 4-17) when a component has an exponential probability of failure. Thus, the Markov model solution verifies the clear relationship between the constant failure rate and the exponential probability of failure over a time period.

Single Repairable Component

A single repairable component has the Markov model of Figure E-2. To develop time-dependent solutions for the state probabilities, assume that the model starts in state 0. To stay in state 0 during the next instant (time = $t + \Delta t$), one of two situations must occur. In the first situation, the model must be in state 0 at time t and stay there. In the second situation, the model must be in state 1 at time t and move to state 0 during Δt. Mathematically, this is written:

$$S_0(t + \Delta t) = S_0(t)(1 - \lambda \Delta t) + S_1(t)(\mu \Delta t)$$

This can be rearranged as:

$$S_0(t + \Delta t) - S_0(t) = -[\lambda S_0(t) + \mu S_1(t)]\Delta t$$

Dividing both sides of the equation by Δt and taking the limit as Δt goes to zero results in:

$$\frac{dS_0(t)}{dt} = -\lambda S_0(t) + \mu S_1(t) \tag{E-9}$$

In a similar manner:

$$\frac{dS_1(t)}{dt} = \lambda S_0(t) - \mu S_1(t) \tag{E-10}$$

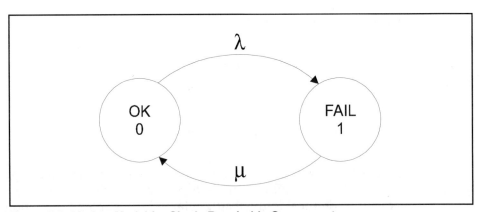

Figure E-2. Markov Model for Single Repairable Component.

Equations E-9 and E-10 are first-order differential equations. Again, using the Laplace transform solution method:

$$sS_0(s) - S_0(0) = -\lambda S_0(s) + \mu S_1(s)$$

Rearranging:

$$(s + \lambda)S_0(s) = \mu S_1(s) + S_0(0)$$

With further algebra:

$$S_0(s) = \frac{\mu}{s+\lambda}S_1(s) + \frac{1}{s+\lambda}S_0(0) \tag{E-11}$$

In a similar manner:

$$S_1(s) = \frac{\lambda}{s+\mu}S_0(s) + \frac{1}{s+\mu}S_1(0) \tag{E-12}$$

Substituting Equation E-11 into Equation E-12:

$$S_1(s) = \frac{\lambda}{s+\mu}\frac{\mu}{s+\lambda}S_1(s) + \frac{\lambda}{s+\mu}\frac{1}{s+\lambda}S_0(0) + \frac{1}{s+\mu}S_1(0)$$

Collecting the terms in a different form:

$$\left(1 - \frac{\lambda}{s+\mu}\frac{\mu}{s+\lambda}\right)S_1(s) = \frac{1}{s+\mu}\left[S_1(0) + \frac{\lambda}{s+\lambda}S_0(0)\right]$$

Creating a common denominator for the left half of the equation yields:

$$\left[\frac{(s+\mu)(s+\lambda) - \lambda\mu}{(s+\mu)(s+\lambda)}\right]S_1(s) = \frac{1}{s+\mu}\left[S_1(0) + \frac{\lambda}{s+\lambda}S_0(0)\right]$$

If both sides of the equation are divided by the first term, the $S_1(s)$ term is isolated.

$$S_1(s) = \left[\frac{(s+\mu)(s+\lambda)}{(s+\mu)(s+\lambda) - \lambda\mu}\right]\frac{1}{s+\mu}\left[S_1(0) + \frac{\lambda}{s+\lambda}S_0(0)\right]$$

Multiplying the denominator of the first term and canceling out equal terms:

$$S_1(s) = \frac{1}{s(s+\lambda+\mu)}[(s+\lambda)S_1(0) + \lambda S_0(0)] \tag{E-13}$$

Continuous-Time Markov Modeling

To move further with the solution, we must arrange Equation E-13 into a form that will allow an inverse transform. A partial fraction expansion of $S_1(s)$ where:

$$S_1(s) = \frac{A}{s} + \frac{B}{s+\lambda+\mu} \qquad (E-14)$$

will work. This means that

$$\frac{A}{s} + \frac{B}{s+\lambda+\mu} = \frac{1}{s(s+\lambda+\mu)}[(s+\lambda)S_1(0) + \lambda S_0(0)] \qquad (E-15)$$

This can be algebraically manipulated into the form:

$$[(s+\lambda)S_1(0) + \lambda S_0(0)] = A(s+\lambda+\mu) + B(s)$$

This relation holds true for all values of s. Therefore, to solve for A and B, we should pick a value of s that will simplify the algebra as much as possible. To solve for A, a value of $s = 0$ is the best choice. At $s = 0$,

$$[\lambda S_1(0) + \lambda S_0(0)] = A(\lambda + \mu)$$

Therefore:

$$A = \frac{\lambda}{\lambda+\mu}[S_1(0) + S_0(0)] \qquad (E-16)$$

Solving for B gives the result:

$$s = -(\lambda+\mu)$$

Substituting for s:

$$[-\mu S_1(0) + \lambda S_0(0)] = -B(\lambda+\mu)$$

Rearranging,

$$B = \frac{1}{\lambda+\mu}[\mu S_1(0) - \lambda S_0(0)] \qquad (E-17)$$

Substituting Equations E-16 and E-17 into Equation E-14:

$$S_1(s) = \frac{\lambda}{\lambda+\mu}\frac{1}{s}[S_1(o) + S_0(o)] + \frac{1}{\lambda+\mu}\frac{1}{s+\lambda+\mu}[\mu S_1(o) - \lambda S_0(o)]$$

Using a similar method for state 0:

$$S_0(s) = \frac{\mu}{\lambda+\mu}\frac{1}{s}[S_1(0)+S_0(0)] + \frac{1}{\lambda+\mu s}\frac{1}{s+\lambda+\mu}[\lambda S_0(0) - \mu S_1(0)]$$

Taking the inverse Laplace transform:

$$S_0(t) = \frac{\mu}{\lambda+\mu}[S_0(0)+S_1(0)] + \frac{e^{-(\lambda+\mu)t}}{\lambda+\mu}[\lambda S_0(0) - \mu S_1(0)]$$

and

$$S_1(t) = \frac{\lambda}{\lambda+\mu}[S_0(0)+S_1(0)] + \frac{e^{-(\lambda+\mu)t}}{\lambda+\mu}[\mu S_1(0) - \lambda S_0(0)]$$

Since the system always starts in state 0:

$$S_0(0) = 1$$

and

$$S_1(0) = 0$$

Substituting,

$$S_0(t) = \frac{\mu}{\lambda+\mu} + \frac{\lambda e^{-(\lambda+\mu)t}}{\lambda+\mu} \tag{E-18}$$

and

$$S_1(t) = \frac{\lambda}{\lambda+\mu} - \frac{\lambda e^{-(\lambda+\mu)t}}{\lambda+\mu} \tag{E-19}$$

Equations E-18 and E-19 provide the time-dependent analytical formulas for state probability. In this case, equation E-18 is the formula for availability since state 0 is the success state. If t is set equal to infinity in Equations E-18 and E-19, the second term of each goes to zero. The results are

$$S_0(\infty) = \frac{\mu}{\lambda+\mu} \tag{E-20}$$

and

$$S_1(\infty) = \frac{\lambda}{\lambda+\mu} \tag{E-21}$$

The limiting state probability is the expected result at infinite time. Thus, Equations E-20 and E-21 provide this information.

These methods can be used to solve for analytical state probabilities for more complex models; however, the mathematics become quite involved for realistic models of several states. The use of numerical techniques in combination with a discrete-time Markov model is rapidly becoming the method of choice when time-dependent state probabilities are needed.

Limiting State Probabilities

Analytical formulas for availability and unavailability can be derived using limiting state probability techniques. For the case of a single repairable component (Figure E-2), the **P** matrix is

$$\mathbf{P} = \begin{bmatrix} 1-\lambda & \lambda \\ \mu & 1-\mu \end{bmatrix} \tag{E-22}$$

Limiting state probabilities are obtained by multiplying the matrices:

$$\begin{bmatrix} S_0^L & S_1^L \end{bmatrix} \begin{bmatrix} 1-\lambda & \lambda \\ \mu & 1-\mu \end{bmatrix} = \begin{bmatrix} S_0^L & S_1^L \end{bmatrix}$$

This yields:

$$(1-\lambda)S_0^L + \mu S_1^L = S_0^L$$

and

$$\lambda S_0^L + (1-\mu)S_1^L = S_1^L$$

Algebraic manipulation of both equations yields the same result:

$$S_1^L = \frac{\lambda}{\mu} S_0^L$$

Substituting the above into the relation

$$S_0^L + S_1^L = 1$$

yields:

$$S_0^L + \frac{\lambda}{\mu} S_0^L = 1$$

Solving and noting that state 0 is the only success state,

$$A(s) = S_0^L = \frac{\mu}{\lambda + \mu} \tag{E-23}$$

State 1 is the failure state; therefore, unavailability equals:

$$U(s) = S_1^L = 1 - S_0^L = \frac{\lambda}{\lambda + \mu} \tag{E-24}$$

Fortunately, Equations E-23 and E-24 agree completely with Equations E-20 and E-21. Both methods yield an analytical solution for limiting state probability. These results can be used to show one of the classic equations of reliability theory. Remember that for a single component with a constant failure rate:

$$MTTF = \frac{1}{\lambda}$$

Since a constant repair rate is assumed, MTTR is defined as:

$$MTTR = \frac{1}{\mu}$$

Substituting into Equation E-23, the well-known equation

$$A(s) = \frac{MTTF}{MTTF + MTTR} \tag{E-25}$$

is obtained.

Multiple Failure Modes

Continuous analytical solutions can also be obtained for Markov models with multiple failure states. Figure E-3 shows a single component with two failure modes.

Again, assume that the model starts in state 0. The model will be in state 0 in the time instant only if it stays in state 0. This can be expressed mathematically as:

$$S_0(t + \Delta t) = S_0(t)(1 - \lambda^S \Delta t - \lambda^D \Delta t)$$

This can be rearranged as:

$$S_0(t + \Delta t) - S_0(t) = -\lambda^S S_0(t) \Delta t - \lambda^D S_0(t) \Delta t$$

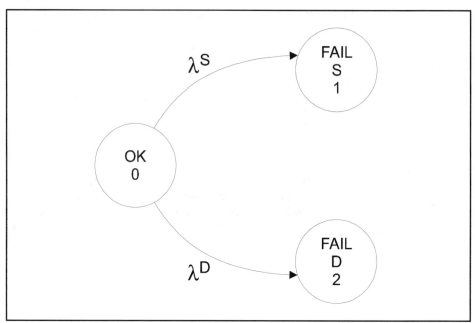

Figure E-3. Markov Model for Single Component with Multiple Failure Modes.

Dividing both sides by Δt:

$$\frac{S_0(t+\Delta t)-S_0(t)}{\Delta t} = -\lambda^S S_0(t) - \lambda^D S_0(t) \tag{E-26}$$

The left side of Equation E-1 is the deviation with respect to time. Taking the limit as Δt goes to zero results in:

$$\frac{dS_0(t)}{dt} = -\lambda^S S_0(t) - \lambda^D S_0(t) \tag{E-27}$$

Using a similar process:

$$\frac{dS_1(t)}{dt} = \lambda^S S_0(t) \tag{E-28}$$

and

$$\frac{dS_2(t)}{dt} = \lambda^D S_0(t) \tag{E-29}$$

Equations E-27 to E-29 are first-order differential equations with constant coefficients. Using a Laplace transform to convert from the time domain (t) to the frequency domain (s):

$$sS_0(s) - S_0(0) = -\lambda^S S_0(s) - \lambda^D S_0(s)$$

$$sS_1(s) - S_1(0) = \lambda^S S_0(s)$$

and

$$sS_2(s) - S_2(0) = \lambda^D S_0(s)$$

For the initial conditions $S_0(0) = 1$, $S_1(0) = 0$, and $S_2(0) = 0$, the equations reduce to:

$$sS_0(s) - 1 = -\lambda^S S_0(s) - \lambda^D S_0(s) \quad \text{(E-30)}$$

$$sS_1(s) = \lambda^S S_0(s) \quad \text{(E-31)}$$

and

$$sS_2(s) = \lambda^D S_0(s) \quad \text{(E-32)}$$

Rearranging Equation E-30:

$$(s + \lambda^S + \lambda^D)S_0(s) = 1$$

Therefore:

$$S_0(s) = \frac{1}{s + \lambda^S + \lambda^D} \quad \text{(E-33)}$$

Substituting Equation E-33 into Equations E-31 and E-32 and solving:

$$sS_1(s) = \frac{\lambda^S}{s + \lambda^S + \lambda^D}$$

A little algebra gives:

$$S_1(s) = \frac{\lambda^S}{s(s + \lambda^S + \lambda^D)} \tag{E-34}$$

and

$$S_2(s) = \frac{\lambda^D}{s(s + \lambda^S + \lambda^D)} \tag{E-35}$$

Taking the inverse transform of Equation E-33 results in:

$$S_0(t) = e^{-(\lambda^S + \lambda^D)t} = R(t) \tag{E-36}$$

Partial fractions must be used to proceed with the inverse transform of Equations E-34 and E-35. Using Equation E-34:

$$S_1(s) = \frac{A}{s} + \frac{B}{s + \lambda + \mu} \tag{E-37}$$

This means that:

$$\frac{A}{s} + \frac{B}{s + \lambda^S + \lambda^D} = \frac{\lambda^S}{s(s + \lambda^S + \lambda^D)}$$

This can be algebraically manipulated into the form:

$$\lambda^S = A(s + \lambda^S + \lambda^D) + B(s)$$

This relation holds true for all values of s. Therefore, to solve for A and B, we should pick a value of s that will simplify the algebra as much as possible. To solve for A, a value of $s = 0$ is the best choice. At $s = 0$,

$$\lambda^S = A(\lambda^S + \lambda^D)$$

Therefore:

$$A = \frac{\lambda^S}{\lambda^S + \lambda^D} \tag{E-38}$$

Continuous-Time Markov Modeling

Solving for B, choose:

$$s = -(\lambda^S + \lambda^D)$$

Substituting for s:

$$\lambda^S = -B(\lambda^S + \lambda^D)$$

Rearranging,

$$B = \frac{-\lambda^S}{\lambda^S + \lambda^D} \tag{E-39}$$

Substituting Equations E-38 and E-39 into Equation E-37:

$$S_1(s) = \frac{\lambda^S}{s(\lambda^S + \lambda^D)} - \frac{\lambda^S}{(\lambda^S + \lambda^D)(s + \lambda^S + \lambda^D)}$$

Taking the inverse Laplace transform:

$$S_1(t) = \frac{\lambda^S}{\lambda^S + \lambda^D} - \frac{\lambda^S e^{-(\lambda^S + \lambda^D)t}}{\lambda^S + \lambda^D}$$

This can be rearranged as:

$$S_1(t) = \frac{\lambda^S}{\lambda^S + \lambda^D}(1 - e^{-(\lambda^S + \lambda^D)t}) \tag{E-40}$$

Similarly:

$$S_2(t) = \frac{\lambda^D}{\lambda^S + \lambda^D}(1 - e^{-(\lambda^S + \lambda^D)t}) \tag{E-41}$$

Equations E-36, E-40, and E-41 provide the time-dependent analytical formulas for state probability. Figure E-4 shows a plot of probabilities as a function of time. Note that at long periods of time, failure state probabilities begin to reach steady-state values.

494 Continuous-Time Markov Modeling

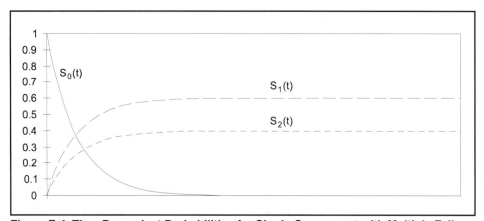

Figure E-4. Time Dependent Probabilities for Single Component with Multiple Failure Modes.

APPENDIX F

Answers to Exercises

Chapter 1

1.1 Yes, S84.01 describes the concept of safety integrity levels and presents example methods on how to determine the safety integrity level of a process. It defines three levels for the process industries.

1.2 Yes, availability is defined as the probability of successful operation at any moment in time. This is equivalent to percent uptime.

1.3 Yes, in both ISA S84.01 and IEC61508.

1.4 Yes, if quantitative targets (typically a SIL level and required reliability) are defined as part of the safety requirements.

1.5 Not in the opinion of the author. Qualitative techniques also are required in order to properly understand how the system works under failure conditions. Remember that the data used in quantitative techniques have an amount of uncertainty, and that stress conditions vary considerably from site to site. Qualitative guidelines should be used in addition to quantitative analysis.

Chapter 2

2.1 Software failures, 9; hardware failures, 10; maintenance failures, 5; operations failures, 1. Histogram is shown in Figure F-1. Venn diagram is shown in Figure F-2.

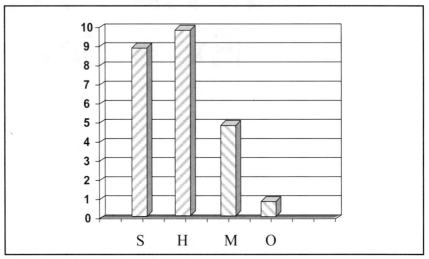

Figure F-1. Histogram of Failure Types.

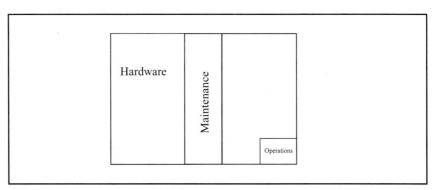

Figure F-2. Venn Diagram of Failure Types.

2.2 The chances that the next failure will be a software failure are $\frac{9}{25}$.

2.3 Figure F-3 shows the pdf. Figure F-4 shows the cdf.

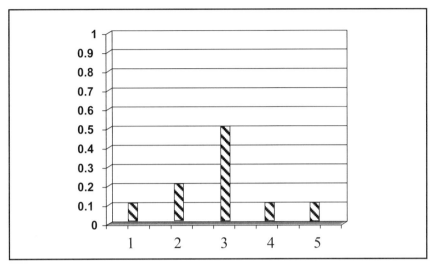

Figure F-3. Pdf for Problem 2.3.

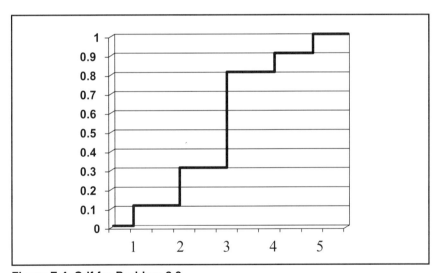

Figure F-4. Cdf for Problem 2.3.

2.4 The mean is calculated by multiplying the values times the probabilities per Equation 2-8.

$$\text{Mean} = 1 \times 0.1 + 2 \times 0.2 + 3 \times 0.5 + 4 \times 0.1 + 5 \times 0.1 = 2.9$$

2.5 The variance is calculated using Equation 2-12. Answer: 1.09.

2.6 Mean of the data equals 427.94 days. Average failure time in hours equals 427.94 × 24, or 10,271 hours. Median failure time in days equals (183 + 228)/2 = 205.5, or 4,932 hours.

2.7 Using a standard normal distribution, z = (80 − 40)/10 = 4.00. From Appendix A, at z = 4.00, 0.999969. Probability of failure then equals 1 − 0.999969 = 0.000031.

Chapter 3

3.1 The answer depends on a particular plant site.

3.2 Yes, software design faults can cause control systems to fail. This risk may go up as the quantity of software in systems increases.

3.3 Manufacturing faults.

3.4 Stress screening during manufacture, "burn-in."

3.5 Wearout occurs when the strength of a device decreases with time.

3.6 Using a standard normal distribution with x = 2,000, μ = 1,500, σ = 200, and z = 2.50. From Appendix A, probability of failure = 1 − 0.993791 = 0.006209.

3.7 Using a standard normal distribution with z = 2.7, probability of failure = 1 − 0.996533 = 0.003467.

3.8 Yes, HALT is used to identify weakness in a design. When the weak areas are removed, the design is strengthened as part of the design process.

3.9 Contact lubricant seals electronic connectors against humidity and corrosive atmospheres. It also provides strength against dirt and dust.

Chapter 4

4.1 Reliability.

4.2 Availability.

4.3 Availability = 0.999.

4.4 The formula MTTF = $1/\lambda$ applies to single components with a constant failure rate or series systems with a constant failure rate.

4.5 Unavailability = 0.001.

4.6 The failure rate equals 0.000001903 = 1,903 FITS.

4.7 Availability = 0.9999847.

4.8 Reliability = 0.9917.

4.9 Availability = 0.997.

4.10 Safety availability = 0.999.

Chapter 5

5.1 The steps for an FMEA are:
1. List all components.
2. For each component, list all failure modes.
3. For each component/failure mode, list the effect on the next higher level.
4. For each component/failure mode, list the severity of effect.

5.2 An FMEA does not identify combinations of failures since each component is reviewed individually. Operational and maintenance failures may be missed. All failure modes of components must be known, or else they will not be included.

5.3 Yes, within practical limits the diagnostics can be verified using by simulating the faults with short circuits, open circuits, parallel components, etc. It is not practical to physically generate all component failures.

5.4 Yes, like most human engineering activities, an FMEA or an FMEDA can benefit greatly by review. Different persons see things differently and can help find errors.

5.5 Answer depends on process and plant.

Chapter 6

6.1 0.142625.

6.2 0.15; the error is almost 5%.

6.3 0.000000944.

6.4 Fault tree analysis is a top-down approach that is capable of identifying combinations of failure events that cause system failure. It is systematic and allows the review team to focus on a specific failure. It is limited, however, in that it depends greatly on the skill of the reviewers. A fault tree can only show one failure (or failure mode) on a single drawing. This sometimes obscures interaction among failure states.

6.5 Answer depends on specific circumstances.

Chapter 7

7.1 This is a series system. The system availability is
$$0.8 \times 0.7 \times 0.75 = 0.42$$

7.2 System availability equals 0.6195.

7.3 System availability equals 0.187, not very impressive for components with such high availabilities. Note that the system availability is always much lower than component availabilities. (Did you get 0.487? Don't forget there are four sensors and two valves.)

7.4 System availability now equals 0.206, not much of an increase.

7.5 MTTF = 2,000,000 hr.

7.6 System MTTF = 3,000,000 hr.

7.7 High success probabilities are given, so use a failure approximation method.

7.8 Success + one failure mode = two modes, six components; therefore, $2^6 = 64$.

7.9 Success + two failure modes = three modes, four components; therefore, $3^4 = 81$.

Chapter 8

8.1 $2^3 = 8$.

8.2 The **P** matrix is

$$\mathbf{P} = \begin{array}{c} \\ 1 \\ 2 \end{array} \begin{bmatrix} 0 & 1 & 2 \\ 0.9898 & 0.01 & 0.0002 \\ 0.05 & 0.945 & 0.005 \\ 0 & 0.05 & 0.95 \end{bmatrix}$$

Limiting state probabilities are calculated to be

$$\begin{bmatrix} 0.814067 & 0.16607 & 0.019863 \end{bmatrix}$$

Steady-state availability = 0.814067 + 0.16607 = 0.980137.

8.3 The problem was solved using a spreadsheet with matrix multiplication. A row vector starting in state 0 was set up:

$$S = \begin{bmatrix} 1 & 0 & 0 \end{bmatrix}$$

This is matrix-multiplied (=mmult(XX:YY,WW:ZZ). (Don't forget CNTL-SHIFT-ENTER if using Excel.) This row matrix is repeated until steady state is reached. In the solution spreadsheet, steady state was reached somewhere in the 270 to 310 hr range, depending on the criteria used for steady state.

8.4 The limiting state probability row is

$$[0.649351 \quad 0.193836 \quad 0.033727 \quad 0.123085]$$

$$A = 0.649351 + 0.193836 = 0.843187$$

8.5 The **P** matrix is

$$\mathbf{P} = \begin{array}{c} \\ 1 \\ 2 \\ 3 \end{array} \begin{bmatrix} 0 & 1 & 2 & 3 \\ 0.973 & 0.02 & 0.002 & 0.005 \\ 0.05 & 0.933 & 0.002 & 0.015 \\ 0.05 & 0 & 0.95 & 0 \\ 0.05 & 0 & 0 & 0.95 \end{bmatrix}$$

The **Q** matrix is

$$\mathbf{Q} = \begin{bmatrix} 0.973 & 0.02 \\ 0.05 & 0.933 \end{bmatrix}$$

The **I − Q** matrix is

$$(\mathbf{I} - \mathbf{Q}) = \begin{bmatrix} 0.027 & -0.02 \\ -0.05 & 0.067 \end{bmatrix}$$

The **N** matrix is

$$\mathbf{N} = \begin{bmatrix} 82.81829 & 24.72188 \\ 61.8047 & 33.37454 \end{bmatrix}$$

The MTTF is 107 hr.

8.6 The MTTF equals $1/\lambda$.

Chapter 9

9.1 The repair rate for failures detected by on-line diagnostics is 0.25. The repair rate for failures detected by periodic inspection is 0.0002281.

9.2 System MTTF = 6,881 hr.

9.3 System MTTF = 18,671 hr.

9.4 Thermocouple burnout results in an open circuit. One technique would supply several volts through a high-ohm resistor so that the voltage in the input circuit would go to several volts (instead of the normal millivolts).

9.5 In normal operation, the switches would be open-closed or closed-open, depending on valve position. During valve movement it would be reasonable to accept a period of time where the two switches are open-open. If the condition closed-closed appears, at least one switch has failed closed. If open-open occurs after a reasonable period of time, it could be assumed that either the valve is stuck or one of the switches has failed open, depending on previous valve position. There is no clean, direct way to detect an open switch. If the switches had multiple contacts, perhaps extra inputs could be compared for better diagnostics.

9.6 Shorted wiring to a 4-20 mA transmitter could be detected by checking if the current goes above the 20 mA range. (Suggest at least 21 mA to avoid a false diagnostic trip.)

9.7 A frozen analog output could be detected by checking to see if the output remains the same for several readings. (We normally assume that the output changes at least a small amount during normal operation.)

9.8 For most switches there is no electrical difference between a normally closed switch and a short-circuit failure, making it extremely difficult to detect these failures. Many control system designers are replacing switches with analog sensors to add diagnostic capability to the system.

9.9 SPDT contacts can be used for diagnostics by knowing that one contact is open when other is closed. These could be wired to two inputs. The inputs should always be 0/1 or 1/0 (except for a brief period of time when the switch changes state). This technique will only detect certain failure modes of the switch contacts. Other switch failure modes, including contact actuator failures, are not detected.

Chapter 10

10.1 Design errors, manufacturing errors, maintenance faults, and operational errors. Another source is environmental stress, which includes electrical events (power spikes, lightning, and high current levels); mechanical stress (shock and vibration); chemical stress (corrosive atmospheres, salt air, and humidity); physical stress (temperature); and heavy usage (high data rates and high memory usage).

10.2 A software fault can fail two redundant controllers if both machines are running the same software and if the two units are subject to the same software stress (e.g., identical inputs, identical timing, and identical machine state).

10.3 Diversity is the use of different-design redundant components in order to reduce common-cause susceptibility. Diversity works best when the "diverse" components respond differently to a common stress.

10.4 Hardware diversity can be best achieved when a mechanical component is redundant with an electrical component. Some level of diversity is achieved when programmable electronic circuits are redundant with nonprogrammable circuits. Software diversity is achieved through two different programs to accomplish the same task. Software diversity can also be achieved by having redundant computers execute the same program differently (different operating "modes").

10.5 Assume full physical separation; therefore, add 0.002 (equivalent to different cabinets) + 0.002 (fully isolated) + 0.01 (no diversity) = 0.014.

10.6 Using the beta model, the dangerous common-cause failure rate is $0.025 \times 0.00005 = 0.0000125$ failures per hour.

Chapter 11

11.1 The answer can vary according to experience. Author's list: crash during word processor usage that destroyed Chapter 4, PC hung up during e-mail file transfer, PC hung up during print with three applications open, PC crashed after receiving e-mail message....

11.2 Software failures can be modeled statistically because the failure sources create stress that can be characterized as random variables, much like hardware.

11.3 Software strength is increased when fewer design faults are present, when "stress rejection" is added to the software, when software execution is consistent (no multitasking, few or no interrupts, little or no dynamic memory allocation), and when software diagnostics operate on-line to detect and report faults.

11.4 Stress on a software program include inputs (especially unexpected inputs), the timing of the inputs, the contents of memory, and the state of the machine.

11.5 Software complexity appears to have grown three orders of magnitude.

11.6 $e = 32$, $n = 21$; therefore, $32 - 21 + 2 = 13$.

11.7 No, some faults occur when certain input data occur.

11.8 $e = 33$, $n = 11$; therefore, $33 - 11 + 2 = 24$. Compare that to the number in Figure 11-6! An interrupt adds many possible control paths that must be tested to get good test coverage.

11.9 Test time = 700 hr, $k = 0.01$, $N_0 = 372$, $\lambda = 0.00339$.

11.10 Approximately 1,055 hr of testing are required.

11.11 $\lambda(1,000) = 0.0328$.

11.12 Approximately 334,000 hr of testing would be required.

11.13 Not likely; perhaps the program should be simplified.

Chapter 12

12.1 PFD = 0.0003499, RRF = 2,858, PFD_{avg} = 0.0001596, RRF = 6,265.

λ^S	0.00005	1oo2, Fault Tree with Diagnostics, Common Cause			
λ^D	0.00005				
RT	8				
TI	4,380				RRF
SD	24		PFD_{ft}	0.0001284	7,790
μ_P	0.00045496		PFD_{avgft}	0.0000443	22,587
μ_{SD}	0.04166667		$PFD_{avgftap}$	0.0000400	25,020
μ_O	0.125				
CS	0.9		PFD_{dcc}	0.0003499	2,858
CD	0.95		PFD_{avgdcc}	0.0001596	6,265
β	0.02	λ^{DDN} 0.00004655			
		λ^{DDC} 0.00000095			
λ^{DD}	0.0000475	λ^{DUN} 0.00000245			
λ^{DU}	0.0000025	λ^{DUC} 5E-08			

12.2 Example 12-20 (no common cause) had a PFD of 0.0001284 (RRF of 7,790). Exercise 12.1 (common-cause β = 0.02, smaller value) has a PFD of 0.0003499 (RRF of 2,858). The difference is −63%, showing that smaller numbers for β are still very significant.

12.3 PFD = 0.048, RRF = 21, PFD_{avg} = 0.016, RRF = 62.

λ^S	0.00005	1oo2, Fault Tree with Diagnostics		
λ^D	0.00005			
RT	8			
TI	8,760			RRF
μ_P	0.00022789	PFD_{ft}	0.0480486	21
μ_O	0.125	PFD_{avgft}	0.0160308	62
CD	0.5			
λ^{DD}	0.0000250			
λ^{DU}	0.0000250			

12.4   PFD$_{avg}$ approximation = 0.01598. Results are close.

λ^S	0.00005	1oo2, Fault Tree with Diagnostics		
λ^D	0.00005			
RT	8			
TI	8,760		RRF	
μ_P	0.00022789	PFD$_{ft}$	0.0480486	21
μ_O	0.125	PFD$_{avgft}$	0.0160308	62
CD	0.5	PFD$_{avgftap}$	0.0159870	63
λ^{DD}	0.0000250			
λ^{DU}	0.0000250			

Chapter 13

13.1   An FMEA (or equivalent) is necessary to understand how the system components fail and how these failures affect the system.

13.2   Many different degraded modes of operation exist. In some cases, the response time of the system slows. This should be considered a failure if the system response time becomes slower than the requirement. Other degraded modes of operation include "loss of redundancy." In fault-tolerant systems, one unit failure is often called "degraded mode." This is not a system failure. As one considers the needed performance of a system, other degraded modes of operation may become clear.

13.3   One failure mode, no diagnostics, and no common cause: One failure rate should be modeled.

13.4   Common-cause failures should be modeled when multiple (redundant) components exist in a system.

13.5   It is important to distinguish detected failures from undetected failures in two circumstances. First, detected failures may control how a redundant system operates. When this occurs, detected failures must be separated from undetected failures. Second, when detected failures result in a different repair situation.

13.6   Using a first-order approximation (see Equations 12-1 and 12-19), probability of the switch failing dangerously is

$$PFD_{SW} = PFD_{SW,DD} \times RT + PFD_{SW,DU} \times TI$$
$$= 100 \times 10^{-9} \times 8 + 100 \times 10^{-9} \times 8760$$
$$= 0.0008768$$

The probability of the valve failing dangerously is

$$PFD_{VLV} = 100 \times 10^{-9} \times 8760$$
$$= 0.000876$$

The probability of two controllers failing dangerously is determined by first calculating basic fault probabilities:

$$PFD_{C,DDC} = 0.0000000227 \times 8$$
$$= 0.0000001816$$

$$PFD_{C,DUC} = 0.0000000003 \times 8760$$
$$= 0.000002628$$

$$PFD_{C1,DDN} = 0.0000002050 \times 8$$
$$= 0.00000164$$

$$PFD_{C1,DUN} = 0.0000000020 \times 8760$$
$$= 0.00001752$$

$$PFD_{C2,DDN} = 0.0000002050 \times 8$$
$$= 0.00000164$$

$$PFD_{C2,DUN} = 0.0000000020 \times 8760$$
$$= 0.00001752$$

Next, calculate combinations of two normal-mode controller failures:

$$PFD_{C,DD/DD} = 0.00000164 \times 0.00000164$$
$$= 2.7 \times 10^{-12}$$

$$PFD_{C,DU/DU} = 0.00001752 \times 0.00001752$$
$$= 3 \times 10^{-10}$$

$$PFD_{C,DU/DD} = 0.00000164 \times 0.00001752$$
$$= 2.88 \times 10^{-11}$$

$$PFD_{C,DD/DU} = 0.00001752 \times 0.00000164$$
$$= 2.88 \times 10^{-11}$$

Approximating the two controllers failing dangerously can be done by adding the probabilities from the OR gate:

$$PFD_C = PFD_{C,DDC} + PFD_{C,DUC} + PFD_{C,DD/DD}$$
$$+ PFD_{C,DU/DU} + PFD_{C,DU/DD} + PFD_{C,DD/DU}$$

$$PFD_C = 0.0000001816 + 0.000002628 + 2.7 \times 10^{-12} + 3 \times 10^{-10}$$
$$+ 2.88 \times 10^{-11} + 2.88 \times 10^{-11}$$
$$= 0.0000028$$

The system PFD can be approximated by adding the PFDs of the switch, dual controller, and valve:

$$PFD_{SYSTEM} = PFD_{SW} + PFD_{VLV} + PFD_C$$
$$= 0.0008768 + 0.000876 + 0.0000028$$
$$= 0.00176$$

13.7 Substituting the numeric values into the **P** matrix gives:

$$\mathbf{P} = \begin{bmatrix} 0.999996 & 3.31\text{E-}06 & 4.1\text{E-}07 & 4\text{E-}09 & 1.23\text{E-}07 & 2\text{E-}07 \\ 0.041667 & 0.958333 & 0 & 0 & 0 & 0 \\ 0.125 & 2.45\text{E-}06 & 0.874997 & 0 & 5.3\text{E-}07 & 0 \\ 0 & 2.45\text{E-}06 & 0 & 0.999997 & 3.28\text{E-}07 & 2.02\text{E-}07 \\ 0.125 & 0 & 0 & 0 & 0.875 & 0 \\ 0 & 0 & 0 & 0 & 0 & 1 \end{bmatrix}$$

Solving by matrix multiplication gives the value:

$\text{PFD}_{\text{SYSTEM}}$ @ 1 year = 0.001754

Chapter 14

14.1 The total failure rate is

$4 \times 277 + 2 \times 183 + 2{,}555 + 1{,}800 + 2 \times 212 + 2 \times 290 = 6{,}633$ FITS

14.2 $\lambda_{\text{Detected}} = 6{,}326.71$ FITS

$\lambda_{\text{Undetected}} = 306.29$ FITS

14.3 $MTTF = \dfrac{1}{6633 \times 10^{-9}} = 150{,}761$ hr

14.4 The software failure rates are

$\lambda_S^C = 1800 \times 0.2 = 36$ FITS

$\lambda_S^N = 1800 \times (1 - 0.2) = 1764$ FITS

The hardware failure rates are

$\lambda_H^C = 4833 \times 0.05 = 241.65$ FITS

$\lambda_H^N = 4833 \times (1 - 0.05) = 4591.35$ FITS

14.5 1oo2, 1oo2D, followed by 2oo3.

14.6 2oo2D, 2oo3, 1oo2D.

14.7 Yes, previously undetected failures will be detected and repaired. This results in higher safety and higher availability in redundant systems.

Chapter 15

15.1 Death to multiple persons is classified as a C3 consequence. The frequency of exposure is continuous, F2. Alarms give a possibility of avoidance, P1. Since no protection equipment is installed, the probability of unwanted occurrence is high, W3. The risk graph indicates SIL3. The RRF should be in the range of 1,000 to 10,000.

15.2 The necessary RRF is 250,000/25,000 = 10. This is classified as SIL1.

15.3 SIS reliability and safety analysis should consider all components — from sensor process connection to valve process connection. Typically, this include impulse lines, manifolds, sensors, controllers, power supplies, solenoids, air supplies, valve actuators, and valve elements. If communications lines are required for safe shutdown of an SIS, then they must be included in the analysis.

15.4 All failures are dangerous undetected. The dangerous undetected failure rate is

$$\frac{1}{8760} = 0.00011415 \text{ failures per hour}$$

15.5 Using Equation 14-1, the approximate PFD is calculated as 0.5. At this high failure rate, the approximation method is expected to have considerable error. A more accurate method would be to use Equation 4-16. In this case, PFD = 0.3935, a considerable difference.

15.6 Using Equation 14-1, the approximate PFD is calculated as 2.0. Since probabilities must be a number from 0 to 1, this is obviously a bogus number. Using Equation 4-16, PFD = 0.865, a more reasonable number.

Chapter 16

16.1 Yearly failure costs equal 0.01 × $1,000 per hour × 8760 hr per year = $87,600. The totals are

Availability	0.99		
Procurement Costs	$100,000		
Yearly Risk/Failure Cost	$87,600		
Yearly Operational Cost	$5,000		
Discount Rate	0%		
Total Yearly	$92,600	Cumulative	$100,000
1	$92,600	Year 1	$192,600
2	$92,600	Year 2	$285,200
3	$92,600	Year 3	$377,800
4	$92,600	Year 4	$470,400
5	$92,600	Year 5	$563,000

16.2

Availability	0.99497487		
Procurement Costs	$110,000		
Yearly Risk/Failure Cost	$44,020		
Yearly Operational Cost	$5,000		
Discount Rate	0%		
Total Yearly		$49,020	Cumulative	$110,000
1	$49,020	Year 1	$159,020
2	$49,020	Year 2	$208,040
3	$49,020	Year 3	$257,060
4	$49,020	Year 4	$306,080
5	$49,020	Year 5	$355,101

The cost is justified.

16.3

Availability	0.99950025		
Procurement Costs	$210,000		
Yearly Risk/Failure Cost	$4,378		
Yearly Operational Cost	$6,000		
Discount Rate	0%		
Total Yearly		$10,378	Cumulative	$210,000
1	$10,378	Year 1	$220,378
2	$10,378	Year 2	$230,756
3	$10,378	Year 3	$241,133
4	$10,378	Year 4	$251,511
5	$10,378	Year 5	$261,889

The cost is justified.

16.4 $FV = \$100,000 \times (1.04)^5$

$= \$121,665.$

16.5 $117,528.

16.6

Availability	0.99		
Procurement Costs	$100,000		
Yearly Risk/Failure Cost	$87,600		
Yearly Operational Cost	$5,000		
Discount Rate	5%		
Total Yearly		$92,600	Cumulative	$100,000
1	$88,190	Year 1	$188,190
2	$83,991	Year 2	$272,181
3	$79,991	Year 3	$352,173
4	$76,182	Year 4	$428,355
5	$72,555	Year 5	$500,910

At a discount rate of 5%, the life-cycle cost dropped from $563,000 to $500,910.

APPENDIX C

C.1 Each coin must land heads-up. The probability of that event is one-half. The combination of all three heads events is

$$½ × ½ × ½ = 1/8$$

C.2 P(Control loop success) = P(Transmitter success) ×

P(Controller success) × P(valve success)
= (1 − 0.01) × (1 − 0.005) × (1 − 0.05)
= 0.99 × 0.995 × 0.95 = 0.9358

P(Control loop failure) = 1 − 0.9358 = 0.0642

Note that an approximation could be obtained by adding the failure probabilities. The approximation would be

P(Approx. control loop failure) = 0.01 + 0.005 + 0.05 = 0.065

This method is not exact because the failures are not mutually exclusive. However, it is usually faster and produces a conservative result.

C.3 P(At least one heads up) = 1 − P(No heads up) = 1 − 1/8 = 7/8

Alternatively, the problem could be solved by creating a list of all combinations of three coin toss outcomes. There will be eight possible combinations. Each combination will be mutually exclusive. It will be seen that seven of the eight combinations will have at least one heads.

C.4 P(Going broke) = P(6 or 8) = P(6) + P(8) since outcomes are mutually exclusive.

P(Going broke) = 5/36 + 5/36 = 10/36

P(Passing GO) = P(9 or more) = P(9) + P(10) + P(11) + P(12)
= 4/36 + 3/36 + 2/36 + 1/36 = 10/36

C.5 All combinations of no loops operating, one loop operating, and two loops operating are system failures. Combinations of three or four loops operating are system success. Combinations of one loop operating are

$$\frac{4!}{1!(4-1)!} = 4$$

Combinations of two loops operating are

$$\frac{4!}{2!(4-2)!} = 6$$

There is one combination of no loops operating. A total of 11 combinations of successful/failed controllers represent system failure.

Answers to Exercises

C.6   Combinations given by the basic counting rule:
$$3 \times 2 \times 1 = 6$$

C.7   Combinations of letters are given by the basic counting rule. In this case,
$$5 \times 4 \times 3 \times 2 \times 1 = 120$$

C.8   Using the formula for combinations,
$$\frac{3!}{2!(3-2)!} = 3$$

C.9 P(One successful controller and one failed controller) = $2 \times 0.95 \times 0.05 = 0.095$

Remember, there are two combinations of one failed and one successful controller.

C.10 P(One successful controller and two failed controllers) = $3 \times 0.95 \times 0.05 \times 0.05 = 0.007125$

Remember, there are three combinations of two failures and one successful unit.

P(Three failed controllers) = 0.000125

P(System failure) = 0.007125 + 0.000125 = 0.00725

C.11 P(Software failures) = 0.60

P(Maintenance failures) = 0.20

P(Operational failures) = 0.10

P(Hardware failures) = 0.1

C.12   The system operates safely only if all components operate safely.

P(System operating safely) = $1 - P$(Sensor fail dangerous) $\times$ $1 - P$(Sensor fail dangerous) $\times 1 - P$(Logic solver fail dangerous) $\times$ $1 - P$(Valve fail dangerous) = $0.999995 \times 0.999995 \times 0.99999 \times 0.99995 = 0.9999300011$

P(System fail dangerously) = $1 - 0.9999300011 = 0.0000699989$

Note, an approximation can be obtained by adding the probabilities of dangerous failure of each element. For this case,

P(Approx system failure) = $0.000005 + 0.00005 + 0.00001 + 0.00005 = 0.00007$

C.13 P(Boiler explosion) = P(dangerous condition) $\times P$(protection system failing to respond) = $0.00001 \times 0.000001 = 1\text{E-}11$

C.14 $\quad P(\text{Safety system fail dangerously}) = \dfrac{P(\text{Explosion with safety system})}{P(\text{Explosion without safety system})}$

$$= \dfrac{1\text{E}10-7}{2\text{E}10-5}$$

$$= 0.005$$

The inverse of this number is known as the risk reduction factor (RRF). In this case,

$$\text{RRF} = \dfrac{1}{0.005} = 200$$

Index

1oo1 architecture, 344, 411
1oo1 PLC, 196
1oo1D architecture, 358, 362
1oo2 analog voting, 108
1oo2 architecture, 197, 348, 351, 406, 412, 413
1oo2 series, 272, 300
1oo2D architecture, 198, 381, 384, 385, 409
1oo3 analog voting, 109
2oo2 architecture, 356
2oo2D architecture, 376, 379
2oo3 architecture, 364, 370, 408
2oo3 voting, 5
4oo5 system, 129

A. C. Brombacher, 45
absorbing failure state, 170
accelerated stress testing, 39
acceptable risk, 398
active review process, 39
annuity, 438
asynchronous functions, 250
availability, 61, 76, 119

basic execution time (BET) model, 256
basic faults, 103
basic process control system (BPCS), 1
bathtub curve, 71
beta factor, 312
bounding approximations, 142
burner management system (BMS), 214

censored data, 481
checksum, 205
common cause, 298, 312, 348, 351, 370
common-cause defense rules, 223

common-cause failures, 90, 275
common-cause stress, 213
comparison coverage factor, 207
comparison diagnostics, 86, 204, 389
compliance voltage, 36
compound interest, 433
constant failure rate, 65, 70, 73, 485
continuous-time Markov models, 483
corrosive atmospheres, 53
coverage factor, 313
cumulative distribution function (cdf), 15
curve-fitting, 258

dangerous coverage factor, 86, 200, 293
data mismatch, 223
degraded mode, 187
degraded states, 148
design faults, 38
diagnostic coverage factor, 85, 96, 187, 200
diagnostics, 358, 410
discount rate, 434
discrete-time models, 151
diversity, 223, 225
divide by zero failure, 238
dynamic memory allocation, 238

electrolytic capacitor, 35
electrostatic discharges, 54
environmental failure sources, 42
event-based failure costs, 428
expected value, 18
exponential distribution, 23
external failure sources, 37
external watchdog, 358

failure databases, 44
failure modes, 78
failure modes and effects analysis (FMEA), 89, 307
failure modes, effects, and diagnostic analysis (FMEDA), 93, 200, 292, 313
failure on demand, 428
failure rate, 68, 478
failure simulation testing, 210
failure sources, 31
failure types, 31
false trip, 79, 440
false trip rate, 348
fast-transient, 54
fault tree analysis (FTA), 101
fault tree review team, 101
fault-tolerant system, 213
FIT, 68, 309
FM, 242
FMEDA, 93, 200, 292, 313
foolproofing techniques, 225
fully repairable systems, 149, 160, 173
functional failures, 32, 35, 252

hardware diversity, 224
hazard analysis, 427
hazard rate, 68
hazardous event severity matrix, 402
highly accelerated life testing (HALT), 54, 68
histogram, 10
human reliability analysis, 90
humidity, 37
humidity testing, 53

ideal dual architecture, 187
IEC68-2, 52
imperfect inspection, 271
imperfect repair, 282
incorrect repair procedures, 43
instantaneous failure rate, 150
internal failure sources, 37
interrupt, 250
ISO 9000-3, 241

language of money, 423
Laplace transforms, 484
life test data, 477
lightning, 34
limiting state probability, 157, 158, 176
logarithmic Poisson (LP), 260
lognormal distribution, 26

maintenance policy, 326, 341
manual inspections, 192
manufacturing faults, 39

matrix inversion, 175, 191
McCabe complexity metric, 246
mean, 18
mean dead time (MDT), 65
mean time between failures (MTBF), 66
mean time to fail dangerously (MTTFD), 83
mean time to fail safely (MTTFS), 83, 199
mean time to failure (MTTF), 59, 64, 73, 123, 128, 149, 188, 346
mean time to repair (MTTR), 65, 281
mechanical shock, 53
median, 20
MIL-STD-1629, 89
minuteman missile program, 101
MTTFS, 83, 199
MTTR, 65, 281
multiple failure modes, 296

N matrix, 175
nonrepairable component, 150
NORAD, 38
normal distribution, 23

on-line diagnostics, 271, 292, 296, 298
on-line repair rate, 189
operational profile, 263
output readback, 205

parallel system, 124
partial fraction, 488
path space, 244
periodic inspection, 285, 292, 342, 427
periodic inspection interval, TI, 272
periodic maintenance, 189, 427
PFD average (PFDavg), 81
physical failures, 32, 34
physical separation, 225
plausibility assertions, 242
PLC input circuit, 96
power system, 106, 111
pressure transmitters, 108
priority AND gate, 105
probability density function (pdf), 11, 477
probability of failing safely (PFS), 80
probability of failure on demand (PFD), 78, 80, 149, 196
program flow control, 242
programmable electronic systems (PES), 335
programmable logic controllers (PLC), 335

Q matrix, 177

R matrix, 177
random failures, 34
random variable, 9, 59

reference diagnostics, 86, 204
regular Markov models, 160
reliability, 60, 119, 477
reliability network, 118
repair rate, 76
resulting fault, 103
risk, 397
risk analysis, 428
risk cost, 397, 426, 439
risk graph, 400
risk reduction factor (RRF), 78, 196, 399
roller coaster curve, 72

S71.04, 53
S84.01, 2, 78, 402
safe coverage factor, 86, 200, 293
safety availability, 82
safety functional verification, 426
safety instrumented system (SIS), 1, 214
safety integrity levels (SIL), 399, 426
safety PLC, 80, 96, 200, 204
safety-critical software, 242
sample mean, 18
sample variance, 21
self-diagnostic detection times, 341
series system, 119
shock rate, 220
simplification techniques, 330
simulation, 48
single repairman model, 188, 279
single-board controller (SBC), 335
single-loop controllers (SLC), 335
smart transmitters, 207
software common-cause strength, 224
software diagnostics, 242
software diversity, 224
Software Engineering Institute, 241
software error, 224
software fault-avoidance, 240
software maturity model, 241
software strength, 241
software stress rejection, 242
software testing, 242
standard deviation, 22
standard normal distribution, 25
starting state, 157, 287
state diagrams, 147
state merging, 330
state space, 244
statistics, 9
steady state probabilities, 280
steady-state availability, 164
stress rejecter, 248
stress rejection, 242

stressors, 45
stuck signal detector, 206
success paths, 118
surge, 54
system architectures, 335
system failure, 308
system FMEA, 425
systematic failures, 33

temperature stress testing, 52
test coverage, 245
test interval, 342
thermocouples, 109
time-based failure costs, 428
time-dependent availability, 167
time-dependent failure rate, 184
time-dependent probabilities, 282
training, 425
transition matrix, 152
trigger events, 103
TUV, 242

U.S. space program, 187
U.S. Space Shuttle, 225
unavailability, 63, 491
uncensored data, 481
uniform distributions, 13
unknown failure modes, 202
unreliability, 61, 477
useful approximation, 75

variance, 21
vibration, 53
virtual infinity, 261
voting circuit, 207

watchdog timers, 204
wearout, 51, 71, 184
windowed watchdog, 204